THE PRIMARY BATTERY

THE ELECTROCHEMICAL SOCIETY SERIES

THE CORROSION HANDBOOK
Edited by Herbert H. Uhlig

ELECTROCHEMISTRY IN BIOLOGY AND MEDICINE
Edited by Theodore Shedlovsky

ARCS IN INERT ATMOSPHERES AND VACUUM
Edited by W. E. Kuhn

MODERN ELECTROPLATING, THIRD EDITION
Edited by Frederick A. Lowenheim

FIRST INTERNATIONAL CONFERENCE ON ELECTRON AND
ION BEAM SCIENCE AND TECHNOLOGY
Edited by Robert Bakish

THE ELECTRON MICROPROBE
Edited by T. D. McKinley, K. F. J. Heinrich and D. B. Wittry

CHEMICAL PHYSICS OF IONIC SOLUTIONS
Edited by B. E. Conway and R. G. Barradas

VAPOR DEPOSITION
Edited by C. F. Powell, J. H. Oxley and J. M. Blocher, Jr.

HIGH-TEMPERATURE MATERIALS AND TECHNOLOGY
Edited by Ivor E. Campbell and Edwin M. Sherwood

ALKALINE STORAGE BATTERIES
S. Uno Falk and Alvin J. Salkind

THE PRIMARY BATTERY (*In Two Volumes*)
Volume I Edited by George W. Heise and N. Corey Cahoon
Volume II Edited by N. Corey Cahoon and George W. Heise

ZINC-SILVER OXIDE BATTERIES
Edited by Arthur Fleischer and J. J. Lander

THE CORROSION MONOGRAPH SERIES
R. T. Foley, N. Hackerman, C. V. King, F. L. LaQue, H. H. Uhlig, Editors

THE STRESS CORROSION OF METALS, *by Logan*
CORROSION OF LIGHT METALS
by Godard, Bothwell, Kane, and Jepson

THE CORROSION OF COPPER, TIN, AND THEIR ALLOYS
by Leidheiser

CORROSION IN NUCLEAR APPLICATIONS, *by Warren C. Berry*
HANDBOOK ON CORROSION TESTING AND EVALUATION,
Edited by William H. Ailor

THE PRIMARY BATTERY

VOLUME II

Edited by

N. COREY CAHOON

BATTERY DEVELOPMENT LABORATORY
BATTERY PRODUCTS DIVISION
UNION CARBIDE CORPORATION (RET.)

and

The late GEORGE W. HEISE

Sponsored by

THE ELECTROCHEMICAL SOCIETY, INC.
New York, New York

A WILEY-INTERSCIENCE PUBLICATION

JOHN WILEY & SONS

NEW YORK · LONDON · SYDNEY · TORONTO

Copyright © 1976 by John Wiley & Sons, Inc.

All rights reserved. Published simultaneously in Canada.

No part of this book may be reproduced by any means, nor transmitted, nor translated into a machine language without the written permission of the publisher.

Library of Congress Cataloging in Publication Data
Main entry under title:

The Primary battery.

1 (Electrochemical Society series)
   Includes bibliographical references.
   1. Electric batteries.   I.   Heise, George W., ed.
II.   Cahoon, N. Corey, 1904-   ed.
TK2921.P75   621.35'3   73-121906
ISBN 0-471-12923-2 (v. ii)

Printed in the United States of America

10 9 8 7 6 5 4 3 2 1

George W. Heise, 1888–1972.

Contributors

DAVID BELITSKUS, *Senior Scientist, Alcoa Laboratories, Alcoa Center, Pa.*

RALPH J. BRODD, *Technical Research Manager, Research Laboratory, Battery Products Division, Union Carbide Corp., Parma, Ohio.*

N. COREY CAHOON, *Senior Scientist, (ret.) Battery Development Laboratory, Battery Products Division, Union Carbide Corp., Cleveland, Ohio.*

JOSEPH S. DERESKA, *Staff Development Engineer, Battery Development Laboratory, Battery Products Division, Union Carbide Corp., Cleveland, Ohio.*

DONALD J. DOAN, *Staff Scientist, Eagle-Picher Industries, Inc., Couples Depat., Joplin, Missouri.*

WALTER J. HAMER, *Chemical Consultant and Science Writer, 3028 Dogwood Street, N. W., Washington, D. C.*

CHARLES W. JENNINGS, *Organization 2433, Sandia Laboratories, Albuquerque, New Mexico.*

JOHN W. PAULSON, *Senior Engineer, Ray-O-Vac Division, ESB, Inc., Madison, Wisconsin.*

JOHN L. ROBINSON, *Engineering and Metals, The Dow Chemical Co., Midland, Michigan.*

ROBERT C. SHAIR, *Manager of Energy Products, Motorola, Inc., Fort Lauderdale, Florida.*

JOHN J. STOKES, JR., *Scientific Associate, Joining Division, Alcoa Laboratories, Aluminum Co. of America, Alcoa Center, Pa.*

RICHARD M. WILSON, *Group Leader, Research Laboratory, Battery Products Division, Union Carbide Corp., Parma, Ohio.*

Advisory Board

Foreword

The Battery Division of The Electrochemical Society, Inc., was organized at the 1947 Fall Meeting of the Society in Boston. Its purpose included the stimulation of research, publication, and the exchange of information relating to primary or secondary cells for the production of electrical energy. What propitious timing! The 1967 Census of Manufactures indicates that the total annual product of the battery industry in the United States has steadily increased to the present annual value of about one billion dollars, about triple that reported for 1947.

The growth in volume of the battery industry has been accompanied by a demand in military and consumer markets to extend the utility of batteries. In the field of primary batteries, the one of particular interest here, the batteries must operate usefully over a wide range of ambient temperatures and environmental conditions, must have a wide range of energy and power densities, and often must be available in unusual packaging arrays. On one hand, development and application of transistorized circuits have led to miniaturization as in the case of hearing aids and their battery requirements. On the other hand, there is a trend toward larger sizes for stationary batteries used in railway signaling installations.

It should be no wonder then that all possible materials and their combinations are scrutinized and examined constantly in striving to create and invent improved cells. The technology of manufacturing at sustained high quality economically and the requisite knowledge become increasingly complex with time. There are no reasons at this time to justify any belief that the tasks in this field will become any easier.

This conclusion is indicated with certainty by the results of rather extensive research and development programs over the past decade. In particular, there may be cited the work in the fields of fused salt, solid-state electrolyte, organic electrolyte, and fuel-cell systems.

A partial answer to the problem of an explosively expanding literature is its periodic review and appraisal. Such compilations will be of great value to those concerned with all aspects of the industry, to the users and consumers who must understand the battery and its capabilities for its proper application and the avoidance of maintenance, and to those who have and seek a general background and interest in scientific and technologic achievements. The scope of the primary battery field was found to be broad enough to justify more than one published volume. This book is an authoritative review of primary battery technology.

ARTHUR FLEISCHER

Orange, New Jersey
January 1970

Preface

The objective for Volumes I and II is to present the science involved in primary batteries in a comprehensive manner. It is recognized that batteries originated as an art and much art is still present in battery manufacture. However, the art is considered largely proprietary and is therefore not appropriate for inclusion in these volumes.

Both volumes of *The Primary Battery* are directed to several groups of readers. They contain much of value for the scientists in the battery field who, through their continuing research and production efforts, are improving the performance of many commercial battery systems and are developing new types. We believe that technologists in fields related to batteries will find that this volume provides interesting and useful information which can be utilized in applying batteries to their products. Students in scientific and engineering schools, as well as those individuals interested in science, are a third group, and we feel that the many examples of applied electrochemistry, present in some detail, will be useful in their particular fields.

Each of the 11 chapters in Volume II presents a description of either a battery system or some important field closely related to battery technology. Many of the chapters offer, some for the first time, a comprehensive analysis of the subject and present examples of the applied science involved. Volume I covered the alkaline primary cells, fuel cells, solid electrolyte and nonaqueous systems, as well as standard cells. Volume II describes the Leclanché cell and its analogs, the zinc chloride cell, the aluminum and magnesium cells, the organic depolarized cells, low-temperature cells, and thermal and water-activated cells. In addition, it covers subjects important to battery technology, such as the nomenclature and testing procedures, internal resistance measurement, the reversibility of battery systems, and a survey of energy sources other than batteries. Thus, Volume II is planned to complement Volume I with the objective that the two-volume work will provide a comprehensive coverage of the primary battery field. Each of the contributors is a well-qualified expert in his specialty.

Thanks are due to many individuals who helped to make this volume possible: the contributors and their organizations who provided much time and effort in preparing manuscripts for the individual chapters; the members of the advisory board and the reviewers who helped, by thorough and painstaking reviews to insure that the data presented were both well organized and correct; the industrial and government organizations that released the chapters written by members of their staff; the publishers, as acknowledged in the text, who permitted the use of copyrighted material; the officers of The Electrochemical Society and the Battery Division who have furnished support for this project.

No preface to this volume would be complete without a tribute to the coeditor, George W. Heise. From the inception of the plans for these two volumes until shortly before his death in 1972, he contributed a prodigious amount of time to the planning, writing, and editing operations. Throughout our joint effort, he applied his scientific and literary talents to achieve our objective of making these volumes a comprehensive and authoritative reference work in primary battery technology. It is a privilege to acknowledge George W. Heise's many important contributions to the Battery Division of The Electrochemical Society as coeditor of these volumes. George W. Heise was President of the Electrochemical Society in 1948 and was an active and staunch supporter of the Battery Division since its organization in 1946.

Fairview Park, Ohio   N. COREY CAHOON
August, 1975

Contents

R. C. SHAIR

1.

Leclanché and Zinc Chloride Dry Cells

N. Corey Cahoon

This chapter describes both the conventional Leclanché dry cell system and the newer zinc chloride cell which has appeared during the last few years. In many respects, the zinc chloride cell is very similar to the conventional cell but, since NH_4Cl is absent from the electrolyte, it has certain advantages. In particular, the zinc chloride cell is highly resistant to leakage on heavy discharge tests. The literature describing this new system is so limited that a separate chapter is not warranted. Instead, it is described in a separate section at the end of this chapter.

As every serious worker in the field is aware, the literature on the Leclanché battery system is quite extensive, as would be expected in an art that is now over 100 years old. In this review no effort has been made to include in the bibliography every reference available. Instead, selected references have been chosen to lead the interested reader to the original work referred to in the text.

An apparent omission from this review is a detailed presentation of the formulations used for commercial cell types. Very little data of this type have appeared in the literature, as battery manufacturers consider such information proprietary. Thus the treatment of the Leclanché cell system presented herewith emphasizes the science of cell technology rather than the art of formulation.

Of the many electrochemical systems adapted for use in primary batteries during the 19th century, very few are in commercial use today. The Leclanché system is the most important of this group, and the reasons for its continued existence and growth are both technical and economic.

One of the original advantages of the Leclanché system, pointed out by the inventor[1],[2] was the lower cost of the raw materials as compared with those for the mercury sulfate cell then in use for the newly installed telegraph system. A major factor in the growth of the Leclanché system has been the low-cost materials that are commercially available in adequate purity. Equally important has been the stability or keeping quality of the finished batteries. Although this may not have been important in the early days of the fluid electrolyte or wet cells, today batteries are often stored for considerable periods without serious loss in capacity. This factor has materially helped establish the commercial success of this system.

Another important item has been the relatively noncritical construction that has developed, which is applicable to many sizes of unit cells. This factor, coupled with the ready adaptability of these unit cells for use in multicell batteries, has made it possible for the industry to meet a wide range of applications.

One of the most attractive features of the Leclanché system has been the portability of the unit cells and most batteries made from them. Furthermore, such batteries are capable of producing some electrical output on the heaviest of current drains even though their efficiency under such conditions may be quite low. Often, even a short service life may be quite acceptable to the customer, especially when weight and portability are considered as in the demand for batteries for cable testing and similar applications.

Many of the important developments that followed the invention of the Leclanché cell have been covered in the Historical Introduction in Volume I(3). Here, the emphasis is placed on the technology, which in more recent times, has provided the foundation on which the modern battery manufacture and operation is based.

Although the Leclanché system originated as a wet cell, it was only when the so-called "dry" or "spillproof" units became available that the basis for commercial acceptance was developed. It is not surprising that dry cell manufacture, which developed as an art, was for many years far ahead of the science. During the period 1930–1970, rapid developments in the technology were made. The fundamental chemistry and electrochemistry involved were given concentrated attention by several groups of workers, and the industry was not slow in applying the new findings to commercial units. The results of this concentrated technological effort have, in retrospect, been amazing. Without any essential change in cell constituents the battery capacity attainable from standard units in 1920 was essentially quadrupled by 1950. In fact, the modern "D" size cell can be formulated to deliver as much energy, that is, 8–10 amp-hr, as the No. 6 size unit, approximately eight times larger, of the early 1900s. Such a "D" size cell is an experimental unit and not commercially available. Figure 1.1 shows the progress in increasing the cell output over a 50-year period(3).

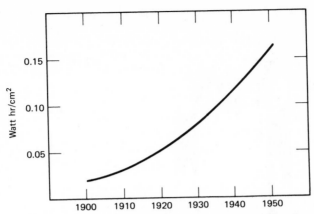

Figure 1.1 Fifty years of progress in the Leclanché cell.

DESCRIPTION OF TYPICAL LECLANCHÉ CELLS

The original Leclanché cell utilized an MnO_2-carbon cathode, a zinc anode, and an aqueous solution of ammonium chloride as an electrolyte. The evolution from the original invention to the modern dry cell has been described elsewhere(3). Present-day dry cells, often referred to as carbon-zinc cells, basically utilize the same system, although the electrolyte now contains zinc chloride in addition to the original ingredients. Three examples of construction are described below.

Figure 1.2 shows a cutaway illustration of one type of a modern unit cell. Here, the zinc can anode serves as the mechanically strong container for the unit, although a jacket consisting of layers of plastic and paper provides the external covering. The cathode bobbin consists of an intimate mixture of powdered manganese dioxide and carbon wet with electrolyte and molded under pressure onto a central carbon rod. This assembly is placed inside the zinc can. The narrow, intervening space is filled with a cereal paste electrolyte which is often cooked to solidify it by the application of heat during the assembly of the cell. An insulating disc at the bottom, the "Star Bottom" in Fig. 1.2, prevents electronic contact between the cathode and anode and thus prevents short circuits. Insulating washers at the top of the bobbin maintain the latter centrally in the zinc can. The zinc can is the anode or negative electrode of the unit, although a tinplated steel bottom disc makes the external contact and serves as the negative terminal. The carbon rod extending from the top of the cathode provides the positive current collector. A metal cap on this or

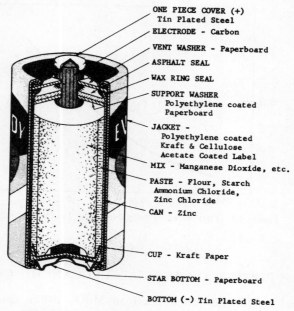

ONE PIECE COVER (+)
Tin Plated Steel

ELECTRODE - Carbon

VENT WASHER - Paperboard

ASPHALT SEAL

WAX RING SEAL

SUPPORT WASHER
Polyethylene coated
Paperboard

JACKET -
Polyethylene coated
Kraft & Cellulose
Acetate Coated Label

MIX - Manganese Dioxide, etc.

PASTE - Flour, Starch
Ammonium Chloride,
Zinc Chloride

CAN - Zinc

CUP - Kraft Paper

STAR BOTTOM - Paperboard

BOTTOM (-) Tin Plated Steel

Figure 1.2   Cross section of standard round cell.

a complete metal top cover for the cell contacts the carbon electrode and serves as the positive terminal of the cell.

Figure 1.3 shows an alternative construction for a unit cell. Here, a paper jacket carrying a positive metal cover is lined with an electronically conductive carbon layer by injection molding. A premolded cylinder of cathode mix is placed in this container and the zinc anode element is then pressed into the cathode mix. The zinc anode element is made from a zinc sheet cut and formed into an X-shaped cross section and then covered with a separator sheet of paper and plastic. After this unit is pressed into the cathode mix, the separator absorbs electrolyte from the cathode and places the cell in condition to operate. The cell is closed with suitable insulation parts and a metal cover which effectively contacts the protruding end of the zinc anode as shown in the illustration. Because of its unique construction, this cell is often referred to as the external cathode or the "inside-out" round cell(4). Since zinc is not used as a container in this cell, the construction provides a high degree of leakage resistance on severe tests.

The third example utilizes the anode and cathode elements arranged in a modern version(185)(186)(187) of the Voltaic Pile(5). The elements are usually rectangular in shape and the resulting cell is designated a flat cell.

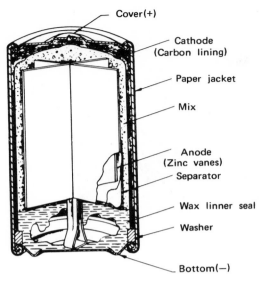

Figure 1.3   Cross section of external cathode or "inside-out" round cell.

Each cell is partially enclosed in an insulating elastic envelope as shown in Fig. 1.4a. Suitable contacts to adjacent cells and to battery terminals are made through the openings in this envelope when the cells are piled together to become battery assemblies as shown in Fig. 1.4b and c. This construction utilizes a larger percentage of the space (185) in a rectangular package than when cylindrical unit cells are used. The flat cell construction is therefore used for batteries where the space factor is important, as in highly portable units.

Figure 1.5 shows a variety of styles and sizes of batteries and unit cells that are commercially available. Although the examples cited here are taken from only one manufacturer, interested readers are referred to the Engineering Data Books published by the various battery manufacturers for the details of a particular battery construction in which they might be interested.

TECHNICAL CONSIDERATIONS

To many electrochemists, the Leclanché cell system probably appears to be a simple combination of a manganese dioxide redox system, an aqueous electrolyte of inorganic salts, and a zinc anode. In spite of the apparent simplicity, no substantial progress was made for many years in explaining the well-established behavior of a battery in terms of the

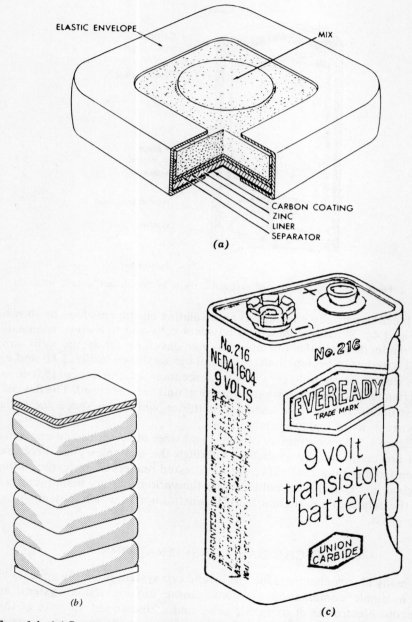

Figure 1.4 (a) Cross section of "mini-max" flat cell. (b) Assembly of flat cells. (c) Battery of flat cells.

Figure 1.5 A variety of Leclanché cell and battery types.

chemistry and electrochemistry of its elements. In 1906, Professor Crocker(6) said "One is at a loss to point out any radical or even important advance in primary batteries . . .; the types now on the market . . . are not substantially better than what might have been obtained 60 years ago." The development of the battery continued at a slow pace for many more years during which time technologists were more concerned with manufacturing problems than with scientific aspects.

By the early 1930s, battery technologists were asking important questions to which no satisfactory answers had been available. For example, why was it not possible to calculate the open circuit voltage of a Leclanché cell from thermodynamic data and obtain a better agreement between the actual and calculated values, 1.56 and 1.2 to 1.86, respectively? Moreover, there appeared to be no simple relationship between the amount of ammonium chloride placed in the cell initially and the service output of many common tests. What was the real function of the zinc chloride added to the electrolyte? It greatly influenced the shape of the discharge curve and made necessary a mass of empirical evaluations on every specific cell formulation examined. These and other technical questions were responsible for the rise in technological interest in this battery system and provided the impetus to examine many factors affecting dry cell performance. Fortunately, new research tools and new and improved raw materials, such as synthetic manganese dioxides and acetylene black, were also available.

The individual components of the cell, as originally assembled, are discussed under the headings: the electrolyte; the cathode; the anode; and cell voltage. As cell discharge occurs, both chemical and electrochemical reactions result in changes in the various parts of the cell system. These reactions are discussed under the headings: reactions in the electrolyte; reactions in the cathode; reactions at the anode; shelf reactions· and a comparison of the proposed reaction mechanisms for the cathode reaction.

THE ELECTROLYTE

The electrolyte in the original Leclanché cell(1) was an aqueous solution of ammonium chloride, often with the addition of solid salt beyond the saturation level. Zinc chloride was found to be beneficial and its addition to the ammonium chloride solution became general about the time the dry or spillproof units of Gassner(7) were introduced. In this discussion most of the data are limited to electrolytes containing both salts although significant data on individual salts are not omitted.

The solubilities of zinc and ammonium chlorides in water at 10°, 20°, and 30°C reported by Meerburg(8) in 1903 (Fig. 1.6) provided useful basic

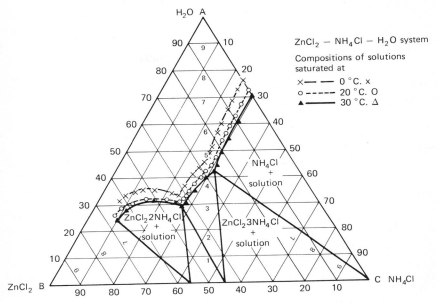

Figure 1.6 The ZnCl$_2$–NH$_4$Cl–H$_2$O system after Meerburg (8).

data for the preparation of electrolyte solutions for battery manufacture. Subsequent determinations (9) of the solubilities of these salts in water adequately confirmed the Meerburg values. Figure 1.7 shows a comparison of Meerburg data at 20°C with recent data at 21°C. The agreement is quite satisfactory particularly since the 21°C data were obtained with commercial salts such as those used in battery manufacture. The compositions of the two solid phases, ZnCl$_2$3NH$_4$Cl and ZnCl$_2$2NH$_4$Cl, obtained by Meerburg were satisfactorily confirmed. Table 1.1 shows the data obtained in the later work. Confirming the findings of Meerburg, at 21°C, the saturated solutions containing up to 25 percent zinc chloride are in equilibrium with solid ammonium chloride, those containing 25.0 percent to 41.4 percent zinc chloride with ZnCl$_2$3NH$_4$Cl, and those above 41.4 percent zinc chloride with ZnCl$_2$3NH$_4$Cl. Practical experience has shown that electrolytes containing from 25 percent to 41.4 percent zinc chloride may be unsatisfactory for dry cell use. Such electrolytes permit the growth of crystals of ZnCl$_2$NH$_4$Cl with a resultant development of high resistance in the depolarizer mass and actual mechanical disintegration of the latter. Table 1.2 shows the compositions of the two solid phases.

At the time the above confirmatory work was done the glass electrode had just become commercially available, and the pH was determined on

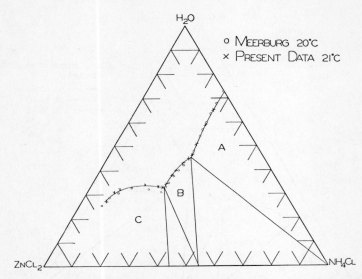

Figure 1.7 A comparison of Meerburg data at 20°C with Cahoon data at 21°C for the ZnCl₂–NH₄Cl–H₂O system.

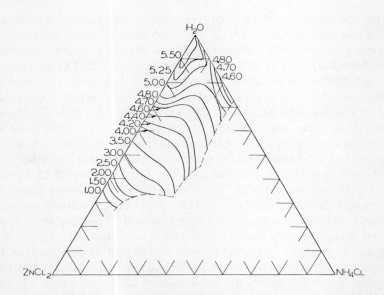

Figure 1.8 The relation between pH and solution composition for the ZnCl₂–NH₄Cl–H₂O system at 21°C.

TABLE 1.1. COMPOSITIONS AND pH VALUES OF SATU-
RATED SOLUTIONS IN THE $ZnCl_2$-NH_4Cl-H_2O SYSTEM
AT 21°C

Composition Percent by Weight		
$ZnCl_2$	NH_4Cl	pH
—	27.15	4.50
6.91	27.35	4.82
13.60	27.66	4.71
17.87	28.19	4.61
20.89	28.67	4.54
24.56	29.02	4.43
25.44	28.96	4.50
26.54	28.30	4.46
29.77	27.79	4.25
33.91	26.73	3.99
38.03	26.04	3.64
39.85	25.99	3.51
41.91	25.26	3.31
43.28	23.64	3.11
47.05	18.83	2.71
52.63	14.26	2.24
57.12	10.54	1.70
59.75	9.72	1.50
64.36	8.55	0.91

TABLE 1.2. COMPOSITIONS OF THE SOLID PHASES IN THE $ZnCl_2$-
NH_4Cl-H_2O SYSTEM AT 21°C

Original Solution		Solid Phase		Formula
$ZnCl_2$	32.7%	Zn	21.80%	
NH_4Cl	27.6%	Cl	58.98%	$ZnCl_2 3NH_4Cl$
H_2O	39.7%	NH_3	16.81%	
$ZnCl2$	49.0%	Zn	26.77%	
NH_4Cl	18.0%	Cl	56.71%	$ZnCl_2 1.88NH_4Cl$
H_2O	33.0%	NH_3	13.00%	

the series of electrolytes saturated with respect to ammonium chloride.
Table 1.1 presents these data while Table 1.3 offers similar results on a
series of zinc chloride solutions. These values are combined with other
determinations to provide the data shown in Fig. 1.8 where iso-pH lines
are drawn through the area of true solutions on the triangular diagram. It

TABLE 1.3. COMPOSITIONS AND pH OF
AQUEOUS SOLUTIONS OF ZNCL$_2$ AT 21°C

ZnCl$_2$ % by Weight	pH
0.935	6.09
4.73	5.75
9.56	5.44
19.20	5.02
29.40	4.60
40.64	3.80
49.09	3.00
59.02	1.80
66.96	0.58

is clear that a wide range of pH from 1.00 to 5.50 is represented by the solutions shown on this figure. Takahashi, Nakauchi and Sasaki(10) have studied the electrolyte system in considerable depth and their papers should be consulted by the reader who may be particularly interested in these aspects.

Electrical conductivity measurements for the system ZnCl$_2$-NH$_4$Cl-H$_2$O at 21°C are presented in Fig. 1.9. The data for the individual salts agree within 3 percent with published values. It is apparent that the

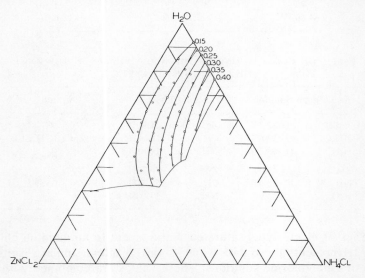

Figure 1.9 The relation between electrolyte conductivity and solution composition for the ZnCl$_2$–NH$_4$Cl–H$_2$O system at 21°C.

conductivity of the saturated solutions generally decreases in value with decreasing ammonium chloride concentration, although not in arithmetic proportion. From this figure, it is evident that the electrolyte is an excellent conductor of electricity and therefore well chosen for use in a battery.

THE CATHODE

A cathode for a primary battery may be defined as an oxygen-yielding compound which, by reaction with the electrolyte, becomes an electron acceptor and provides a half-cell capable of sustained current production at a useful EMF level. The cathode in the Leclanché cell is the MnO_2 electrode. A considerable amount of carbon is needed to make an adequate electrical contact to the particles of MnO_2 which itself has a low degree of electrical conductivity. All types of the MnO_2 mineral occurring in nature are not equally effective in a battery, as Leclanché[1] recognized.

The problem of how to determine the effectiveness of MnO_2 from various mines around the world troubled the industry for many years. Although much careful work was done to evaluate and compare MnO_2 samples, the final test was a trial in actual cells and a comparison of the test results with those attained from other sources. When World War II shut off the supply of Caucasian MnO_2 ore, the type generally used by the industry, alternative sources were examined. Holler and Ritchie[11] at the Bureau of Standards tried to use the potential of the sample of MnO_2 in electrolytes of various pH levels as a means of differentiating these materials. Their work, confirmed some years later by Martin and Helfrecht[12], gave S-shaped curves when the potential was plotted against the electrolyte pH, significantly at variance with the straight line predicted from the Nernst equation. Cahoon[13] examined the potential-pH relationship of MnO_2 using electrolytes containing only zinc and ammonium chlorides and ammonia. He obtained a straight line relationship over the pH range of 0–12. It was believed that the lack of straight lines in the work of Holler and Ritchie and Martin and Helfrecht resulted from the use, by both groups of hydrochloric acid to acidify the ammonium chloride electrolytes. Cahoon[13] employed zinc chloride for this purpose, but when hydrochloric acid was added, curves similar to the earlier type were obtained. The straight line obtained by Cahoon using an African natural ore had a slope within 2 percent of that predicted from the Nernst equation. Figure 1.10 illustrates these data.

An extension[14] of this study to other natural MnO_2 samples showed results that deviated from the straight line described above. Figure 1.11 gives the data obtained from Java and Caucasian ores as curve C,

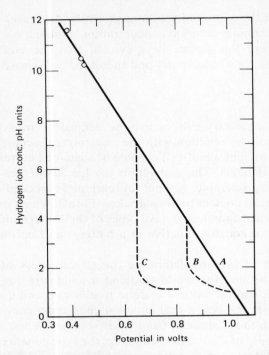

Figure 1.10 The potentials of manganese dioxide versus pH in dry cell electrolyte. *A*, data obtained in the absence of hydrochloric acid. *B*, data obtained in the presence of hydrochloric acid. *C*, Martin and Helfrecht, curve *F*.

Montana ore as curve *M*, and the African ore as curve *A*. From an examination of these curves and associated battery service data, it appeared that the 4 ohm continuous service delivered by "D" size cells made from the three ores was in the same order as the potentials developed at pH 9. Moreover, the service of No. 6 size cells on a representative telephone test was in the same order as the potentials of the three ores at pH 4. Table 1.4 shows the data on which these observations were based. At one time, such curves were helpful in selecting ore samples for further tests, but more sophisticated methods have been devised.

The pH-potential method was applied to a large number of samples including synthetic and beneficiated types of MnO_2. Figure 1.12 shows the curves obtained from electrolytic MnO_2 as curve *E*, a synthetic precipitated type imported from Germany before World War II as curve *S*, a

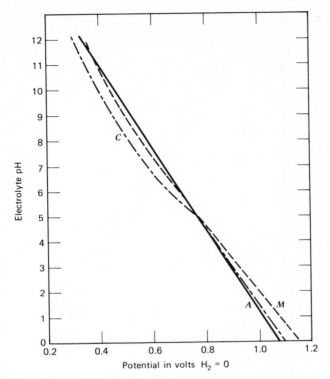

Figure 1.11 The potentials of natural MnO$_2$ from four sources versus pH of dry cell electrolyte. *A*, African ore; *M*, Montana ore; *C*, Java and Caucasian ore.

TABLE 1.4. A COMPARISON OF BATTERY PERFORMANCE WITH SELECTED POTENTIAL VALUES FOR THREE TYPES OF DEPOLARIZERS

| | Depolarizer Ore Source | | |
Battery Test	Africa	Montana	Java
Minutes service on 4 ohm cont test on a "D" size cell	539	465	359
Depolarizer potential at pH 9	0.52	0.50	0.46
Days service on 20 ohm light intermittent test on a No. 6 cell	370	440	425
Depolarizer potential at pH 4	0.83	0.86	0.84

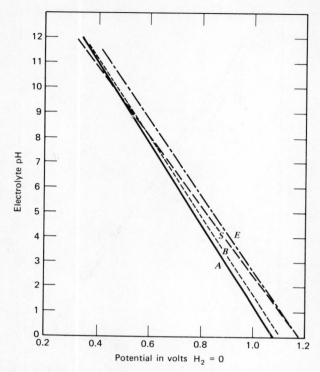

Figure 1.12   The potentials of three synthetic types of MnO_2 versus African ore in dry cell electrolytes. *A*, African ore; *E*, electrolytic; *S*, synthetic chemical; *B*, activated MnO_2.

beneficiated[1] natural ore as curve *B*, all compared to the African natural ore data as curve *A*. All the curves in Fig. 1.12 are straight lines with a slope approximating 0.06 similar to that of the natural African ore. The fact that both the electrolytic and the imported synthetic MnO_2 were used successfully for many years in commercial dry cell production has adequately established them as high-quality depolarizers. The beneficiated ore, as suggested by the position of curve *B*, has been found to possess a depolarizer value for many applications intermediate between that of the truly synthetic and the natural ore.

A number of workers have studied various aspects of the pH-potential relationship of MnO_2. McMurdie, Craig, and Vinal(16) obtained slopes of -0.12 and -0.06 for pH ranges of 2–6 and 6–12, respectively. They

---

[1]The beneficiated or activated ore was prepared by acid leaching Mn_2O_3, prepared from African ore, to form MnO_2 and a manganese salt. The MnO_2 was washed free of residual acid and salt and then dried(15).

recognized the role of the manganous ion in providing the -0.12 slope. Sasaki[17] examined MnO_2 from 24 different sources and found relationships with slopes from -0.064 to -0.089. Johnson and Vosburgh[18] offered the important suggestion that the slope was related to a surface phenomenon, the ion exchange property of the MnO_2. Kozawa[19] examined the surface properties of electrolytic gamma, as well as the alpha and beta varieties of MnO_2 in their original state and after heating at temperatures from $160°$ to $400°C$. He found that the potential was reduced by heating and concluded that an oxide of manganese, lower in valence than MnO_2, was formed on the surface and was responsible for the loss in potential. These findings confirmed earlier work by Kozawa and Sasaki[20] and Maxwell and Thirsk[21]. Kozawa found a similar reduction in potential when $MnSO_4$ solution was reacted with the surfaces of the various oxides.

Muller, Tye, and Wood[22] investigated the ion-exchange properties of various types of natural and synthetic MnO_2 using methods developed for organic ion-exchange resins. All samples of MnO_2 were first converted to the hydrogen form by ball milling the powdered MnO_2 with dilute acid, HCl or H_2SO_4, for periods of 23 days, although some tests were continued for as long as 56 days. This process insured that the particles were smaller than 45 microns in diameter. The product was washed with demineralized water until no acid reaction was observed. Potassium chloride solutions were used to measure the ion-exchange properties and it was found that the reaction was complete in 28 days. Table 1.5 shows the data obtained. The two samples prepared electrolytically and the activated sample (which is equivalent to the beneficiated MnO_2 described in an earlier paragraph) showed the highest cation exchange as expressed as m-equiv. g^{-1}. The authors related the ion-exchange capacity to the combined water content of the MnO_2 sample, with the higher ion-exchange capacity associated with the higher water contents.

Benson, Price, and Tye[23] extended this study to include the pH-potential relation of a sample of gamma MnO_2 of electrolytic origin which was converted to the acid form as described above. Measurements of the potentials developed by this sample in dilute electrolytes of HCl of various pH levels gave a slope of -0.058, in good agreement with theory. In electrolytes containing the cations Li, K, Ba, Ni, La, Cu, and Co, the slope increased from -0.069 to -0.104 in that order. In electrolytes containing manganous ions the slope was -0.122, as expected from theory. These authors conclude that a slope numerically greater than that required by the reaction

$$MnO_2 + H^+ + e^+ \rightarrow MnOOH \tag{1}$$

TABLE 1.5. CATION EXCHANGE IN 1.0 N KCL UNDER VARIOUS CONDITIONS

Material	pH	Cation Exchange (m-equiv. g^{-1})			
		23 Days Acid Treatment		46 Days Acid Treatment	
		28 Days Exchange	56 Days Exchange	28 Days Exchange	56 Days Exchange
Moroccan ore	12.7	0.05	0.05	0.05	0.05
	12.3	0.04	0.05	0.05	0.05
Synthetic	12.6	0.10	0.09		
	12.3	0.09	0.09		
Caucasian ore	12.4	0.21	0.22	0.22	0.22
	11.5	0.20	0.20	0.20	0.20
Electrolytic (source 2)	12.4	0.35		0.36	
	11.6	0.33		0.34	
Electrolytic (source 1)	12.3	0.42		0.43	
	11.0	0.39		0.40	
Activated ore	12.3	0.64	0.67	0.66	0.66
	11.3	0.60	0.61	0.62	0.62

will always be obtained when cations other than hydrogen are present in the external solution because of the ion-exchange properties of the sample. Their work indicates that ion-exchange is a surface phenomenon and that the amount is proportional to the surface area as proposed by Kozawa(19). Caudle, Summer and Tye(24) have studied the pH-potential relationships of the acid forms of both gamma and beta types of MnO_2 in both hydrochloric and sulfuric acid electrolytes. A slope of -0.059 V/pH was found in every case. However the potential values for the two types of MnO_2 were not identical, the relation for the gamma type being about 0.15 V higher than the corresponding value for the beta type at the same pH level. The presence of other cations in the electrolyte did influence the slope of the relationship, but the extrapolation of the potentials developed back to zero time of contact with the solution minimized these effects. In solutions containing $MnCl_2$ the acid forms of both the gamma and beta types of MnO_2 gave the slope of -0.118 V/pH in accordance with theory and the reaction

$$MnO_2 + Mn^{++} + 2H_2O \rightarrow 2MnOOH + 2H^+ \qquad (2)$$

However, extrapolation of the measured potentials back to zero time of contact with the electrolyte gave a slope of -0.059 V/pH unit. These authors concluded that the surface of the solid MnO_2 particles could be a factor, but the nature of the MnO_2 itself also influenced the rate of proton release in accord with earlier data by Korver, Johnson, and Cahoon[25].

In view of the extensive studies on the MnO_2 electrode, it seems certain that the electrochemical behavior of the electrode in the undischarged Leclanché cell operates in accordance with the reaction

$$MnO_2 + H^+ + e^- \rightarrow MnOOH \qquad (3)$$

The reader may wonder how, in view of all the factors (e.g., ion-exchange and interfering cations), did the early work of Cahoon[13] obtain a slope so nearly approaching the theoretical. One possible explanation would require that (a) the particular sample of African natural MnO_2 used was substantially in the acid form, and (b) that neither the ammonium nor zinc cations greatly influence the slope of the pH-potential curve.

The fact that the pH-potential curves for the various samples of MnO_2 do not intercept the potential axis at zero pH all at the same point, as shown in Figs. 1.11 and 1.12, still requires some explanation. Kozawa[19] has tried to relate the difference in potential between the gamma and beta types of MnO_2 to their difference in the onset of the major decomposition temperature on the differential thermal analysis test, that is, 480°C for gamma versus 580°C for beta. He calculates that this difference in decomposition temperature should account for a potential difference of 44 mV. In Fig. 1.13, a difference in the intercepts between electrolytic (gamma) and synthetic pyrolusite (beta) amounts to 160 mV. Based on the finding that the intercept of the pH-potential curve for synthetic pyrolusite (beta) at 1.01 V agrees so well with the calculated $E°$ for beta MnO_2 of 1.015 V, Cahoon[26] has reasoned that the differences in the intercepts of the curves for the various MnO_2 samples may well be the result of different free energies of formation. Thus, he has calculated the values given in Table 1.6 for seven MnO_2 samples studied. It is interesting to note that the equivalent free energies of formation so obtained, are not greatly different from that of the synthetic pyrolusite, the greatest difference being 3.5 kcal.

The behavior of the MnO_2 electrode in solutions containing $MnCl_2$ is definitely associated with cell discharge and will be discussed in some detail later in this chapter under the heading "Reactions in the Cathode."

THE ANODE

The anode of the early Leclanché cells consisted of a rod of metallic zinc arranged so that the upper end provided the negative terminal of the cell and the lower end contacted the electrolyte. In the modern counterpart, zinc is still used as the anode, though usually in the form of a sheet of

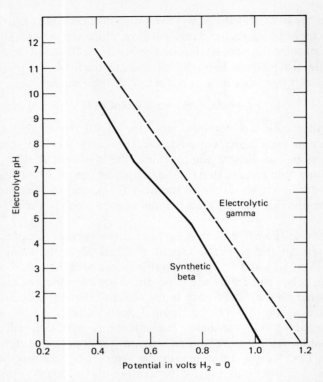

Figure 1.13 A comparison of the pH-potential curves of electrolytic gamma and synthetic beta type MnO_2.

TABLE 1.6. THE EQUIVALENT FREE ENERGIES OF FORMATION FOR VARIOUS SAMPLES OF MANGANESE DIOXIDE

		E_0 Values		ΔF^0 Values	
MnO_2 Sample	Crystal Type	Calc'd	From pH-E Curve	Literature	Calc'd from pH-E Curve
Synthetic Pyrolusite	Beta	1.015	1.01	−111.1 kcal	—
Electrolytic	Gamma	—	1.175	—	−107.6 kcal
Synthetic chemical	Delta	—	1.175	—	−107.6 kcal
Beneficiated	Alpha	—	1.095	—	−109.3 kcal
Ghana	Rho	—	1.070	—	−109.8 kcal
Caucasian	Beta	—	1.085	—	−109.5 kcal
Montana	Rho	—	1.128	—	−108.5 kcal

metal shaped to conform to the design of the remainder of the cell components. In spite of competition from other metals, chiefly magnesium and aluminium, zinc is still the most widely used anode metal in primary cells. The reasons for this are the desirable electrochemical properties, low cost, and suitable physical characteristics of zinc. In a comparison of the electrochemical properties of the three metals, zinc has a number of advantages. Most important of these is the high hydrogen overvoltage, which is actually increased by amalgamation. Moreover, the wasteful corrosion of zinc is usually only a fraction of that generally found with the other materials. The anodic attack is uniform and the operating potential is reasonably high. In making any such comparison, one must recognize that zinc has been used commercially in primary cells for over a century whereas both aluminum and magnesium are relative newcomers to this field.

The potential of zinc in zinc chloride solutions has been reviewed by a number of workers, for example, Getman(27). Unfortunately for the battery industry, most such studies are limited to concentrations below 1.5 molar and do not extend into the region of concentrated electrolytes often encountered in battery operation.

The potentials developed by an amalgamated zinc electrode in a series of zinc chloride solutions, ranging in concentration from 5.04 to 223.21 g $ZnCl_2/100$ g H_2O, are shown in Table 1.7 and Fig. 1.14(28). These values are useful in any consideration of the system in which only zinc chloride solution is used as an electrolyte. They are also applicable to the Leclanché system when the anolyte, after severe discharge, is being studied.

The electrolyte used in Leclanché cell construction is an aqueous solution of zinc chloride in which a considerable amount of ammonium chloride has been dissolved. It is usually near or at saturation with respect to the latter salt. In addition, some undissolved ammonium chloride may be present. In such solutions the zinc electrode develops a potential that is more anodic than that found when ammonium chloride is absent. Takahashi et al.(10) investigated the potentials of an amalgamated zinc electrode in both saturated and unsaturated electrolytes of this system. Cahoon(26) also measured a similar set of potentials in a series of solutions containing from 5.2 to 183 g $ZnCl_2/100$ g H_2O with the data shown in Table 1.8. The potential data are shown in Fig. 1.15 along with those of Takahashi. All values are in satisfactory agreement.

The data given in Fig. 1.15 may be charted as in Fig. 1.16 where the pH of the electrolyte is plotted against the potential of the zinc electrode. The relationship between these properties, when the potential is expressed on

TABLE 1.7. THE RELATION BETWEEN THE POTENTIAL OF AN AMALGAMATED ZINC ELECTRODE AND THE CONCENTRATION AND pH OF A SERIES OF ZINC CHLORIDE SOLUTIONS

| | | Potential of Zinc Electrode | | | |
| | | Measured | Calculated | | |
Solution No.	Concentration of ZnCl₂ g/100 g H₂O	Versus Sat'd. Calomel	Versus Standard H₂ᵃ	pH of Electrolyte	Molality of Electrolyte
1	5.04	1.03078	0.7871	5.85	0.37
2	10.00	1.01520	0.7715	5.65	0.734
3	20.99	0.99627	0.7526	5.20	1.54
4	30.11	0.98363	0.7399	4.85	2.21
5	41.22	0.96696	0.7233	4.45	3.03
6	52.14	0.95061	0.7069	4.20	3.83
7	62.14	0.93466	0.6910	3.95	4.56
8	70.76	0.92378	0.6800	3.45	5.19
9	80.38	0.90441	0.6607	3.15	5.90
10	90.16	0.88764	0.6439	2.80	6.62
11	103.16	0.86898	0.6253	2.42	7.58
12	111.75	0.85676	0.6131	2.15	8.20
13	121.78	0.84039	0.5967	1.85	8.94
14	144.26	0.80675	0.5631	1.32	10.595
15	164.89	0.78574	0.5420	0.82	12.09
16	182.49	0.76775	0.5241	0.50	13.40
17	212.84	0.73493	0.4912	0.13	15.61
18	223.21	0.73383	0.4901	0.	16.40

ᵃOn the basis that the value of the saturated type calomel half-cell is 0.2437 V positive to the standard Hydrogen electrode at 27°C.

the basis of the standard Hydrogen electrode is

$$E_{Zn} = -0.465 - 0.0733 \text{ pH in the range pH 1.33 to 3.85} \tag{4}$$

$$E_{Zn} = -0.392 - 0.0916 \text{ pH in the range pH 3.85 to 5.00} \tag{5}$$

It is unfortunate that the potential of the zinc electrode in battery electrolytes cannot be calculated easily from thermodynamic data, presumably because of the formation of complex ions such as $ZnCl_4^=$ in the zinc chloride solutions and $Zn(NH_3)^=$ in the zinc and ammonium chloride electrolytes.

Electrolytes have been specifically formulated for cells to operate at temperatures below the freezing point (−6°F; −21°C) of the conventional $ZnCl_2$-NH_4Cl-H_2O system. The reader is referred to Chapter 10 where this aspect is discussed in detail.

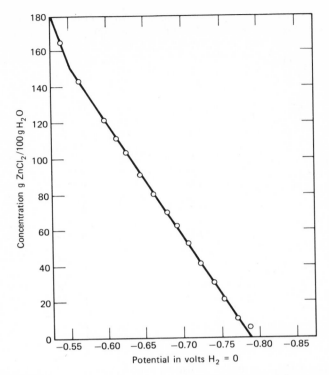

Figure 1.14   The relation between the potential of a zinc electrode and the concentration of a zinc chloride solution.

CELL VOLTAGE

Many battery technologists have attempted to calculate the open circuit voltage of a Leclanché cell from thermodynamic data and the cell reaction. Drucker and Heuttner(29) made such a calculation based on the reaction

$$\text{Zn} + 2\text{MnO}_2 + 2\text{NH}_4\text{Cl(aq.)} \rightarrow 2\text{Mn}_2\text{O}_3 + (\text{ZnO}, 2\text{NH}_4\text{Cl(aq.)}) \qquad (6)$$

They obtained values ranging from 1.32 to 1.41 even when the temperature coefficient was included. Thompson(30) reported calculations made by Kaneke(31) which gave 1.605 for the equation

$$\text{Zn} + 2\text{MnO}_2 + 2\text{NH}_4\text{Cl} \rightarrow \text{Zn(NH}_3)_2\text{Cl}_2 + \text{Mn}_2\text{O}_3\text{H}_2\text{O} \qquad (7)$$

In the absence of data some estimates were used. Thompson, using a slightly different approach, obtained a value of 1.30, but he recognized

TABLE 1.8. THE POTENTIALS OF AMALGAMATED ZINC IN SOLUTIONS OF
ZINC CHLORIDE OF VARIOUS CONCENTRATIONS WHICH ARE SATURATED
WITH AMMONIUM CHLORIDE

Solution No.	g $ZnCl_2$ per 100 g H_2O	pH	Measured EMF	EMF^a $H_2 = O$	Molality $ZnCl_2$
1	5.24	4.94	1.09944	0.8538	0.385
2	11.00	4.86	1.08899	0.8433	0.808
3	17.45	4.84	1.08129	0.8356	1.28
4	24.72	4.77	1.07350	0.8278	1.81
5	32.93	4.69	1.06743	0.8217	2.42
6	42.27	4.61	1.06136	0.8157	3.10
7	53.12	4.54	1.05527	0.8096	3.91
8	65.77	4.40	1.04122	0.7955	4.82
9	80.28	4.20	1.02095	0.7753	5.89
10	98.30	3.89	0.99501	0.7493	7.22
11	119.92	3.52	0.96524	0.7195	8.80
12	146.19	2.47	0.88511	0.6394	10.75
14	110.00	3.65	0.97794	0.7322	8.08
15	135.50	2.92	0.92592	0.6802	9.95
16	163.00	1.98	0.85329	0.6076	11.96
17	183.00	1.33	0.80742	0.5617	13.44

[a]The value of the saturated type calomel electrode was taken at 0.2457 V at 23°C (21).

that important discrepancies in the thermodynamic data left some questions unanswered. Drotschmann(32) tried to correct the equations used by Thompson and Drucker to show $ZnCl_2 2NH_3$ as the reaction product. He obtained a value of 1.41 V. Latimer(33) used the reaction

$$MnO_2 + NH_4^+ + 2H_2O + 2e \rightarrow Mn(OH)_3 + NH_4OH \qquad (8)$$

to arrive at a value of 1.2 V. In view of the fact that none of these workers obtained realistic values, that is, in the 1.5 to 1.8 V range characteristic of actual fresh cells, a somewhat different approach was indicated(26).

Previously, a complete discussion of the relationship between the potential of MnO_2 and the electrolyte pH was presented. Table 1.9 shows appropriate equations for this relationship in the range of pH zero to pH 5 for seven different types of MnO_2.

In the discussion of anodes the measurement of the zinc anode potential and its dependence on the concentration of both zinc and ammonium chlorides were described. These data were graphically shown in Figs. 1.14, 1.15 and 1.16. Both the cathode and anode potentials can be conveniently plotted against the electrolyte pH as shown in Fig. 1.17 where data for Ghana ore have been chosen for the example. The

---

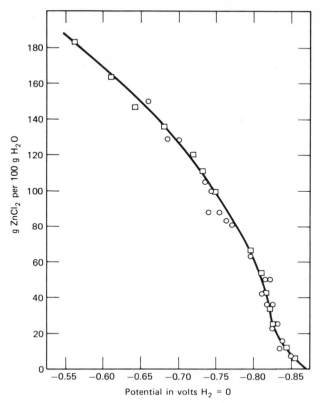

Figure 1.15 The relation between the potential of a zinc electrode and the zinc chloride concentration in aqueous solutions saturated with ammonium chloride. Data from Takahashi, ○(10); Cahoon, □(26).

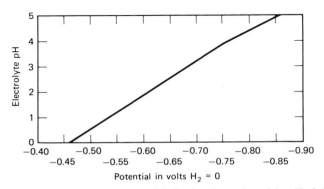

Figure 1.16 The relation between the potential of a zinc electrode and the pH of zinc chloride solutions which are saturated with ammonium chloride.

TABLE 1.9. THE pH-POTENTIAL RELATION FOR SEVEN
TYPES OF MnO₂

Manganese Oxide Type	Equation for pH Range 0–5
Synthetic pyrolusite	$E = 1.015–0.054\,pH$
Electrolytic MnO_2	$E = 1.175–0.06\,pH$
Synthetic chemical	$E = 1.175–0.072\,pH$
Beneficiated	$E = 1.095–0.063\,pH$
African or Ghana ore	$E = 1.070–0.0608\,pH$
Caucasian ore	$E = 1.085–0.063\,pH$
Montana ore	$E = 1.128–0.068\,pH$

horizontal lines at unit pH intervals show the expected open circuit voltage of a freshly made cell utilizing the electrolyte at that particular pH value. Table 1.10 shows a comparison of the actual potentials measured in experimental cells with the initial open circuit voltages for a group of MnO_2 types. The total of the two potentials in any individual cell agrees reasonably well with the actual measured voltage.

The early efforts to calculate the initial cell voltage appeared to have failed for two reasons:

1. They did not take into account the dependence of the cathode potential on the electrolyte pH;
2. They used a calculated value rather than a measured one for the anode potential and did not recognize its dependence on the concentration of salts in the electrolyte.

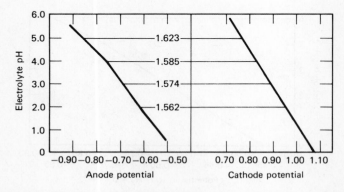

Figure 1.17 A graphical method of estimating cell voltage from the separate potential-pH relations for Ghana ore and zinc electrodes.

TABLE 1.10. A COMPARISON OF THE ACTUAL OPEN CIRCUIT VOLTAGES WITH THOSE OBTAINED BY ADDING THE POTENTIALS OF THE TWO ELECTRODES

| MnO₂ Type | Electrolyte pH | Electrode Potentials | | | Actual OCV |
		Zn	MnO₂	Total	
Synthetic pyrolusite	4.45	−0.800	0.775	1.575	1.55
Electrolytic MnO₂	3.90	−0.748	0.976	1.724	1.74
Caucasian ore	4.45	−0.800	0.800	1.600	1.61
Montana ore	4.45	−0.800	0.815	1.615	1.61–1.63
Beneficiated MnO₂	4.45	−0.800	0.820	1.620	1.66
Ghana ore	4.08	−0.796	0.910	1.697	1.68

REACTIONS IN THE ELECTROLYTE

It is usually assumed that the dissolved electrolyte in a reasonably fresh cell is the same composition throughout the system. This composition may be represented by a point D on the triangular diagram as shown in Fig. 1.18. Generally, this electrolyte contains undissolved ammonium chloride as indicated by the appearance of point D in the area of solution

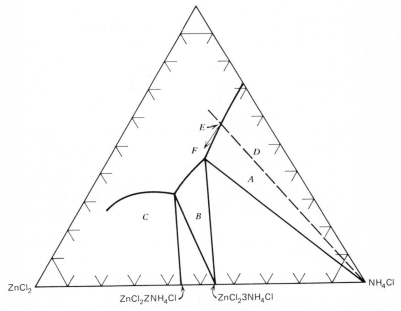

Figure 1.18   The electrolyte system $ZnCl_2$–NH_4Cl–H_2O at the beginning of cell discharge.

plus NH_4Cl salt. The composition of the liquid electrolyte is then shown by point E, where the dotted straight line drawn through the NH_4Cl apex of the diagram and through point D intercepts the saturated solution line.

Once this cell is placed on discharge the electrolyte compositions in the various parts of the cell are changed because of the by-products of the reactions involved and by the consumption of some electrolyte salts. This section deals with examples of both types.

It is easy to assume that ammonium chloride is the active electrolyte salt in the system and that it dissociates as

$$NH_4Cl \rightarrow NH_4^+ + Cl^- \tag{9}$$

At the anode the electrochemical reaction is simple; zinc is formed electrochemically to produce zinc chloride.

$$Zn + 2Cl^- \rightarrow ZnCl_2 + 2e^- \tag{10}$$

The electrolyte at the anode surface, originally of composition E in Fig. 1.18, starts to shift in the direction indicated by the arrow to some point F, that is, toward the $ZnCl_2$ apex of the diagram. The amount of divergence from point E depends on the intensity of the current, the time of discharge, and the extent of diffusion of the $ZnCl_2$ produced into the separator and even into the cathode. Point F can be established by an actual analysis of the electrolyte at the anode surface. However, it is clear that the system $ZnCl_2$-NH_4Cl-H_2O can show the compositions of both the initial electrolyte in the cell and that generated at the anode during discharge.

The electrolyte reactions at the cathode are more complex in that ammonia is generated as a cathodic by-product. Tomassi(34) refers to still earlier work, for example, Priwoznik(35), which identified the ammine, $ZnCl_2 2NH_3$ and the oxychlorides, $ZnCl_2 3ZnO 4H_2O$ and $ZnCl_2 6ZnO 10H_2O$, as products of the cathode reactions. Daniels(36) reviewed the literature with special emphasis on the electrolyte equilibrium and its effect on cell behavior. Friess(37) identified the oxychloride as $ZnCl_2 4ZnO$ and the ammine as $ZnCl_2 2NH_3$ and showed that the composition of the $ZnCl_2$-NH_4Cl-H_2O electrolyte to which ammonia was added influenced the product composition. He indicated the pH levels attained in such reactions.

The generation of ammonia in the cathode during cell discharge is shown in the reaction

$$2MnO_2 + 2NH_4^+ + 2e^- \rightarrow 2MnOOH + 2NH_3 \tag{11}$$

When ammonium hydroxide is added slowly in relatively small amounts to aqueous solutions containing zinc chloride or zinc and ammonium chlorides, a point is soon reached at which a permanent precipitate occurs. Continued addition of ammonia causes first an increase and then a decrease in the amount of precipitate and, finally, complete solution of the precipitated phase. The mechanism of the change occurring on the addition of ammonium hydroxide can be studied by following the changes in pH of the suspension. The pH changes obtained on adding ammonia to a typical zinc chloride-ammonium chloride solution are shown in Fig. 1.19. The amounts of ammonia added to produce two inflection points or cusps in the curve correspond approximately to the calculated volumes required for complete conversion of all the zinc chloride present to di-ammino and tetra-ammino zinc chloride, $ZnCl_22NH_3$ and $ZnCl_24NH_3$. Thus the compound first precipitated by ammonia is the slightly soluble $ZnCl_22NH_3$ which by further reaction with ammonia is converted to the more soluble $ZnCl_24NH_3$. The formation of these compounds may be represented thus;

$$ZnCl_2 + 2NH_4OH \quad\quad \rightarrow ZnCl_22NH_3 + 2H_2O \quad\quad (12)$$

$$ZnCl_22NH_3 + 2NH_4OH \rightarrow ZnCl_24NH_3 + 2H_2O \quad\quad (13)$$

Depending on the composition of the solutions used, a further addition of ammonia may be required for complete solution of the precipitate. The above data suggest that the ammine salts formed by such reactions serve as the buffering agents to reduce the pH rise in the cell electrolyte as ammonia is developed during cell discharge.

Table 1.11 shows the amounts of ammonia required for saturation and complete solution end points, and significant pH values for all solutions.

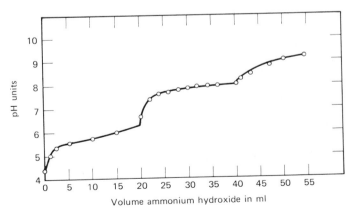

Figure 1.19 The change in pH of a typical aqueous zinc chloride-ammonium chloride solution on the addition of ammonium hydroxide.

TABLE 1.11. THE $ZnCl_2$-NH_4Cl-H_2O-NH_3 SYSTEM AT 20°C

At Saturation				At Complete Solution				pH		
% $ZnCl_2$	% NH_4Cl	%H_2O	g. NH_3 per 100 g. Sol.	% $ZnCl_2$	% NH_4Cl	%H_2O	g. NH_3 per 100 g. Sol.	Orig. Sol.	At Saturation	At Complete Solution
54.16	—	43.84	1.01	30.15	—	69.85	17.44	2.00	2.81	10.70
				15.37	—	54.63	9.27	5.08	—	10.54
17.91	9.09	73.00	1.43	14.67	7.44	77.89	7.77	5.57	5.92	8.25
16.56	17.10	66.34	0.90	13.58	14.03	72.39	7.31	4.82	5.68	8.06
15.29	23.82	60.89	0.69	12.77	19.89	67.34	6.60	4.40	5.45	7.91
13.78	24.52	61.71	0.70	11.73	20.87	67.40	6.04	4.57	5.70	7.95
11.38	20.77	67.84	0.62	9.98	18.22	71.80	5.13	4.75	5.76	7.91
8.85	16.08	75.07	0.56	8.00	14.55	77.45	3.99	4.75	5.81	7.87
19.07	26.62	54.31	0.82	15.09	21.06	63.85	8.28	4.55	5.41	8.20
25.63	28.25	45.12	0.76	17.63	19.43	62.94	11.91	4.38	5.39	9.21
40.52	24.13	35.35	1.70	26.13	15.56	58.30	14.04	3.35	4.78	8.63
				36.91	—	63.09	22.98		—	10.81
28.06	26.24	45.70	1.16	19.79	18.50	61.71	11.56	4.00	5.06	8.65
				24.35	—	75.65	14.14	3.86	—	10.63
44.59	13.39	42.02	2.65	28.89	8.68	62.43	14.71	2.73	4.69	8.50
32.47	9.75	57.77	2.25	23.48	7.05	69.47	11.73	3.84	5.38	8.31
9.27	9.65	61.07	0.94	8.50	8.85	82.65	3.89	5.03	6.23	8.00
				8.53	2.66	88.50	4.97	5.44	5.58	9.32
4.70	26.73	68.58	0.38	4.47	25.44	70.09	2.12	4.96	6.16	7.70
9.43	26.77	63.80	0.52	8.30	23.56	68.14	4.67	4.89	5.91	8.14
				20.66	—	79.34	12.18	4.51	—	10.77

23.69	19.12	57.19	1.25	17.97	14.51	67.53	9.65	4.25	5.46	8.24
23.55	9.29	67.18	1.48	18.33	7.24	74.44	9.11	4.78	5.65	8.34
33.34	19.35	47.31	2.89	23.00	13.35	63.66	12.02	3.94	4.88	8.51
36.85	9.50	53.66	2.22	25.42	6.55	68.02	12.71	3.52	5.04	8.31
45.96	9.69	44.35	2.32	28.88	6.09	65.03	14.86	2.55	4.34	8.31
42.76	18.48	38.82	2.16	27.21	11.73	61.07	14.49	2.72	4.43	8.18
53.71	9.25	37.04	3.38	31.70	5.46	62.84	16.75	1.27	3.73	8.49
32.45	25.03	42.52	1.14	22.50	17.38	60.16	11.85	3.91	4.98	8.03
4.79	19.66	75.55	0.55	4.56	18.73	76.71	2.22	5.20	6.22	7.59
4.78	11.75	83.47	0.60	4.57	11.23	84.20	2.15	5.00	6.45	7.65
4.84	4.99	90.16	0.09	4.49	4.65	90.88	2.73	5.31	5.93	8.84
19.27	4.97	75.76	0.44	14.93	3.85	81.22	6.47	4.82	5.44	9.06
28.67	4.94	66.39	0.80	19.91	3.43	76.66	11.60	4.55	5.15	9.23
				26.58	15.98	57.44	16.80	—	—	9.15
				25.47	17.06	57.47	20.95	—	—	9.62

The term "saturation" is used to indicate the point at which a permanent precipitate first forms, and the term "complete solution" for the point at which the last traces of precipitate redissolve.

The amounts of ammonia required for saturation and the type of complex formed at each precipitation point are shown in Fig. 1.20. The results are in satisfactory agreement with those of Friess. Analyses have shown that the basic compound identified by Friess as $ZnCl_24ZnO$ has a composition approximating $ZnCl_24Zn(OH)_2XH_2O$.

Zinc oxide is slightly soluble in zinc chloride solutions and its presence in the zinc and ammonium chloride solutions may be considered the same as an equivalent amount of ammonia. Thus from Table 1.11 and Fig. 1.20 it appears that zinc oxide would react in two ways, depending on the concentration of the dissolved chlorides. In the range of solutions in which $ZnCl_22NH_3$ is the precipitated phase, the reaction probably is

$$ZnO + 2NH_4Cl \rightarrow ZnCl_22NH_3 + H_2O \tag{14}$$

In the range in which $ZnCl_24Zn(OH)_2$ is the precipitated phase the reaction probably is;

$$4ZnO + 4H_2O + ZnCl_2 \rightarrow ZnCl_24Zn(OH)_2 \tag{15}$$

Microscopic examination of the precipitate formed at saturation furnished a ready means of determining whether it was the crystalline $ZnCl_22NH_3$ or the more amorphous $ZnCl_24Zn(OH)_2$. The data of

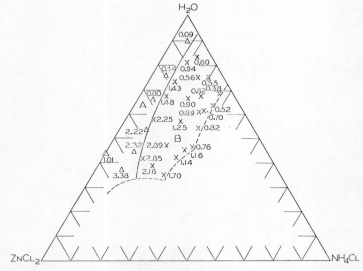

Figure 1.20 The system $ZnCl_2–NH_4Cl–H_2O–NH_3$ at 20°C. The solid phases are indicated by areas A and B, corresponding respectively to the compounds $ZnCl_24Zn(OH)_2$ (experimental points designated by triangles), and $ZnCl_22NH_3$ (crosses). The numbers indicate the amounts of ammonia, (g NH_3 per 100 g solution) required for saturation.

McMurdie(38) on the precipitated phases of the system $ZnCl_2$-NH_4Cl-ZnO-H_2O agree with the data in Table 1.11 so as to indicate the correctness of the above reactions.

The solubility of $ZnCl_2 2NH_3$ at a variety of temperatures has been reported by Takahashi(39). The solubility increases with rising temperature, a phenomenon possibly linked to the well-known fact that batteries provide a greater output when the temperature is raised. Additional data on the formation mechanism of the zinc ammines and their heats of formation have also been reported by Takahashi(40).

Higher ammines of zinc chloride have been described(41) with formulas of $ZnCl_2 6NH_3$ and $ZnCl_2 10NH_3$, but there is little evidence that these compounds are developed in dry cells under normal discharge conditions.

REACTIONS AT THE CATHODE

Electrochemical theory, particularly regarding the development of electromotive force, was well advanced by 1900, yet the early battery technologists were slow to apply its teachings to Leclanché cells. Polarization was frequently considered loosely in terms of the cathodic liberation of hydrogen(42), though how the latter could be visualized as proceeding at the potential of the MnO_2 electrode or the operating voltage of the cell is not immediately clear. Depolarization, as defined by Leblanc(43), is at best only a means of preventing evolution of hydrogen or oxygen; it has all too often been considered, not as the primary electrode process, but as a chemical reaction between MnO_2 and previously liberated hydrogen. There was little basis for this view(44), which regrettably is still found in many books on general chemistry.

Tomassi(34), in his text on primary batteries, refers to earlier work in which various reaction products of the Leclanché wet cell were identified, among them the ammine $ZnCl_2 2NH_3$(45), and oxychlorides of the type $ZnCl_2 3ZnO4H_2O$ and $ZnCl_2 6ZnO10H_2O$. He wrote the equation for the overall cell reaction essentially in its modern form:

$$Zn + Mn_2O_4 + 2NH_4Cl \rightarrow ZnCl_2 2NH_3 + Mn_2O_4H_2 \text{ (i.e., } Mn_2O_3H_2O) \qquad (16)$$

and cited conditions under which Mn_2O_4Zn (hetaerolite?) might be the reaction product.

It should be recognized that the efforts at identifying individual products of the cell reaction must have been based on isolating these materials and subjecting them to analysis, probably of the gravimetric type. Techniques such as x-ray diffraction, which has been used extensively in the study of the cathode, were generally not available until about

1940. In view of this situation, we must acknowledge that shrewd reasoning probably played an important role in the early efforts. However much as the more recent work has served to confirm earlier findings, it has also extended the impact of chemistry and electrochemistry on the practical aspects of battery technology.

It is well known that the discharge of a dry cell cathode at a relatively heavy drain produces ammonia which can be readily identified by its odor. From the studies of the electrolyte system(9) it was apparent that a strong odor of ammonia was associated with an electrolyte pH of about 8 or higher and the presence of $ZnCl_2 4NH_3$. When the rugged type of glass electrode for measuring pH became available, it was immediately applied(46) to determine the pH of the dry cell cathode during discharge. A series of "D" size cells were placed on a 4 ohm drain at 21°C. Periodically representative cells were removed, the tops of the cells dismantled to expose the cathode, and the central carbon electrode withdrawn. The central section of the cathode, I, concentric with the zinc can, was removed with a $\frac{1}{2}$-inch diameter (1.27 cm) cork borer, and a middle section, II, with a $\frac{3}{4}$-inch diameter (1.90 cm) corkborer, both of which had been chrome plated to protect against corrosion. The zinc can was then split and the remaining part of the mix, III, and the paste layer, IV, were removed as separate samples. Measurements of the pH of the three mix samples and the paste layer were made as rapidly as possible. The data are shown in Fig. 1.21. The data show the surprising fact that the innermost and middle layers of the cathode were the first to show a rise in pH, thus indicating that the reaction probably began at this location in the cathode. The pH in the outermost layer of the cathode soon followed along in the pH rise, indicating that all parts of the mix bobbin participated in the discharge. As the test proceeded the pH levels of all three sections of the cathode rose appreciably. These findings are significant in that they tended to dispel the previously held belief that the outermost part of the cathode reacted first due to its location nearest the anode. The paste layer pH fell appreciably during the test as a result of the accumulation of acidic $ZnCl_2$.

In 1919 French and MacKenzie(47, 14) discovered that the cathode discharge in a dry cell proceeded in two steps as represented by the reactions

$$MnO_2 + 2H^+ + 2e^- + 2NH_4Cl \rightarrow MnCl_2 + 2NH_3 + 2H_2O \qquad (17)$$

$$MnO_2 + MnCl_2 + NH_4OH \rightarrow Mn_2O_3 + H_2O + 2NH_4Cl \qquad (18)$$

The first reaction is the electrochemical reduction of Mn^{4+} to Mn^{2+} and since it is electrochemical it operates only when current is drawn from the

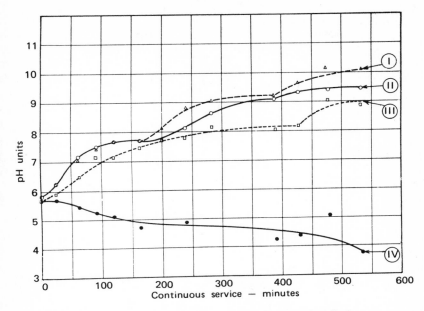

Figure 1.21 The pH changes in a "D" size cell during discharge.

system. The second reaction is chemical and it may operate during both discharge and rest periods. French and MacKenzie also found that the first reaction could be studied separately from the second by discharging an MnO_2 sample while at the same time supplying it with a continuous supply of fresh electrolyte to wash out the $MnCl_2$ and the $ZnCl_2 2NH_3$ formed in the reaction. This reaction was later used by Cahoon(14) as the basis for an evaluation test for MnO_2 and a wide variety of samples were studied. The apparatus is shown diagramatically in Fig. 1.22. The cathode sample consisted of 0.7100 g of the MnO_2 under test and 0.1130 g of acetylene black, ground together in a mortar, and placed in the apparatus at K. A continuous supply of electrolyte was provided by a pump which gave an output of 2.0 cm^3/min. The electrolyte selected was an aqueous solution containing both $ZnCl_2$ and NH_4Cl which had a pH of 3.85. The potential of the cathode was measured periodically during the test by reading the potential against a reference electrode immediately after the discharge circuit was momentarily opened. This method effectively eliminated the contribution of the IR often included in cell readings. The test that was designated the "utilization test" provided a discharge curve as shown B in Fig. 1.23 in comparison to A obtained on a "D" cell discharge on a 4 ohm LIF test. In both instances African ore was used.

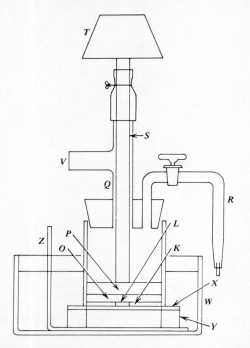

Figure 1.22 A cross-sectional diagram of the cell used in the utilization test for MnO$_2$. K, the depolarizer sample mixed with carbon; L, an insulating button; O, a layer of powdered graphite; P, a disc of graphite with vertical holes for electrolyte flow; Q, glass tee; R, side neck for connection to reference cell; S, carbon rod for exerting mechanical pressure on to the cathode; T, weight applied to rod S; V, connection to electrolyte supply; W, glass dish to hold electrolyte; X, nichrome screen; Y, insulating ring; Z, zinc electrode. A cellulosic separator layer was placed between the cathode mix layer K and the nichrome screen X. At first, readings were made manually but when interrupter equipment became available, the latter was used. No significant difference between results resulted from this change.

Since the same weight of contained MnO$_2$, that is, 0.7100 g, was used in each utilization test the theoretical capacity was 244 min at the constant current of 0.108 amp used. A cutoff potential of 0.26 V was used, a value that appeared to be quite close to that in an actual cell when it reached the cutoff. The result was expressed as a "utilization factor," that is, the percentage of the theoretical capacity actually obtained. Table 1.12 shows some representative values determined with this test. It is clear that a wide range of results are obtained with different MnO$_2$ samples. Although this test was extremely useful in studying various aspects of the behavior of MnO$_2$ under a wide variety of electrolyte and other conditions, it did not become more than a test to evaluate the first of the two cathode reactions. Some MnO$_2$ samples were found which gave high utilization

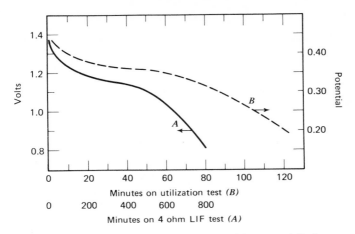

Figure 1.23 A comparison of a utilization test, curve *B*, with an actual discharge curve for a "D" size cell, curve *A*.

factors but poor performance in actual cells. Invariably these materials gave low recuperation rates as described in the following section.

Having established that the pH-potential relationship for the undischarged cathode follows, at least in a general way, that predicted from the Nernst equation, the question may well be raised regarding what develops in this relationship during cell discharge. The development of the rugged type of glass electrode made pH measurements possible on cathodes taken from operating cells. When these measurements were combined with potential data, it became possible to follow this equilibrium.

A series of such measurements(48) were made on "D" size flashlight units, taken at intervals of a few minutes, from a standard 4 ohm continuous test. The data are presented in Fig. 1.24 and Table 1.13. The figure compares the pH-potential curve for the undischarged Ghana ore with that obtained in this series of cells during discharge. It is clear that

TABLE 1.12. A COMPARISON OF THE UTILIZATION
FACTORS OF FOUR REPRESENTATIVE MnO_2
SAMPLES

Sample	Utilization Factor, %
African ore	40
Electrolytic	70–76
Imported chemical MnO_2	78
Activated African	50

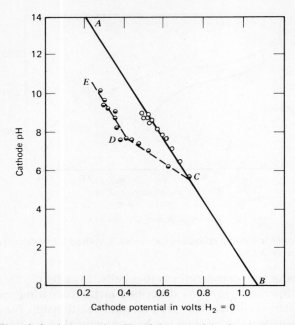

Figure 1.24 The relation between the pH and the potentials of cathode samples taken from cells discharged on 4 ohm continuous test. Samples taken initially, line *CDE* ◐, those after 1 week storage ○, line *AB* shows the relation for the undischarged African ore.

the catholyte pH rose from 5.7, characteristic of the fresh cell, to 10.1, found in the cells at the end of the discharge. Furthermore the potentials found in the cathodes of the discharged cell are lower than those in the undischarged cathode by as much as 0.17–0.21 V in the pH range 8.1 to 10.1. Thus, in a cell discharge in which the working voltage falls from 1.40 to 0.75 V, a drop of 0.65 V occurs; of this amount, 0.44 V or 67.8 percent of this decrease is accounted for in the cathode. Furthermore, 0.26 V or 40 percent of the total drop in voltage is due to a rise in cathode pH alone, while 0.18 V or 27.7 percent of the total must be caused by another factor. It was well known to French and MacKenzie and their associates(47) that MnCl₂ appeared in the cathode of a discharged cell, though this finding was not general knowledge for some time(49). Measurements of the pH-potential relationship of a Ghana ore in a series of dry cell electrolytes containing added MnCl₂(48) gave a clue to the important effect this material had on the potential. The data are shown in Fig. 1.25. It is clear from this figure that the presence of the manganous ion in the electrolyte reduces the potential of the MnO₂ electrode, especially in the pH range above 5.4. Thus, in that pH range the manganous ion is a chemical

TABLE 1.13. POTENTIAL AND pH VALUES AND THE COMPOSITIONS OF SAMPLES OF CATHODE MIX TAKEN DURING CELL DISCHARGE

Minutes on Test	Sample Composition[a]	Immediately after Opening Cell		1 Week Later	
		pH	Potential[b]	pH	Potential[b]
0	1.950	5.75	0.720	5.75	0.720
25	1.941	6.20	0.619	6.55	0.690
63	1.929	7.06	0.525	7.22	0.648
92	1.914	7.42	0.473	7.84	0.612
121	1.904	7.67	0.437	7.88	0.590
165	1.903	7.69	0.403	8.06	0.570
200	—	8.12	0.363	8.62	0.547
240	1.877	8.80	0.357	8.54	0.529
285	1.873	9.05	0.355	8.89	0.514
390	1.840	9.22	0.315	8.62	0.511
430	—	9.64	0.297	8.62	0.501
480	1.836	9.40	0.291	8.62	0.504
540	1.818	10.10	0.284	9.00	0.480

[a] Value tabulated is the value of X in the formula MnO_x.
[b] Potential values are reported with reference to the standard hydrogen electrode.

polarizing agent and is responsible for the 0.26 V drop in potential discussed above and shown in Fig. 1.24.

French and MacKenzie(47) discovered that the cathode reaction occurs in two steps. The first is the electrochemical reduction of Mn^4 to Mn^2 as represented by the reaction

$$MnO_2 + 2H^+ + 2e^- + 2NH_4Cl \rightarrow MnCl_2 + 2NH_3 + 2H_2O \qquad (19)$$

The second reaction is a chemical one between the $MnCl_2$ produced above and the remaining MnO_2 thus:

$$MnO_2 + MnCl_2 + NH_4OH \rightarrow Mn_2O_3 + H_2O + 2NH_4Cl \qquad (20)$$

Since the availability of x-ray diffraction the product of this reaction has been identified as $Mn_2O_3H_2O$ or $MnOOH$ instead of the Mn_2O_3 given above. This theory is confirmed to some extent by the fact that samples of the cathode mix taken from the discharged cells and stored in closed jars for a week after completion of the test resulted in the data shown in Table 1.13 and Fig. 1.24. Here the potentials of the stored samples are approximately the same level as those of the undischarged MnO_2, line AB on Fig. 1.24. It might be thought that the rise in potential could have been due to air oxidation of the manganous ion, but separate measurements on samples of mix stored in a nitrogen atmosphere and in discharged but

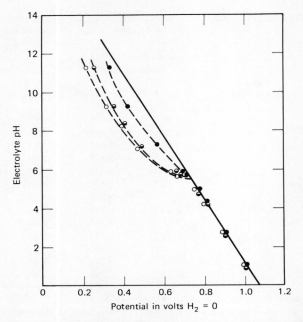

Figure 1.25   The relation between the pH and the potential of African MnO_2 in dry cell electrolytes containing added $MnCl_2$. The full line shows the relation when no $MnCl_2$ is present; ●, ◓, ○ show the relations when 0.14, 1.43 and 7.15 g $MnCl_2 4H_2O$ per liter were added respectively.

unopened cells gave substantially the same results. Thus the rise in potential during a one-week delay appears to have permitted the above reaction to remove the manganous ion from the cathode.

To follow this reaction more closely, analytical methods were devised [Cahoon, Johnson, and Korver(50)], to enable the direct determination of the manganous ion in cathode mix samples taken from discharged cells. Using such methods, a study of "D" size flashlight cells discharged on a 4 ohm HIF[2] test was made. Figure 1.26 shows the discharge curve obtained on the 4 ohm HIF test in comparison with that obtained on the 4 ohm continuous test discussed above. Table 1.14 presents the pH and potential data and the manganous ion contents of the cathodes, measured both at the end of each day's discharge and again after standing overnight. Figure 1.27 illustrates these data graphically, curve *AB* showing the relation for the undischarged ore, curve *EDG* showing the data on

[2]The 4 ohm Heavy Industrial Test (HIF test) consists of a 4-minute discharge out of every 15 minutes for an 8-hour period, repeated daily as described in Chapter 8.

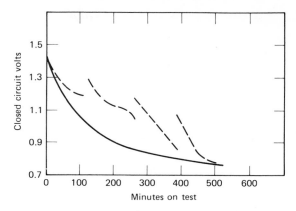

Figure 1.26 A comparison of the discharge curves of a "D" size cell on 4 ohm continuous test, (solid line), and the heavy industrial flashlight test (dashed line).

samples immediately after the daily discharge, and curve *CF* showing the readings on samples allowed to stand overnight. It is clear from these data that the manganous content of the cathode develops during each day's discharge and it is associated with the rise in pH and the decrease in cathode potential. Similarly during the 18-hour overnight rest period the

TABLE 1.14. POTENTIAL AND pH VALUES AND DIVALENT MANGANESE CONTENTS OF SAMPLES OF CATHODE MIX TAKEN DURING A 4 OHM HIF TEST

Output		Samples Removed at End of Daily Test			Samples after Standing Overnight[c]		
Minutes on Test	Ampere-Hours	pH	Cathode Mix Potential[a]	Mn^{++}[b] Content	pH	Cathode Mix Potential[a]	Mn^{++}[b] Content
0	—	5.60	0.747	—	—	—	—
	—	5.60	0.786	—	—	—	—
128	0.65	5.95	0.534	26.8	5.80	0.534	22.2
	0.64	5.95	0.539	28.3	5.73	0.608	19.6
256	1.26	6.45	0.460	39.3	6.00	0.551	34.0
	1.27	6.10	0.494	35.0	5.90	0.551	37.4
384	1.78	6.88	0.399	35.1	6.20	0.520	25.1
	1.75	6.55	0.432	32.7	6.03	0.524	27.9
480	2.08	7.30	0.392	28.9	6.75	0.502	22.0
	2.15	6.55	0.446	25.5	6.45	0.510	17.9

[a]Potential values are reported with reference to the standard hydrogen electrode.

[b]The manganese content is given in ampere-hours, (1.024 g per ampere-hour), as a percentage of the total output to that point in the discharge.

[c]An 18-hour period separated these analyses from those made immediately after discharge.

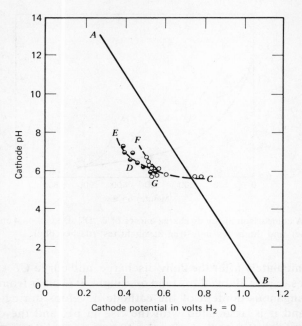

Figure 1.27 The relation between the potential and the pH of cathode samples taken from a "D" size flashlight cell during a 4 ohm HIF test; ◑ initial samples line *EDG*, ○ after 18 hours rest line *CF*. Line *AB* shows data for an undischarged African ore.

manganous content falls somewhat, while the pH falls and the potential rises again. This behavior is reflected in the discharge curve in Fig. 1.26 which begins each morning at a much higher voltage than the cell possessed at the end of the previous day's discharge. This phenomenon has been designated "recuperation" and reaction 20 termed the "recuperation reaction."

These findings agree well with the theory developed by French and MacKenzie but they emphasize that the recuperation reaction may be slow and that the 18-hour period is not enough time to permit complete removal of the manganous ion from the cathode on this HIF test. Thus, on this severe test the cathode operates in a condition that might be termed "continually polarized" by the manganous ion present throughout the test. It should be emphasized that the electrochemical reaction above operates only when current is being taken from the cell, whereas the recuperation reaction can operate both during the discharge and the rest periods.

Studies designed to measure the rates of the recuperation reaction outside of a dry cell and to learn what factors controlled it were reported by Korver, Jonnson, and Cahoon(51). The determination of the reaction rate was made in a flask, held in a thermostatically controlled bath at a selected temperature. One mole of the type of MnO_2 selected for study was suspended in one-half liter of 2N NH_4Cl in this flask and purged with nitrogen. At time zero, a liter of a solution containing 2M NH_4Cl and 1.5M $MnCl_2$ at the same temperature as the contents of the flask was added. As the reaction proceeded the hydroxyl ions were consumed and more were added in the form of measured amounts of an NH_4OH solution of known concentration to maintain a constant pH. Periodic analyses of the supernatant solutions, taken from small samples of the slurry in the flask, for their manganese contents provided a direct way to follow the course of the reaction. Figure 1.28 shows an example of the data obtained from electrolytic MnO_2 studied at a temperature of 43.95°C and a pH of 5.40. Here the line connecting the points shown as circles records the progress of the reaction. Thus, $Z = 1$ when the reaction is completed. The experimental data are represented quite well by the relation

$$\frac{t}{Z} = \frac{t}{b} + \frac{1}{k} \tag{21}$$

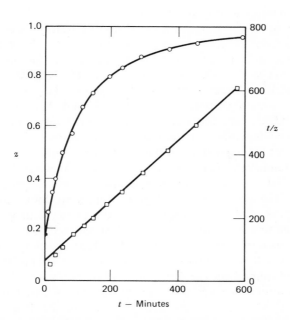

Figure 1.28 The relation between the recuperation reaction and time for an electrolytic MnO_2 sample tested at 43.95°C and pH 5.40.

where b and k are constants selected by conventional methods of least squares. By differentiation, one obtains the relation

$$\frac{dZ}{dt} = k \left(1 - \frac{Z}{b} \right)^2 \tag{22}$$

which is the rate expression for the process for a given starting material at a selected constant pH and temperature. The constant k is the specific rate constant and dZ/dt is the instantaneous rate in terms of moles of manganous ion uptake per mole of MnO_2 initially present per minute. In Fig. 1.28, the t/Z values, shown as squares, obtained in this experiment are plotted as a function of t. It is seen that, except for the initial readings, the data conform to a straight line well within the range of the experimental error. From Eq. 22, one sees that for any given finite time there is a finite, single value for the rate. At zero time, the rate is numerically equal to k, and as time increases the rate becomes slower. As the time becomes infinitely large, the reaction approaches completion and tends toward zero.

When the reaction is half completed, that is, at the half-life ($t = t_{0.5}$), the instantaneous rate is one-fourth of k. However the half-life rate $\left. \dfrac{dZ}{dt} \right]_{t = t_{0.5}}$ is a useful quantity in this study.

A number of MnO_2 samples from different sources were studied at temperatures from 21° to 45°C and in solution pH levels of 4.8 to 6.6. In Fig. 1.29, the logarithm of the half-life is plotted as a function of pH. The lines are isotherms. Starting materials were electrolytic MnO_2, synthetic pyrolusite, African ore, and activated African ore. The apparent energies of activation were calculated to be 26, 17, and 22 kcal/mole for the electrolytic, African ore and activated ore, respectively. The reaction rate was found to be dependent on the type of MnO_2 studied. Table 1.15 shows a comparison of the logarithms of the reaction rates of four representative samples at a temperature of 21°C and a pH of 6.5. As would be expected, a correlation exists between reaction rate and the general order of dry cell performance. However, the true relationship is no doubt complicated.

An effort was made to relate some of the physical properties of the MnO_2 samples to the recuperation rate [Cahoon and Korver(52)]. Other workers, for example, Buser and Graf(53), Kozawa(54), Vosburgh(55), and Euler(56), had studied the effect of the surface area of the powdered MnO_2, as determined by the BET method of Brunnauer, Emmett, and Teller(57). Another measurement was introduced, designated the particle surface area (PSA)(52), as determined by calculation from the particle size spectrum. The latter is found by sieving and measurements on the subsieve portion on the micromerograph. In such a calculation, all

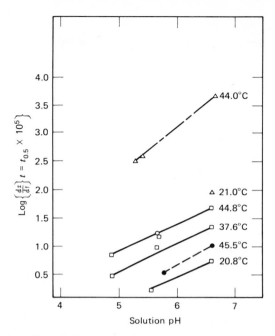

Figure 1.29 The logarithm of the recuperation reaction rate is presented as a function of temperature and electrolyte pH.

particles are assumed to be solid spheres. Table 1.16 shows a comparison of the data obtained by these two methods on a representative group of MnO_2 samples. It is surprising that the PSA values all fall in a narrow range, but it must be recognized that this value is a measure of the surface that may be contacted by the carbon black phase in the cathode manufacture. On the other hand, the BET surface area may all be in

TABLE 1.15. REACTION RATES OF REPRESENTA-
TIVE MnO_2 SAMPLES AT THE HALF-LIFE

Type of MnO_2	$Log \left\{ \dfrac{dz}{dt} \right]_{t=t0.5}^{\times 10^5} \right\}$
Activated African ore	2.38
African ore	0.660
Synthetic pyrolusite	0.013
Electrolytic MnO_2	2.13

Note. Above data determined at 21°C and a solution pH of 6.5.

TABLE 1.16. A COMPARISON OF THE TOTAL AND
PARTICLE SURFACE AREAS OF REPRESENTA-
TIVE MnO_2 SAMPLES

	Surface Area	
Type of MnO_2	Total BET m^2/g	Particle PSA m^2/g
Electrolytic MnO_2	52.8	0.47
Activated African ore	50.5	0.40
African ore	7.4	0.18
Synthetic pyrolusite	1.58	0.47

contact with the electrolyte system. While it is apparent that the more effective types of MnO_2 possess the higher BET surface areas, the relation between this property and the reaction rates given in Table 1.15 is complex. Since the reaction rate given above represented the actual rate at the half-life of the reaction, it was recognized that this might not be the most appropriate value to choose to represent the rate of a reaction that was changing throughout the course of the measurement. Therefore an alternative calculation was made to obtain the reaction rate when the reactants were half consumed, that is, $Z = \frac{1}{2}$, and shown as $\dfrac{dZ}{dt}\bigg]_{Z=Z_{0.5}}$ designated "reaction rate II." Data from earlier rate determinations were recalculated and two additional samples of MnO_2 were studied to obtain a representative series of values for reaction rate II at pH 6.6 and at the temperatures of 21°, 35°, and 45°C. The data obtained are shown in Table 1.17 and graphically in Fig. 1.30. The data were subjected to a statistical study and the equations for the three lines and their correlation values are given in the parentheses. It is indeed encouraging that a definite relationship between the reaction rate II and the surface areas of the powdered MnO_2 has been established. Table 1.18 shows the composition and some of the properties of the six samples of MnO_2 included in this study. The material designated "type M" was supplied by the Manganese Chemicals Corp.(116). The "chlorate" sample was prepared by the American Potash Corp., now a part of the Kerr–McGee Corp., and was prepared by the oxidation of manganous sulfate in solution by sodium chlorate under selected conditions(58).

In view of the fact that the six samples of MnO_2 considered here include such diverse materials as natural ore, synthetic types prepared by four different processes, and an activated African ore sample, it appears

TABLE 1.17. REACTION RATES II DETERMINED AT pH 6.50 AND THREE DIFFERENT TEMPERATURES

	$\text{Log}\dfrac{dz}{dt}\Big]^{\times 10^6}_{z=z0.5}$			BET Surface Area m^2/g
	Temperature			
Sample	45°C	35°C	21°C	
Electrolytic MnO_2	4.70	4.10	3.22	52.8
Activated African MnO_2	5.30	4.43	3.20	50.5
African (Ghana) ore	2.25	1.55	0.505	7.4
Synthetic pyrolusite	1.86	—	—	1.6
Type M synthetic	4.09	3.11	1.57	70
American potash	3.30	2.84	1.59	7.0

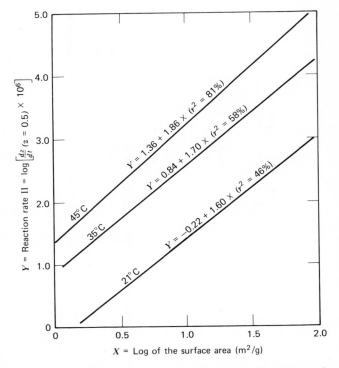

Figure 1.30 The relation between the logarithm of reaction rate II and the total surface areas of six MnO_2 samples.

TABLE 1.18. THE COMPOSITIONS AND PROPERTIES OF SIX REPRESENTATIVE SAMPLES OF MnO_2

Sample	x in MnO_x	% Mn	% Total MnO_2	% Free MnO_2	Total Mn Oxides	Crystal Type	Real Density g/cm^3	BET Surface Area m^2/g
Electrolytic	1.951	58.07	87.43	82.97	91.07	Gamma	4.25	52.8
Activated	1.916	50.83	73.79	67.27	79.30	Alpha	4.09	50.5
African ore	1.976	54.99	84.81	82.61	86.61	Rho	4.49	7.4
Synthetic pyrolusite	1.997	62.12	97.97	97.62	98.11	Beta	5.07	1.6
Type M	1.949	61.0	91.6	86.5	95.7	[a]	4.75	70
Chlorate	1.990	59.5	93.3	92.5	93.9	Rho	4.57	7.0

[a]The crystalline type of this product is difficult to characterize on the accepted basis due to its low order of crystallinity.

particularly significant that the correlation of the data is so satisfactory. Moreover, these samples represent four different crystal allotropes of MnO_2 and again all give concordant results. Thus, as far as the recuperation reaction rate II is concerned, a unit area of any one of the six MnO_2 samples is the equivalent of that of any other member of the group. These data, of course, leave many questions unresolved as to the exact mechanism of the recuperation reaction, such as the role of the adsorption of the manganous ion from the electrolyte. However, the data do suggest that the recuperation reaction is the important reaction that limits the rate of cell discharge on most applications.

The general introduction of x-ray diffraction equipment on a commercial basis early in the 1940s offered the battery technologist a new research tool that would be extremely useful in the identification of raw materials and also the crystalline products formed during cell discharge. McMurdie[38] applied this technique to the differentiation of MnO_2 types, a subject that is discussed more thoroughly in the section entitled "Depolarizers." He also used this method and the then new electron microscope to identify the products formed during cell discharge. He confirmed that the crystals formed in the paste layer of a cell some months old were $ZnCl_2 2NH_3$ and that the basic zinc chloride, $ZnCl_2 4Zn(OH)_2$, was formed in discharged cells. Studying the change in MnO_2 during cell discharge, he showed the presence of $ZnOMn_2O_3$, identical with the mineral hetaerolite, in certain discharged cells prepared with synthetic MnO_2. Later, Copeland and Griffith[59] published their findings on hetaerolite and suggested the battery reaction might be written as

$$Zn + 2MnO_2 \rightarrow ZnOMn_2O_3 \tag{23}$$

They, too, had used synthetic MnO_2 in their experimental cells.

In their study of the products of cell discharge, Cahoon, Johnson, and Korver[50] established an important correlation between the electrical output of experimental cells and that calculated from the products formed during the discharge. They developed a method of analysis for the discharged cell and gave examples of their data as shown in Table 1.19. These results show that both manganite, MnOOH, and hetaerolite, $ZnOMn_2O_3$, are produced although the relative proportions vary with the severity of the discharge and the type of MnO_2 initially used. Thus, electrolytic MnO_2 gave a much greater amount of hetaerolite than did African ore. These authors presented the two-step reaction theory to account for the formation of the two products.

The electrochemical reaction produces manganous ion:

$$MnO_2 + 4H^+ + 2e^- \rightarrow Mn^{++} + 2H_2O \tag{24}$$

TABLE 1.19. A COMPARISON OF THE CHEMICALLY DETERMINED AND ELECTRICAL OUTPUTS OF "D" SIZE EXPERIMENTAL CELLS

Oxide Type	4 ohm Test	Soluble Mn^{++} (1)	Hetaerolite $ZnO:Mn_2O_3$ (2)	Manganite $MnOOH$ (3)	Total of Cols. (1), (2), (3)	Electrical Output
African ore	Continuous	0.06	0.001	1.27	1.33	1.33
African ore	HIF	0.43	0.13	1.74	2.30	2.15
African ore	LIF	1.15	0.61	3.16	4.92	4.91
Electrolytic MnO_2	HIF	2.02	4.41	1.65	8.08	8.19

Note. Data above are expressed in ampere-hours. The abbreviations HIF and LIF indicate the heavy and light industrial flashlight tests, respectively.

The chemical reaction that follows as the second step may be either

$$MnO_2 + Mn^{++} + 2OH^- \rightarrow 2MnOOH \qquad (25)$$

which produces manganite, or

$$MnO_2 + Mn^{++} + 4OH^- + Zn^{++} \rightarrow ZnOMn_2O_3 + 2H_2O \qquad (26)$$

by which the hetaerolite can be formed. Studies of the recuperation reaction by Korver, Johnson, and Cahoon[25] showed that hetaerolite would be produced if zinc were present in the electrolyte and other conditions were satisfactory, thus confirming the above reaction.

The ampere-hour equivalents of the three cathode products, that is, Mn, MnOOH, and $ZnOMn_2O_3$, can be readily calculated from the equations given above. Thus the quantities of each product found in the analysis of the cathode of a discharged cell can be converted to equivalent ampere-hours of output. The sum of the ampere-hour equivalents for the three products, that is, the chemically determined output of the cell, should agree with the ampere hour output determined from the discharge curve of the same cell, or the electrical output. The agreement between the chemical and electrical outputs, as given in Table 1.19 is a powerful argument in favor of the two-step reaction mechanism described herewith.

It will be seen that the sum of reactions 24 and 26 give substantially the reaction

$$Zn + 2MnO_2 \rightarrow ZnOMn_2O_3 \qquad (27)$$

One interesting conclusion from the study of hetaerolite formation in the cathode is that the above reaction consumes zinc ions at the same rate that zinc ions are being added to the electrolyte system anodically. The result is that the cell operates in a nonvariant electrolyte under such conditions. It thus would be expected that the cathode pH would remain constant and the discharge curve reasonably flat. Such results are often obtained with highly active depolarizers on selected tests.

REACTIONS AT THE ANODE

The electrochemistry of zinc in the Leclanché cell appears simple and uncomplicated. It dissolves anodically according to the reaction

$$Zn + 2Cl^- \rightarrow ZnCl_2 + 2e^- \qquad (28)$$

The zinc chloride produced in this reaction is extremely soluble in the electrolyte and, as a result, no passivating film such as develop in some other battery systems occurs. Zinc chloride is an acid salt and the pH of the anolyte decreases as the concentration rises. The effects of the concentration of zinc chloride on the pH of the aqueous solution is discussed at greater length in the electrolyte section of this chapter, with specific data being shown in Table 1.3. Figure 1.21 depicts the decrease in pH of the anolyte in a "D" size flashlight cell during a 4 ohm continuous test. The polarization which develops at the anode surface, chiefly concentration polarization, is often small but may amount to 0.2 V (9) at the current drains used in flashlight service.

A somewhat more sophisticated approach, involving the use of an interrupter (28) measured the polarization of both electrodes in a group of "D" size flashlight cells. The cells were discharged on a pulse current of 0.500 amp and, since the interrupter alternately opened and closed the circuit on a 50 percent duty cycle, the average discharge current amounted to 0.250 amp for the time elapsed during the test. It is important to note that the cell is on drain at a current of 0.500 amp when it is being discharged. With the interrupter technique cell voltages and electrode potentials may be read within 0.1 msec after the circuit is opened to give an "immediate open circuit potential" (IOCE) for the electrode (253).

The discharge curve and the open circuit voltage of the cell are shown in Fig. 1.31. The vertical height, on the voltage scale, between corresponding points on the two curves represents the IR drop in the cell. It is clear that this value is reasonably constant during the first 200 minutes but increases thereafter.

Figure 1.32 shows the IOCE values for the anode and cathode with the cathode curve at the top and the anode curve at the bottom of the figure. The part of the chart below the anode potential line and above −0.83 V, the initial value of the anode potential, represents the polarization of the anode during the test. To study this factor further, a group of cells were simultaneously discharged on this test. Periodically, representative cells

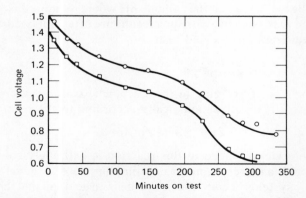

Figure 1.31   The open and closed circuit voltages during a 0.500 amp pulse test on a "D" size cell monitored by an interrupter.

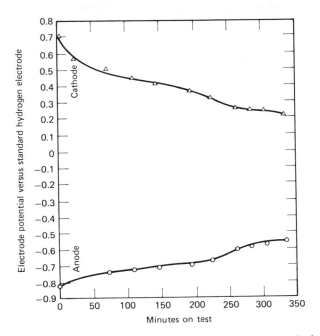

Figure 1.32 The immediate open circuit potentials of the anode and cathode, that is, the IOCE values of both electrodes during the 0.500 amp pulse test.

were removed, opened as quickly as possible, usually in less than one minute, and the paste layer removed. Chemical analyses were made on these samples to provide the results shown in Table 1.20 along with the last values of the IOCE determined just before the cell was opened. The values of the $ZnCl_2$ concentration, calculated from the analyses are plotted against the IOCE values measured at the anode in Fig. 1.33. For comparison curves relating the zinc potential to zinc ion concentration in aqueous solutions of $ZnCl_2$ and in $ZnCl_2$, solutions containing 15, 20 and 25 per cent $NH_4Cl(10)$ as well as saturated solutions are shown. It is quite clear that the IOCE in the undischarged cell fell on the line for an NH_4Cl saturated solution containing 28 g $ZnCl_2/100$ g H_2O. As the discharge progressed the zinc chloride concentration increased and the line relating these values moved diagonally across the chart to reach the potentials of an aqueous $ZnCl_2$ solution at a concentration of about 140 g $ZnCl_2/100$ g H_2O.

Obviously the polarization developing at the anode is concentration polarization due to the accumulation of $ZnCl_2$ at the anode surface. As this product diffuses away from the anode and into the cathode the potential of the anode will return to a more nearly normal level and, in effect, will be less

TABLE 1.20. THE COMPOSITION OF THE ANOLYTE LAYER AND THE POTEN-
TIAL OF THE ZINC ELECTRODE

Minutes on Test	% ZnCl$_2$	% NH$_4$Cl	% H$_2$O	Grams ZnCl$_2$ per 100 g H$_2$O	Immediate Open Circuit Potential of Zinc Electrode
0	15.8	27.9	56.3	28.0	−0.830
45	26.1	18.0	56.0	46.5	−0.772
91–95	37	10	47.4	80.0	−0.731–0.734
128–133	38	8	45.2	84.0	−0.725–0.697
244–251	41	11	38.4	106	−0.669
292–298	56.6	2.1	41.7	136	−0.551–0.569
336–341	57.0	—	40	142	−0.564–0.531
379–388	59.5	1.8	38.7	154	−0.545–0.558

Potentials reported with respect to the standard hydrogen electrode.

polarized. Similarly, if the cell had been discharged on a lighter or on an intermittent drain the additional time available for diffusion of ZnCl$_2$ away from the anode would have shown a smaller anode polarization than that given in the example. It should be noted that, while Fig. 1.32 also shows the polarization developed at the cathode, this part of the analysis would be treated as described in the section on reactions at the cathode and therefore will not be further discussed here.

SHELF REACTIONS

The Leclanché cell contains an active oxidizing agent in the cathode, an aqueous inorganic electrolyte, a cereal gel or equivalent separator, and an anode of zinc—a reducing metal. Most of these materials are reactive components and it is the problem of the battery manufacturer to package the system so that deterioration is kept at a minimum. A summary of the published data on this aspect is presented in the section entitled "Battery Performance." In the present section, the various factors that influence the reactions during shelf storage are discussed under the headings, shelf reactions at the anode, and shelf reactions at the cathode.

Shelf Reactions at the Anode

Although the mechanism of reaction of the zinc electrode is straightforward and progresses with an efficiency of nearly 100 percent, the problems that develop from its use are mainly those associated with the wasteful corrosion of the metal during the period when the cell remains on the shelf, that is, between manufacture and actual use.

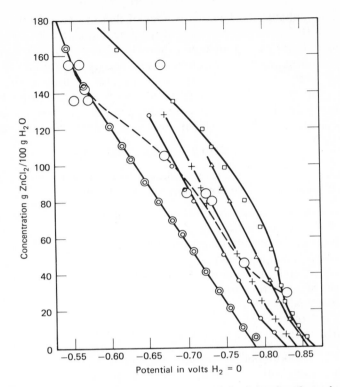

Figure 1.33 The immediate open circuit, IOCE, potentials of the zinc electrode compared with reference curves.

○ Immediate open circuit potentials of Zn under test conditions:

Zinc potentials in $ZnCl_2$ solutions
- □ saturated with NH_4Cl
- △ containing 25% NH_4Cl
- + containing 20% NH_4Cl
- ○ containing 15% NH_4Cl
- ◉ containing 0% NH_4Cl

The zinc anode in the modern dry cell possesses remarkable stability in what might normally be expected to be a very corrosive electrolyte. However, the fact that modern dry cells can be stored for a matter of a few years without loss of more than a small part of their capacity is a tribute to the technical effort expended in this direction by the industry. When a cell is actually placed into service, this condition is radically changed. The anolyte changes in pH and concentration, the loss of mercury from the anode surface(3), together with the accumulation of impurities left after the anodic solution of the zinc from the alloy, and other changes, all introduce complications. The importance of these items may be shown by the fact that the wasteful corrosion of the zinc anode in cells on long-time intermittent service may consume as much as 50 percent of the weight of the metal utilized in generating current. The loss of this amount of metal is significant, but when it is combined with the additional loss of two molecules of NH_4Cl for each gram atom of zinc, the total becomes economically important as well as resulting in a direct loss in the service capacity of the cell. The wasteful corrosion results from the reaction

$$Zn + 2NH_4Cl \rightarrow ZnCl_2 \cdot 2NH_3 + H_2 \tag{29}$$

A number of important factors relating to the general problem of the control of wasteful anode corrosion are listed below.

1. The role of the composition of the zinc alloy,
2. The effects of impurities,
3. The hydrogen overvoltage on the zinc surface,
4. The role of amalgamation,
5. The use of colloids to control wasteful corrosion,
6. The use of inhibitors to reduce wasteful corrosion,
7. Effective sealing of cells,
8. The effect of storage temperature,
9. The measurement of wasteful corrosion.

Generally speaking, all of these factors must be considered simultaneously in designing a dry cell for optimum storage life. They are discussed in the following sections.

The Role of the Composition of the Zinc Alloy   In the early days of dry cell manufacture the cylindrical zinc cans were made by soldering together the edges of a rolled sheet of zinc and soldering a zinc disc on one end as a bottom. The mechanical properties of the metal probably were as important as the electrochemical properties, since this can was the container for the cell. Small amounts of lead and iron were always present in the zinc alloy and a small amount of cadmium was often added to provide increased stiffness to the metal. Burgess(60) long ago advocated the use of zinc of high purity as one means of reducing wasteful corrosion and pioneered in the production of drawn zinc cans to help achieve this purpose. Extruded cans are a more recent development. Today, a large percentage of the smaller size cells are assembled in drawn or extruded zinc cans. In Europe, cans made by impact extrusion predominate(61).

In spite of the trend to eliminate the lead-tin solder seams, which were so common in cells a generation ago, there are many No. 6 cells made today using soldered cans. This raises an important question as to the effect of the exposed solder on the wasteful corrosion of the zinc can in such cells. It appears that the small amount of solder exposed on the inside of such a can does not seriously increase the wasteful corrosion under the conditions normally present in cell storage and use.

Despite the large amount of study that has undoubtedly been given to the selection of the zinc alloy compositions and to the methods of zinc can manufacture, very little has been published. One of the few papers in this field (61) discusses the beneficial effects of lead and cadmium additions to pure zinc from both the metal structure as shown by metallographic examination and the behavior of the alloy in actual cells. The author concludes that the most satisfactory alloy for impact extrusion contains about 0.10 percent of lead and 0.05 percent of cadmium and possesses a relatively fine-grained structure. He states that this composition is not far from that used in making drawn zinc cans.

The Effect of Impurities   Any compound of a metal below zinc in the electromotive series which is soluble in the electrolyte of the Leclanché cell is a potential source of reduced storage quality. Such materials as salts of copper, nickel, lead, arsenic, antimony and other heavy metals are particularly troublesome because they plate out on the surface of the zinc anode and develop local cathodic areas. These, in combination with the original zinc anode, permit "local couples," that is, electrolytic cells, to develop resulting in a higher than normal wasteful corrosion rate. Figure 1.34 shows a diagram of such a local couple with the anodic and cathodic areas indicated. The fact that the hydrogen overvoltage on deposited heavy metals is usually much lower than that of zinc, together with the difference in potential, is given as the basis for the increased rate of attack. This situation permits the purer zinc areas to act as anodes and to electrochemically dissolve, while the less pure cathodic areas permit the evolution of hydrogen in accord with conventional corrosion theory.

It seems relatively straightforward to reduce the heavy metal content of the electrolyte to a low level by the use of salts of adequate purity. However, the extension of this principle to all cell ingredients is complicated by the use of MnO_2 ores as cathodic depolarizers. Most ores contain small amounts of many heavy metals and it becomes the problem of the battery technologist to devise ways of utilizing these ores without suffering serious shelf storage losses in finished cells.

It is surprising that although the battery industry is largely based on the use of MnO_2 ores, the storage quality of the product is acceptable. An important exception is the growing group of batteries that are made from synthetic manganese dioxide which, because of its chemical or electrochemical origin, is substantially free from heavy metal impurities.

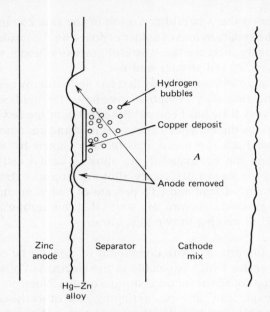

Figure 1.34   The proposed operation of a local couple in producing wasteful corrosion of a zinc anode. The location marked *A* may be visualized as that where the soluble copper salt initially was placed.

One important feature of heavy metal impurities in the Leclanché system is the tendency of certain of these to form treelike growths that may short-circuit the unit and cause rapid failure. Generally, such trees develop from a concentrated area of a soluble heavy metal compound located in the cathode near the separator. For example, the presence of a lump of a soluble copper compound in such a location could result in a metallic copper tree starting at the zinc electrode surface and growing through the separator layer. When it reaches the cathode, short-circuiting of the cell occurs. The rate of discharge of the cell through this short circuit depends on the size and number of the trees present.

The effects of the additions of selected heavy metal impurities to samples of electrolyte containing a zinc specimen have been investigated by many workers including Aufenast and Muller(62). They followed the extent of the corrosion by measuring the volume of hydrogen developed under a set of standard conditions. Their investigation covered additions of antimony, molybdenum, arsenic, copper, nickel, lead, iron, manganese, nitrate, and phosphate. In addition to their work with isolated samples of zinc, they also measured the hydrogen evolution of groups of eight R 20, that is, "D," size cells maintained at 35°C in a special apparatus. Precise amounts of impurities, copper, nickel, iron, both ferrous and ferric, were added to these cells during manufacture. Figure 1.35 shows the data obtained on these cells. The data clearly show how increasing the amounts of copper and nickel present results in increased amounts of hydrogen gas. The authors concluded that the most injurious impurity is antimony with molybdenum following it closely, but only in combination with other metals. Under some conditions, arsenic is equally injurious. Nickel is next in order with iron and copper following it. Manganese and lead appear ineffective as do the nitrate and phosphate.

The Hydrogen Overvoltage on the Zinc Surface   It is well known that zinc possesses a high hydrogen overvoltage and this property is often given a major share of the credit for the stability of the zinc anode in the electrolyte of the Leclanché cell. In spite of this unique property the behavior of unprotected zinc in a dry cell is far from satisfactory because wasteful corrosion proceeds at too high a rate. The problem becomes one of taking advantage of the high overvoltage phenomenon but at the same time reducing the wasteful corrosion loss to the lowest possible level. Three important additional protective methods must be combined with the naturally high overvoltage of the metal to achieve this objective. They are amalgamation, the addition of colloids and inhibitors to the electrolyte, and effective sealing of cells. These are discussed in following paragraphs.

The Role of Amalgamation   The beneficial effect of amalgamation of the zinc electrode was discovered very early in the development of primary cells [Kemp(243)] and was used by Leclanché. In view of this situation, it is surprising that amalgamation of the zinc electrode of the dry cell was not generally adopted until about 1910. Amalgamation operates in a number of ways to reduce the wasteful corrosion of the zinc electrode. The presence of the mercury increases the hydrogen overvoltage above that of untreated zinc. The mercury serves as a solvent for small amounts of certain heavy metal impurities, for example, lead and copper and thus nullifies their effect as cathodes for local couples at the electrode surface. In addition, the mercury provides a smooth surface(63) that discourages wasteful corrosion. Thus the intrinsic properties of the base metal are modified or obscured by the nature of the amalgamated surface(64).

Mercury is generally introduced into the cell system by the addition of a mercury salt. Mercuric chloride has long been used and the mercury deposits from this salt onto the surface of the zinc by replacement. Less soluble mercury salts have been patented(65). The quantity of mercury that can be used effectively is quite small, usually about 0.25 percent, because of embrittlement of the zinc sheet or container. The constitutional diagram of the zinc-mercury series of alloys by Pushin(66) is given in Fig.

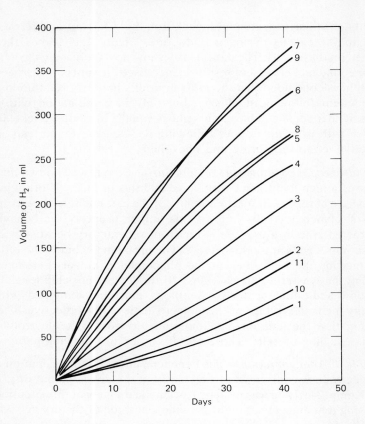

1. Electrolyte
2. Electrolyte plus 1 mg Cu
3. Electrolyte plus 2 mg Cu
4. Electrolyte plus 4 mg Cu
5. Electrolyte plus 1 mg Ni
6. Electrolyte plus 2 mg Ni
7. Electrolyte plus 4 mg Ni
8. Electrolyte plus 1 mg Cu plus 1 mg Ni
9. Electrolyte plus 2 mg Cu plus 2 mg Ni
10. Electrolyte plus 1 mg Fe^2, 2 mg Fe^2, or 4 mg Fe^2.
11. Electrolyte plus 1 mg Fe^3, or 2 mg Fe^3.

All quantities are per eight R 20 cells

Figure 1.35 The effect of impurities on the hydrogen evolution in cells at 35°C. This illustration is reproduced with the permission of the author and publisher (62).

1.36. Figure 1.36 shows the effects of the mercury on the potential of the resulting alloy in a ZnSO₄ solution.

The Use of Colloids to Control Wasteful Corrosion   Every experimenter who has worked with Leclanché cells has been surprised at the reduction in the rate of the wasteful corrosion from the level shown by an amalgamated zinc electrode when contacted by electrolyte to the much lower level characteristic of the same electrolyte in which a colloid has been suspended. Typical colloids or gelling agents such as wheat flour and

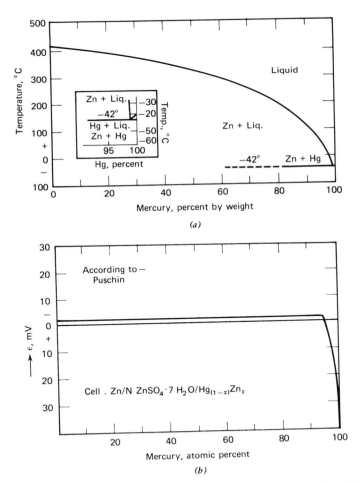

(a)

(b)

Figure 1.36 (*a*) Constitutional diagram of the zinc-mercury series of alloys. (*b*) Relative solution potential of zinc-mercury alloys of different compositions.

various starches have been successfully used by the industry for many years(67) as thickening agents for the electrolyte layer used to contact the zinc anode. Both cooked and cold set(68) varieties of paste have been widely used. The role of the colloid is undoubtedly complex. It may be in part physical since the colloid provides an adhesive, elastic electrolyte-saturated layer at the metal surface, which automatically maintains a continuous electrolyte path between the two electrodes in the cell. Hamer(69) examined wheat flour and found that gliadin, contained in the gluten fraction of wheat flour, had an inhibitory effect on the wasteful corrosion of zinc in the Leclanché cell. This finding prompted an extensive survey of inhibitors used in other corrosion fields and included trials of many of these in actual cells. Unfortunately many of the materials that were tried formed high-resistance deposits on the zinc anodes and prevented the attainment of normal battery life. As a result most of these materials found no useful application in dry cells.

The successful development of the plastic film liner(70) involving alkyl cellulose ethers and similar compounds was found to give even better battery performance than cells made with flour and starch pastes. Part of the advantage is presumably due to improved corrosion inhibition. Additional information on this subject will be presented in the section on separators under the heading "Materials of Construction" later in this chapter.

The Use of Inhibitors to Reduce Wasteful Corrosion   An interesting approach has been to use selected inhibiting materials that are added to the colloid to further reduce the wasteful corrosion. Chromates and/or bichromates(71) have long been recognized as one means of protecting zinc anodes against wasteful corrosion because chromates from a protective film on the surface of the zinc electrode. Chromate additions were actually used in production of No. 6 cells before 1920. The beneficial effect of this film on cells during storage is well established, although its effectiveness under service conditions may be questioned. Complex chromium compounds such as the chrome glucosates have also been proposed.

A group of corrosion inhibitor compounds has been patented by Zimmerman and Powers(72). These compounds generally possess the formula $R-O(R')_n-R''$, where R stands for alkyl, aryl, or aryl alkyl radicals, R'' may be either hydrogen or a grouping similar to R, R' is an alkoxy radical such as ethoxy or propoxy, and n may be any number between 1 and 50. These compounds are essentially linear polymers having wetting characteristics. Their inhibiting properties result from the high hydrogen overvoltage created on the zinc anode surface presumably by the physical and chemical adsorption of the inhibitor compounds. About 13 compounds having inhibiting properties are described. The results obtained from the use of two of these materials, paraphenyl phenoxy polyethylene glycol "(PPG)", and an alkyl phenol polyglycol ether (Neutronix-600) are given as examples in the above patents. It has been found possible to use such inhibitors both alone and in combination with mercury. Table 1.21 taken from U.S. Patent 2,900,434(72) shows some of the characteristics of experimental cells containing one of these corrosion inhibitors.

A proprietary material identified as "Raysol" has been used for some years by the Ray-O-Vac Company as a means of reducing the wasteful corrosion of zinc in certain of the Leclanché cells marketed by this company. Unfortunately, no information on the compound employed or the mechanism by which it operates is available.

Aufenast and Davis(73) have studied two groups of corrosion inhibitors, that is, surface active agents and chelating agents. Of the surface active agents, they found Lissapol N, a nonyl phenol ethylene oxide condensate, to be the most effective at concentrations of 0.5 percent and higher. Eight chelating agents including the sodium salt of ethylene diamine tetra acetic acid were investigated. Both groups were studied in actual cell experiments. In the comparison of the chelating agents, additions of 0.005 percent of each of copper and nickel were added along with the chelating agent to establish effectiveness of the latter. Data presented show that such amounts of impurities, although adequate to ruin the control cells, did not significantly reduce the capacity of the cells containing the chelating agent.

Effective Sealing of Cells   When oxygen from the air has access to the inner part of any Leclanché cell the oxygen effectively reacts with the protective hydrogen film to give the overall reaction

$$Zn + 2NH_4Cl + \tfrac{1}{2}O_2 \rightarrow ZnCl_2 \cdot 2NH_3 + H_2O \qquad (30)$$

Under these conditions the rate of wasteful corrosion increases. Effective sealing of cells to prevent oxygen access to the zinc-electrolyte interface thus becomes an important requirement for long life cells. Effective sealing is, of course, necessary to prevent loss of moisture from the battery during storage. Thus the importance of adequate sealing plays a dual role in extending the useful life of commercial units.

The Effect of Storage Temperature   The wasteful corrosion reaction, like many chemical reactions is temperature dependent that is, it increases in rate as the temperature is raised. This observation becomes particularly important since batteries may be stored at temperatures up to 65°C in such places as warehouses, freight cars, or on shipboard in tropical

TABLE 1.21. THE EFFECT OF CORROSION INHIBITORS ON THE PERFORMANCE OF CELLS

PERFORMANCE OF CELLS AT 21°C

Example No.	Depolarizer	Inhibitor	Zinc Cans Used	Initial		12 Months			18 Months		
				Volt	Amps.	Volt	Amps.	CM[a]	Volts	Amps.	CM[a]
4	Electrolytic manganese dioxide	Hg	Extruded high purity	1.58	9.3	1.47	4.2	45	1.45	2.2	24
6	Electrolytic manganese dioxide	Hg+"PPPG"	Extruded high purity	1.59	7.9	1.42	5.1	65	1.49	4.6	58
7	Electrolytic manganese dioxide	Hg	Soldered, alloy	1.58	8.9	1.50	4.7	53	1.48	3.2	36
9	Electrolytic manganese dioxide	Hg+"PPPG"	Soldered, alloy	1.59	7.9	1.50	4.8	61	1.48	4.1	52
10	African ore	Hg	Extruded high purity	1.64	7.5	1.59	6.3	84	1.58	6.1	81
12	African ore	Hg+"PPPG"	Extruded high purity	1.66	6.4	1.58	4.9	77	1.58	5.5	86
13	African ore	Hg	Soldered, alloy	1.65	7.4	1.59	5.9	80	1.58	5.8	78
15	African ore	Hg+"PPPG"	Soldered, alloy	1.60	6.2	1.58	4.7	76	1.56	4.9	79

PERFORMANCE OF CELLS AT 45°C

Example No.	Depolarizer	Inhibitor	Zinc Cans Used	Initial		12 Months			18 Months		
				Volts	Amps.	Volts	Amps.	CM[a]	Volts	Amps.	CM[a]
4	Electrolytic manganese dioxide	Hg	Extruded high purity	1.57	9.3	1.46	2.2	24	1.47	0.9	10
6	Electrolytic manganese dioxide	Hg+"PPPG"	Extruded high purity	1.58	7.7	1.49	3.9	51	1.49	3.6	47
7	Electrolytic manganese dioxide	Hg	Soldered, alloy	1.59	9.4	1.48	4.1	44	1.48	3.4	36
9	Electrolytic manganese dioxide	Hg+"PPPG"	Soldered, alloy	1.58	7.7	1.45	3.6	47	1.46	3.5	45
10	African ore	Hg	Extruded high purity	1.65	7.5	1.57	5.0	67	1.56	2.9	39
12	African ore	Hg+"PPPG"	Extruded high purity	1.65	6.2	1.57	4.8	77	1.57	4.5	73
13	African ore	Hg	Soldered, alloy	1.64	7.3	1.55	4.7	64	1.57	4.6	63
15	African ore	Hg+"PPPG"	Soldered, alloy	1.65	6.2	1.55	4.4	71	1.53	4.2	68

[a]CM is used to designate current maintenance in percent.

latitudes. The effect of such higher temperatures on the storage quality of Leclanché cells has been of particular interest to the military organizations and considerable money and effort have been spent on evaluating the extent of battery depreciation under such conditions. The effect of shelf storage temperature is reported in the section on battery performance later in this chapter.

The Measurement of Wasteful Corrosion   The determination of the extent of wasteful corrosion of the anodes in experimental cells stored under various shelf conditions is relatively simple. Cells intended for this purpose are assembled with individually weighed and identified zinc anodes. After the completion of the storage period, the cells are dismantled, and the anodes cleaned, dried, and reweighed. The loss in weight is the result of wasteful corrosion.

When such experimental cells are discharged, the combined wasteful corrosion that develops during the discharge and any preceding shelf period can be measured in a similar manner. However, the amount of zinc dissolved electrochemically must be calculated from the discharge curve and used to make a suitable correction to the above calculation.

Unfortunately, the main effect of wasteful anode corrosion is the decrease in the service life of the battery. Since so many factors contribute to the wasteful corrosion process, it is very difficult to assign to each factor its correct importance in a particular instance. Under such circumstances the trained battery technologist can often recognize specific indications, such as characteristic deposits of corrosion products on or near the anode, which suggest that specific factors have predominated.

Shelf Reactions at the Cathode

Although most research regarding shelf reactions has been concentrated on reactions involving the anode, some studies (74)(62)(217)[3] have shown that a limited amount of deterioration at the cathode may also occur. Hamer's work (74) was undertaken because of the need to store batteries for use in World War II under optimum conditions. Hamer investigated the development of gases evolved from batteries stored at 55°C using the mass spectrometer for analyzing the gases. In addition to hydrogen evolved from the anode reactions described above, some carbon dioxide and traces of ammonia and carbon monoxide were found. He suggested that the carbon dioxide resulted from some reaction between the manganese dioxide and carbon or other battery components. The amount of CO_2

---

[3]Further data on shelf properties of Leclanché cells will be found in the section entitled "Keeping Qualities," later in the chapter.

found in the gas developed by BA-38 experimental batteries ranged from 1.0 to 6.9 mole percent whereas the amount of hydrogen ranged from 41.3 to 55.9 mole percent of the total gas evolved. Although every effort was made to make this study quantitative, it is difficult to extrapolate the amount of CO_2 found into the corresponding amount of MnO_2 consumed.

Storage at temperatures as high as 55°C is a severe requirement for Leclanché cells and not usually found in commercial storage and use of batteries. However, for military applications, high temperature storage is a common requirement.

Aufenast and Muller(62) studied the reaction between starches and flour and various types of manganese dioxide. They measured the volume of carbon dioxide generated from a mixture of 10 g of paste and 10 g of MnO_2 material at both 20°C and 35°C in a special apparatus over a period of up to 21 days. On the 35°C test the electrolytic MnO_2 produced the largest amount of gas, 50 ml, followed fairly closely by the activated MnO_2, 33 ml, both samples providing more gas than the group of natural ores studied. Of the natural ores, Moroccan gave the most, 24 ml, Chinese an intermediate level, about 8 ml, and West African and Indian ores the least amounts, 5 and 4 ml, respectively, of gas on this test. Control experiments with D-glucose, sucrose, maltose, melibiose, and raffinose, all sugars that could be present in flour in small amounts, showed gas development when mixed with MnO_2 under the test conditions.

The authors point out that in contrast to hydrogen the solubility of carbon dioxide in cell electrolyte is quite large and thus the latter may diffuse readily through the cell without forming bubbles or dislodging the paste layer.

A COMPARISON OF THE PROPOSED REACTION MECHANISMS FOR THE
CATHODE REACTION

In approaching any comparison of the proposed reaction mechanism, it must be realized that definite differences in viewpoint exist between the various authors who have studied this subject intensively. Thus, some have based their conclusions on work with bobbins in actual cells while others have used tiny experimental electrodes under carefully controlled laboratory conditions. Since the bobbin of an actual cathode is of finite thickness, much greater than that of a tiny electrode, the actual reaction might be different from one place to another in the bobbin, particularly at high current densities. In spite of the differences in technique that have been used, the following comparison will offer a brief summary of several viewpoints on this important subject.

In order to satisfactorily explain the cathode reaction, any proposed theory must be able to include the following items:

1. The drop in potential of the MnO_2 electrode during discharge;
2. The rise in electrode potential during recuperation periods;
3. The amounts of the products of the reaction must be directly converti-
   ble into ampere-hours of output, the total being in agreement with that
   actually obtained.

Early theory by LeBlanc(43) proposed that the primary discharge
product was hydrogen which accumulated on the surface of the electrode.
This layer was subsequently oxidized by the MnO_2 and the process was
termed "depolarization." In the absence of any direct evidence for this
theory, it is not now seriously considered and the facts concerning the
discharge mechanism can be explained in a more satisfactory manner.

Earlier, Divers(45) had tried to explain the mechanism on the basis of
the accumulation of by-products on the surface of the MnO_2 and he
correctly identified MnOOH and $ZnO \cdot Mn_2O_3$ as reaction products although he had none of the modern x-ray tools to confirm his conclusions.

Coleman(77) proposed a solid state diffusion process whereby the
lower oxide of manganese diffuses into the MnO_2 particle. Vosburgh(55)
and Scott(78) added the concept that the removal of MnOOH could be the
result of the diffusion of the proton and the electron into the interior of
the MnO_2 particle as did Coleman earlier. Daniel-Bek(79) proposed that
the carbon and graphite portions of the cathode played a role since their
rest potentials were lower than that of the MnO_2. Thus, when the
electrode is discharged the potential of the carbon decreases more than
that of the MnO_2 and its surface is partly discharged much as had been
proposed by LeBlanc earlier. The potential of the carbon determines the
measured potential of the electrode as a whole. When the external circuit
is broken the carbon must be recharged by the local cell current before a
steady state is reattained.

Based on diffusion equations, Scott(78) showed that a solution of the
diffusion problem is possible provided the following assumptions are
made:

1. The effect of the gradient of the electrical potential on the migration of
   the ions through the solid is negligible compared to that of the
   concentration gradient;
2. The diffusion coefficient is constant;
3. The diffusion is linear and semiinfinite;
4. The MnOOH concentration is zero at the start of the discharge.

Kornfeil(80) extended these relationships and provided experimental data
that gave surprising correlation between observed and calculated values.

These studies and others [Brouilet et al.(81)] gave a considerable impetus to the proton diffusion theory.

Vosburgh and his students(82)(83)(84)(85)(86) have studied the mechanism in considerable detail and their results are summarized in a paper by Vosburgh(55). He discusses several possibilities which involve the diffusion of protons and electrons and concludes that the mechanism is best represented in four steps.

1. Electrons from the electrode and protons from the electrolyte meet at the surface of the MnO_2 exposed to the electrolyte with many of the protons penetrating the MnO_2 lattice:

$$MnO_2 + H^+ + e^- \rightarrow MnOOH \tag{31}$$

2. When enough MnOOH accumulates on the surface of the MnO_2 the reaction goes farther, that is, to the formation of $Mn(OH)_2$:

$$MnOOH + H^+ + e^- \rightarrow Mn(OH)_2 \tag{32}$$

Then Mn^{++} may appear in the electrolyte through the reaction as at pH 5:

$$2MnOOH + 2H^+ \rightarrow MnO_2 + Mn^{++} + 2H_2O \tag{33}$$

3. Inward diffusion of the lower oxide into the residual MnO_2 takes place but this is a slow process.
4. The discharge terminates with a rapid decrease in the electrode potential. Possibly the lattice hardens and protons can no longer penetrate it as suggested by Naumann and Fink(87).

Era, Takehara, and Yoshizawa(88) have also studied the mechanism of the cathode reactions of MnO_2 in acid, neutral and alkaline electrolytes. They summarize Vosburgh's study by the statement that Vosburgh considers the removal of the MnOOH from the surface of the MnO_2 as the rate-determining step, a statement with which they apparently agree. These authors point out that Kornfeil's equation fits the data more satisfactorily in the pH range of 6–8 than in the alkaline systems. They developed a slightly different equation to apply to the MnO_2 electrode in KOH electrolyte and offer convincing data comparing the recovery of the potential after discharge to show the correlation between observed and calculated values. They recognize the roles played by the protons and electrons in the recovery process.

In Leclanché cells, Brenet(89) emphasizes that the presence of combined water in the MnO_2 promotes the movement of the proton into the solid phase. Brenet and his collaborators have studied the mechanism of the cathode reaction extensively and many papers have been published.

Brenet(90) points out that the cathode reaction mechanism is not a straightforward process and that the crystal structure of the MnO_2 must be considered. Furthermore, he believes that no single reaction can present the complex situation that exists. He emphasizes the role of the combined water and offers the following general equation(91)

$$MnO_x(OH)_{4-2x} + 2(2 - 3y)H^+ + 2(1 - y)e^- \rightleftarrows (1 - 2y)Mn^{++} + yMn_2O_3$$
$$+ (4 - x - 3y)H_2O \qquad (34)$$

or the equivalent equation

$$MnO_x(OH)_{4-2x} + 2(2 - 3y)H^+ + 2(1 - y)e^- \rightleftarrows (1 - 2y)Mn^{++} + 2yMnOOH$$
$$+ (4 - x - 4y)H_2O \qquad (35)$$

Brenet(89) considered that the electrochemical reduction proceeded to Mn^{++} within the MnO_2 lattice. The Mn^{++} passes into solution and forms $Mn(OH)_2$ which is oxidized to $MnOOH$ by the remaining MnO_2. Brenet has emphasized the role of the crystal structure of the MnO_2 throughout his studies. He is given the credit for first suggesting the possibility of lattice dilation without additional evidence for a second manganese oxide phase in the MnO_2 cathode. Brenet has pointed out(92) that this observation has been confirmed by several other workers(250)(81), Laurent and Morignat(93), Huber(94), and Gosh(95). He stresses that the same phenomenon has been found by Feitknecht(96) with chemical reduction of MnO_2.

Bode and Schmier(97) examined chemically reduced samples of both gamma and beta types of MnO_2 in order to determine more about the reaction mechanism. They found that the gamma MnO_2, when reduced with hydrazine hydrate gave potentials that continuously decreased with the extent of reduction, thus indicating a homogeneous reduction. The beta MnO_2, under similar conditions, gave potentials which, after an initial drop, maintained the same level during reduction from $MnO_{1.97}$ to $MnO_{1.6}$, thus showing that a heterogeneous reduction operated.

Kozawa and Powers(98) have examined the discharge mechanism of the MnO_2 cathode in ammonium chloride solutions with and without added zinc chloride. They have focused their attention on that initial part of the discharge, that is, until the MnO_2 is reduced to $MnO_{1.75}$. They have listed four possible reaction mechanisms.

a. Proton-electron:

$$MnO_2 + H_2O + e^- \rightarrow MnOOH + OH^- \qquad (36)$$

b. Two-phase:

$$2MnO_2 + H_2O + 2e^- \rightarrow Mn_2O_3 + 2OH^- \qquad (37)$$

c. Mn(II) ion:

$$MnO_2 + 4H^+ + 2e^- \rightarrow 2H_2O + Mn^{++} \tag{38}$$

d. Hetaerolite:

$$2MnO_2 + Zn^{++} + 2e^- \rightarrow ZnO \cdot Mn_2O_3 \tag{39}$$

On the basis of two criteria:

1. the slope of the pH-potential curve, and
2. the potential change versus the value of x in MnO_x

they have concluded that the proton-electron mechanism must be the reaction that is taking place at the beginning of the discharge for both gamma and beta MnO_2 which contains practically no water or OH^-. They point out that this conclusion agrees with the fact that both Mn^{++} ion and hetaerolite are not produced at the early stage of the discharge. They recognize that more complex reactions develop as the discharge proceeds and produce such products as Mn^{++} and hetaerolite.

A unique approach to the problem of reaction mechanism in the Leclanché cell has been used by Miyazaki(99). He utilized the electron probe and x-ray diffraction to study the distribution of manganese, oxygen, and zinc, and the types of compounds formed in individual particles of MnO_2 taken from cells after discharge. This study followed his parallel experiments with MnO_2 discharged in alkaline MnO_2-zinc cells(100). He found in Leclanché cells that low-quality natural ore showed very little penetration of zinc into the MnO_2 particle. However, with the higher quality synthetic oxides studied, a considerable amount of zinc entered the MnO_2 particle to form $ZnO \cdot Mn_2O_3$. Corresponding losses of oxygen from the MnO_2 particle were observed as a result of the discharge.

Most of the mechanisms presented in the foregoing paragraphs are concerned with the steps in the reaction by which the main trivalent manganese products of the cathode reaction are formed. These theories are not at variance with the electrochemical-chemical mechanism, proposed originally by French and McKenzie and expanded later by Cahoon and his associates. These workers consider that MnO_2 is reduced electrochemically to Mn^{++} which passes into solution either as the simple Mn^{++} ion or as a complex of some type. The Mn^{++} ion or complex then reacts with remaining MnO_2 to give $MnOOH$ or with Zn and MnO_2 to give $ZnO \cdot Mn_2O_3$. The individual steps in this mechanism were described earlier in the chapter. This theory satisfies the three requirements presented at the beginning of this section. The electrochemical-chemical theory offers a basis for the practical cell design of cell formulations, a feature of considerable importance in the production of commercial

Leclanché cells. Most of the other mechanisms described have not been carried that far.

MATERIALS OF CONSTRUCTION

One of the important reasons for the existence and continued growth of the Leclanché cell industry is the ready availability of the raw materials used in its construction. Although the basic materials have not changed materially since the days of Leclanché, there have been many improvements in both their preparation and in the manner in which they have been applied in dry cell manufacture. Some of these improvements were discovered by the empirical method and therefore may be properly classified as a part of the "art of dry cell manufacture." Other improvements resulted from carefuly planned research on the improvement of battery technology. These may be classified as a part of the "science" associated with dry cell improvement. Both of these aspects of dry cell manufacture are discussed in this chapter.

This section on the materials of construction used in dry cells discusses: depolarizers of the manganese type; carbon components; zinc; separators; and packaging and sealing materials.

Depolarizers

In the Leclanché cell system the word depolarizer means the cathode reactant which is some type of manganese dioxide. This material is available from a number of sources and is described under the headings: natural MnO_2 ores; activated MnO_2; chemical synthetic MnO_2; electrolytic MnO_2; and general considerations on MnO_2.

Natural MnO_2 Ores

Manganese ores are found in many parts of the world and occur as carbonates, silicates, oxides, and other forms. Large deposits are known in the U.S.S.R., India, South Africa, Brazil, Morocco, China, Ghana, with smaller deposits in the United States, Australia, Greece, and so on. The only type of manganese ore that is useful in Leclanché cells is the dioxide while the lower oxides are mainly used as a source of manganese in metallurgical operations. Table 1.22 gives a general classification of the manganese oxide and other important ores. Divalent elements, other than manganese, may be incorporated in the complex lattice as is shown there. Barium, potassium, iron, lead, and zinc are the most common elements found. Gouge and Orban(248) have described many aspects of MnO_2 from 1909 to the present.

Natural manganese dioxide is usually assumed to have been formed by

TABLE 1.22. VARIOUS TYPES OF MANGANESE ORES

I MnO$_2$ Types
   Pyrolusite
   Ramsdellite (rare)
   Psilomelane contains Ba and H$_2$O
   Cryptomelane contains K
   Hollandite contains Ba and Fe
   Coronodite contains Pb

II Mn$_2$O$_3$ Types
    Bixbyite often contains Fe
    Manganite MnOOH
    Hetaerolite ZnO·Mn$_2$O$_3$

III Mn$_3$O$_4$ Types
     Hausmanite
     Hydrohausmanite
     Braunite often contains Si
     Jacobsite often contains Fe

IV Carbonate
    Rhodochrosite

the long-time oxidation of manganous carbonate since some MnO$_2$ mines have underlying carbonate deposits with the MnO$_2$ in the layers nearer the surface. However, the residual carbonate content of MnO$_2$ ores is so low that a direct oxidation may be questioned. Zwicker et al.(101) have identified manganoan Nsutite which they believe is a probable intermediate in the formation of the Nsutite MnO$_2$ found in Ghana ore. They found that the manganoan Nsutite is deficient in oxygen with a ratio of MnO$_2$:MnO::5:1. When this ore constituent is heated at 150°C, it is converted to a formula ratio of MnO$_2$:MnO::20:1 and the x-ray pattern agrees with the gamma or rho crystal structure characteristic of the Ghana ore. They have given the name Nsutite to the naturally occurring ore from Nsuto, Ghana, Africa, where it is abundant.

The properties of natural MnO$_2$ ores from different locations are so different from one another that it has become customary to identify an ore by the name of the mine or location from which it came rather than the mineralogical name. The differences in properties are often reflected in the battery behavior, a fact that Leclanché recognized(1). It is difficult to identify the sources of the MnO$_2$ ores used by the early experimenters. Prior to World War I, Caucasian ore, from the deposits in the Caucasus mountains in Russia, was generally used in the United States for battery depolarizers. The international upheaval of World War I shut off the supply of this ore and battery manufacturers turned to sources in North America.

The deposits at Phillipsburg, Montana, contained ore that was softer and lower in MnO_2 content than the Caucasian ore previously used. These handicaps were overcome and the Montana ore came into general use in the industry. After World War I the discovery of natural ore in what is now Ghana opened this source to the battery industry. Ghana ore appealed to the battery technologist because of its higher MnO_2 content and generally greater effectiveness in cells.

Table 1.23 shows the compositions and some of the properties of a representative group of natural ores. The differences in MnO_2 content often mislead the layman into thinking the ore with the highest content should be the most satisfactory. Usually the battery quality of an ore is a complex resultant of many factors, some of which, for example, surface area, have been discussed earlier in this chapter. It is surprising that natural MnO_2 ores from many parts of the world are remarkably free from large amounts of heavy metal impurities and are therefore suitable for direct use in cells without chemical purification. Some ores, that is, those from the southeastern United States, contain copper in small but significant amounts which prevents their use in cells. Iron is usually present in the form of Fe_2O_3 which is not harmful since it is insoluble in dry cell electrolyte at the pH normally employed.

Natural MnO_2 usually occurs in combination with clay and other minerals. The MnO_2 is upgraded by one or more of several methods to eliminate as much as possible of the useless ingredients. Flotation, separation by differences in density in heavy fluid media, magnetic separation, water washing, and even hand picking have been used. the highest grade is thus selected for battery use. The lumps of ore produced in any of the above processes are dried and broken so that they may be fed to

TABLE 1.23. COMPOSITIONS OF NATURAL ORES

	African (%)	Montana (%)	Nonoalco Mexico (%)
MnO_2	77.4	68.6	74.0
Mn	51.8	—	—
Peroxidation	94.4		
Lead	—	0.185	0.02
Copper	0.12	0.023	0.002
Iron	2.52	1.36	7.5
Arsenic	0.04	0.031	0.005
Nickel	0.04	—	0.02
Cobalt	0.01	—	0.013
Antimony	—	—	0.001

a milling operation. Milling is usually done in a ball or roller mill and the product is often a 200 mesh nominal size.

Activated MnO₂

In their efforts to improve the battery quality of natural MnO_2 ores, the early technologists must have considered many possible methods. Leaching of ores with dilute acids to remove acid-soluble materials(102), such as carbonates, was probably used experimentally but no record of its commercial use has been found.

One early attempt to increase the battery value of a natural ore was developed by Burgess(15). His method consisted of several steps.

1. Roasting MnO_2 ore at 600°–700°C to convert it to Mn_2O_3.
2. Leaching the Mn_2O_3 with dilute H_2SO_4 to permit the reaction

$$Mn_2O_3 + H_2SO_4 \rightarrow MnO_2 + MnSO_4. \tag{40}$$

3. Filtering to separate the solid product, plus washing and drying.

This process provided a beneficiated or activated MnO_2 as distinct from a truly synthetic product. Tables 1.15, 1.16, 1.17, and 1.18 present some of the properties and the composition of representative samples of activated MnO_2. Muller, Tye, and Wood(22) have described the ion-exchange properties of this type of material.

The above process is often combined with other processing steps to recover the Mn values in the $MnSO_4$ solution as MnO_2. Tamura(103) shows a schematic for a chemical method of recovery which may be applied.

The activation method described above has many attractive features in the simplicity of both process and equipment needed. It usually produces a more active MnO_2 than the ore initially used. However, since the acid insoluble gangue in the ore remains with the MnO_2 product, this process is not recommended for ores containing large amounts of such materials. Activated MnO_2 was widely used as a commercial depolarizer during World War II when supplies of electrolytic and synthetic MnO_2 were quite inadequate to meet the demands of a wartime economy.

An activated MnO_2 prepared by the Hanshin Yosetsu Kizai Co. of Japan has been described(104). The properties of the product and a discharge curve comparing the performance of AA size experimental cells made with the activated and the natural ore are given. The process of activation is not described and it is assumed that these authors have used the process given earlier in this section. The author emphasizes that the cost of an activated ore of this type is more nearly equal to that of a natural ore and much less costly than a synthetic MnO_2 of either the chemical or electrolytic types. Table 1.24 shows a comparison of data given in Table 1.18 for African ore

TABLE 1.24. A COMPARISON OF COMPOSITIONS AND PROPERTIES OF ACTIVATED AND NATURAL MnO_2 FROM THREE SOURCES

	African Table 1.18		Hanshin		Fiji	
	African Ore	Activated MnO_2	Natural Ore	Activated MnO_2	Natural Ore	Activated MnO_2
Total MnO_2	84.81	73.79	75.0	72.0	79.96	69.30
Total Mn	54.99	50.83	—	—	50.3	45.3
Free MnO_2	82.61	67.27	—	—	—	—
Total Mn Oxides	86.61	79.30	—	—	—	—
Real density	4.49	4.09	—	—	—	—
BET surface area m^2/g	7.4	50.5	14	35	—	—
Crystal type	Rho	Alpha			Beta	Gamma
SiO_2			8.0	13.0	9.56	13.38
Fe			2.0	2.0	1.44	2.00
MgO					0.13	0.10
CaO					0.13	Trace

and activated African ore with the data for the corresponding Hanshin materials. The discharge curve of the AA size cells discharged on 10 ohms at 20°C shows a higher curve for the cells made with activated MnO_2 than for the natural ore. The activated MnO_2 cell gave 159 minutes and the natural ore cell gave 121 minutes to the 0.80 volt cutoff, thus showing a significant advantage for the activated MnO_2.

Tamura(103) presents an excellent review of the work done on the activation process in Japan. His data on Fiji ore before and after activation are included in Table 1.24. A discharge curve showed that the activated Fiji ore performed comparable to the electrolytic MnO_2.

Chemical Synthetic MnO_2

This section is limited to a discussion of synthetic MnO_2 prepared by chemical methods, leaving to the following section the similar discussion of electrolytic MnO_2.

The use of synthetic depolarizer seems to have originated in Germany at a comparatively early date, presumably because of the availability of by-product MnO_2 as obtained, for example, in the manufacture of saccharin(105). In the United States application followed some time later, although, by 1910, materials of local manufacture as well as imported varieties were actively considered. Substantially all known methods for the production of MnO_2 that seemed commercially practical (Dubois)(106) have been examined either here or abroad, including the reduction of permanganate(107), the oxidation of manganous compounds by various reagents, (108), (109), and even the thermal decomposition of salts like the nitrate(110). In general, the gain in depolarizing activity and the improvement in heavy drain service of fresh cells obtained with such depolarizers was offset by nonuniform and generally unpredictable service and poor shelf characteristics.

The demand arose again in the 1930s for a synthetic depolarizer for use in premium quality cells. Many samples were offered to the manufacturers for trial in their product. A synthetic MnO_2, said to have been prepared as a by-product of saccharin manufacture(105) was selected and imported in considerable quantities from Germany for this application. This product is identified as a chemical synthetic MnO_2 in several tables and figures in this chapter. It was characterized by a content of up to 10 percent combined water and up to 10 percent K_2O and or Na_2O.

More recently a limited number of synthetic chemical processes have been considered, such as:

1. Dean process using ammonium carbamate [Dean(111) and Welsh(112)].
2. Bradley Fitch process using ammonium sulfate [(113), Henn et al. (114)].
3. Nitric acid process [Kaplan(109) and Nossen(115)].
4. Chlorate process [Moore(58)].

None of the above processes appear to have been commercialized. The only chemical synthetic MnO_2 commercially available is produced by the Diamond Shamrock Chemical Co. (116) in the United States and by Societé Européenoe des dèrives du Manganése, (Sedema) in Belgium.

Welsh has pointed out(116) that U.S. Pat. 2,956,860(117) describes the general steps used in the preparation of type "M" MnO_2. The process is based on the reaction

$$5\,MnSO_4 + 2\,NaClO_3 + 4\,H_2O \xrightarrow{\,H^+\,} 5\,MnO_2 + Na_2SO_4 + 4\,H_2SO_4 + Cl_2 \uparrow$$

$$(41)$$

The following steps are involved in the process:

a. The preparation of $MnCO_3$;
b. Roasting $MnCO_3$ to give an 80 percent content of MnO_2.
c. Leaching the roasted mixture with H_2SO_4 to extract the soluble manganese as $MnSO_4$.
d. Reacting the resulting slurry, containing the solid particulate MnO_2 with $NaClO_3$ at 80–90°C.
e. Separating the resulting solid product from the liquid phase followed by washing and drying the product.

The patent emphasizes that the above reaction operates most effectively in the presence of a certain amount of "catalytically active MnO_2" added with the ingredients or in the process itself. Thus the MnO_2 formed by roasting $MnCO_3$ serves so that steps c and d above may be combined.

The $MnCO_3$, originally produced by the carbamate process (111), can be duplicated by the controlled precipitation of soluble manganese with a soluble carbonate. The $MnCO_3$ must be made in such a way that it can be readily oxidized. The susceptibility to oxidation is controlled by the precipitation conditions and is independent of those conditions that optimize the air oxidation step. Also, the $MnCO_3$ must possess certain density and particle size characteristics since these properties are carried through to the final product. Thus, rather precise processing conditions must be maintained.

The step of washing the dioxide product after the oxidation step is also important. The "M" material has a high surface area which is active with respect to hydrogen ions. Large quantities of water are therefore required to remove the excess acidity.

Table 1.25 shows representative chemical and physical properties of type "M" MnO_2. As can be seen from the table the product is a high-grade synthetic MnO_2. It has found commercial use in a variety of cells which are specially formulated for specific applications. Figure 1.37 shows a comparison of the curves obtained by differential thermal analysis on the type "M," electrolytic MnO_2, and a sample of beta MnO_2 currently available from the Baker Chemicals Co. The beginning of decomposition of the MnO_2 under such test conditions is shown by the temperature at which the main vertical peak near the middle of the chart starts to form. Thus, for type "M" this occurs at 600°–800°F, for electrolytic at about 1050°F, and for the beta MnO_2 at about 1050°F. The temperatures at which the peaks reach their apex is about 1000°F for type "M," 1000°F for the electrolytic MnO_2,

TABLE 1.25. CHEMICAL COMPOSITION AND
PHYSICAL PROPERTIES OF TYPE "M" MnO_2

Total MnO_2	90.0%
Total Mn	60.0
Iron	0.1
Water	3.0
Alkali metals	0.3
Alkaline earths	0.4
Heavy metals	< 0.01
Insoluble in HCl	0.30
pH	4.5–6.0
Particle size	100% through 200 mesh
Surface area (BET)	80 m^2/g
Apparent density	21–24 g/in.^{3}
Kerosine absorption	36.5 ml/100 g

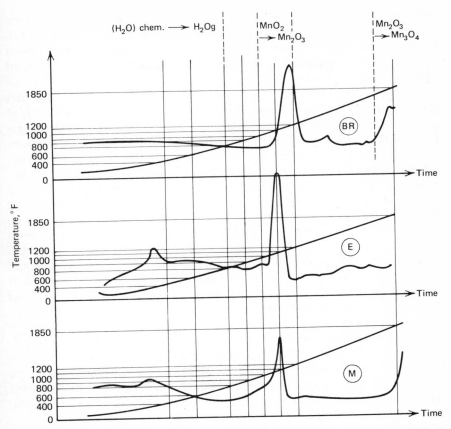

(H_2O) chem. $\longrightarrow H_2Og$ | MnO_2 | Mn_2O_3
| $\longrightarrow Mn_2O_3$ | $\longrightarrow Mn_3O_4$

Figure 1.37 DTA graphs for different types of MnO_2. M above denotes type "M", E electrolytic, and BR beta MnO_2.

and about 1200°F for the beta variety. Although the differential thermal analysis measures properties of an MnO_2 sample quite different from its reduction in a dry cell the method has been used to help characterize MnO_2 from different sources.

The Welsh patent describes two other chemical synthetic MnO_2 varieties, type "W" and type "WB." The processes for these materials differ from that used for the type "M" in that Mn_3O_4 is used as an important ingredient. The Mn_3O_4 could be made by heating an MnO_2 ore. In preparing the type "W" the Mn_3O_4 powder is added slowly to a solution containing $NaClO_3$ and H_2SO_4 at 85°C and maintained at this temperature for from 3–5 hours. The product is a black oxide, one example contained 91 percent MnO_2 and 59.6 percent Mn.

In preparing the type "WB" product the Mn_3O_4 is added to the cold solution containing $NaClO_3$ and H_2SO_4 and the mixture slowly heated to 85°–87°C and maintained there for about four hours. The product is a brown color and one example cited contained 90.2 percent MnO_2 and 60.0 percent Mn.

The types "W" and "WB" are interesting from the point of view that they represent a nearly pure rho crystal structure. They are indeed considerably more crystalline than the type "M" and have correspondingly lower surface areas. They have found limited use in battery applications but the general applicability and economy of the "M" type has, to date, relegated the "W" and "WB" products to a much less important role.

Honda, Mizumaki, and Tanabe(118) have described two types of chemically prepared synthetic MnO_2 made by the Japan Metals and Chemical Co. The manufacturing processes are not given but data on composition, pore volume, electron microscope pictures and experimental cell performance are presented in comparison to that of electrolytic MnO_2. The two products, types A and C differ somewhat in surface area and liquid absorption but are very similar in composition. The type C material is shown to have a gamma crystal structure on x-ray diffraction very similar to that of the electrolytic MnO_2. Differential thermal analyses also show decomposition occurs at very nearly the same temperatures. The discharge curves of the experimental "D" size cells show similar discharge curves to those with electrolytic MnO_2 although the closed circuit voltages levels and the actual total outputs are not identical. These differences may be explained in part by the difference in liquid absorption and the resulting difference in actual MnO_2 content of the cells made with these products. Table 1.26 shows data taken from their paper.

Tamura(103) has reported the process used by the Japan Metals and Chemical Co. in the form of a flowchart. It is shown in Fig. 1.38. There is no clear indication as to how type A differs from type C in preparation. Discharge curves on cells made with these products show the similarity to those of the electrolytic MnO_2. Tamura reports attempts by others to develop processes for chemical synthetic MnO_2. He also suggests a different scheme in another flow diagram in his paper.

Undoubtedly, there are many methods of preparing a synthetic MnO_2 by a chemical process. However, it seems that the economic situation may influence the choice of the process as much as any other factor, except for the performance of the product in dry cells.

Electrolytic MnO$_2$

Electrolytic MnO_2 had long tantalized the battery technologist as an attractive depolarizer for dry cells especially since it was produced in the

TABLE 1.26. COMPOSITIONS AND PROPERTIES OF CHEMICALLY PREPARED SYNTHETIC MnO$_2$

	Chemical Synthetic MnO$_2$		Electrolytic MnO$_2$
	Type A	Type C	
MnO$_2$	90.3	92.4	92.8
MnO	3.0	2.1	1.7
x in MnO$_x$	1.96	1.97	1.98
Combined H$_2$O[a]	1.7	1.9	4.4
SO$_4$	0.44	0.30	1.06
Fe ppm	75	55	190
Ni ppm	12	12	10
Co ppm	4	5	9
Cu ppm	2	3	5
Pb ppm	63	63	840
Sb ppm	3	3	3
Mo ppm	2	2	1
As ppm	1	1	1
pH[b]	4.50	4.20	4.80
Specific density	4.2	4.2	4.5
Surface area m^2/g			
Kozawa method	88	64	44
BET method	86	72	32
Air perm. method	3.05	1.59	0.29
EMF at 5.0 pH	645	663	688
Activity[c]	55	56	55

From Honda, Mizumake, and Tanabe (118).
[a]Water released from a sample dried at 105°C by heating to 750°C.
[b]pH determined by immersing sample in a 20% NH$_4$Cl solution at pH 5.4.
[c]Determined by reaction with hydrazine.

refining of zinc as an anode mud. However, samples of this material contained large amounts of lead and calcium sulfate and all efforts to find an economical process for removing them failed. The process for preparing MnO$_2$ by direct electrolysis had been studied in the laboratory by Van Arsdale and Maier (119) and it appeared to be a direct and straightforward operation. However, when Nichols (120) undertook to prepare pilot quantities for dry cell use, problems arose. Lead anodes were found to add too much lead to the product and so limited the shelf quality of the batteries, although fresh service results were most encouraging. The carbon and graphite anodes then available were oxidized too rapidly to be practical. Unavoidably, the project was dropped. It was only when graphite anodes, resistant to attack in acid electrolytes, were developed by Johnson (121) that a commercial process for electrolytic MnO$_2$ was developed. Fortunately the production was started in 1940 (122) in time to

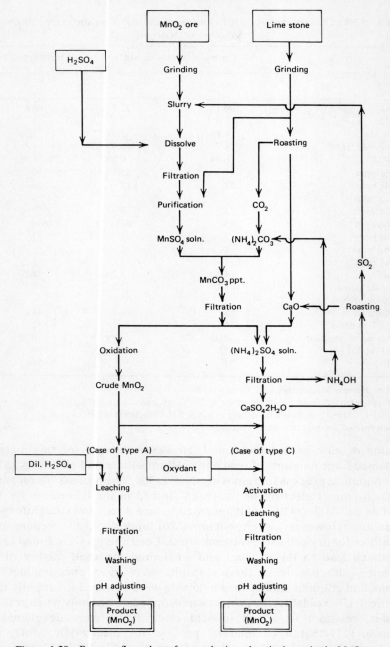

Figure 1.38 Process flow-sheet for producing chemical synthetic MnO_2.

be an important factor in providing high-quality depolarizer for batteries used in World War II. Wartime demand for batteries prompted other manufacturers to start production of electrolytic MnO_2 for their own use. Furthermore, interest in this material by the U.S. Army Signal Corps resulted in the establishment of a large plant by the Western Electrochemical Co., now part of the Kerr–McGee Corp., to prepare electrolytic MnO_2 for the battery industry.

The general process for manufacturing electrolytic MnO_2 is given in Fig. 1.39. It is probable that each manufacturer has found it necessary to adapt the general process to the special conditions presented by the particular ore used as a source of manganese. Few of the details were initially published. However, a U.S. Army Signal Corps contract with Kissin, Georgia Inst. Tech.(123) to develop the process provided many details. These workers used platinum anodes, to which the MnO_2 deposit did not adhere, to eliminate the grinding operation. All commercial plants, however, have used graphite or other anodes to which the anode deposit adheres and must be mechanically removed without breaking the anode. More recently both titanium and lead anodes have been employed on a limited scale.

Recently, a group of Japanese authors have published many details of the electrolytic MnO_2 process in a two-volume work with special emphasis on the developments in Japan. Kozawa(124) has reviewed the process of making electrolytic MnO_2 thoroughly. Miyazaki and Suemoto(125) as well as Kato(126) have discussed particular aspects of the process. Takahashi and Kozawa have discussed the production of electrolytic MnO_2 in Japan for the period 1957 to 1970(127). Koshiba and Nishizawa(128) have published on the crystal structure of electrolytic MnO_2. Sudo(129) presents the properties of a new type of electrolytic MnO_2 prepared by an undisclosed process. In the process shown in Fig. 1.39, both the ore reduction and the stripping of the anodes are so-called batch operations. Araki(130) has attempted to convert these steps in the process to continuous processes. He used a tower for leaching the $MnCO_3$ feed material on a continuous basis. He also devised a continuous electrolytic cell in which the electrolytic product was Mn^{3+} which later disproportionated to form MnO_2. His process appears similar to that described by Welsh(131) which was characterized by the use of much stronger sulfuric acid electrolyte than the general process.

In general the leaching operation shown in Fig. 1.39 is designed to produce an electrolyte which is about 1 M in $MnSO_4$. As pointed out by early workers, Van Arsdale and Maier(119), iron must be removed from the electrolyte in order that a maximum current efficiency may be attained. The iron removal results in an electrolyte that is slightly acid. Sometimes further purification is obtained by the addition of soluble sulfide salts

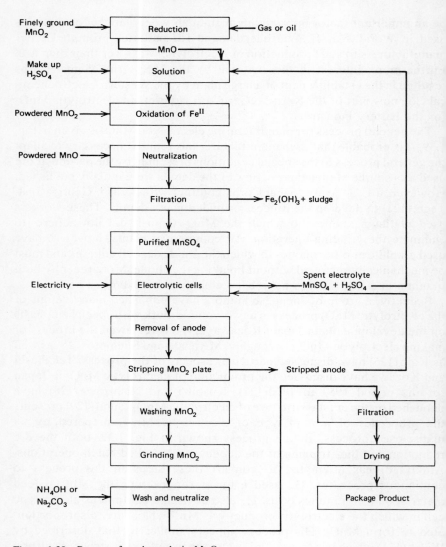

Figure 1.39 Process for electrolytic MnO_2.

followed by another filtration to remove the heavy metals which may be present in the ore used(124).

The purified electrolyte is fed continuously to a series of electrolytic cells and the spent electrolyte emerging from these cells is, as expected, lower in manganese content and proportionately higher in H_2SO_4 content.

As shown in the figure, it is recycled to become regenerated in $MnSO_4$ content. Plating is usually done at 1.5 to 3.0 amp/dm^2 with the electrolyte at about 90°C. A current efficiency of about 60 percent is usually attained(132).

With graphite electrodes plating is allowed to proceed for several days to build up a plated layer thickness of from $\frac{1}{4}$ to 1 inch, (0.6 to 2.5 cm). The anodes are then removed, washed and the MnO_2 deposit removed mechanically, taking precautions that the graphite anode is not broken in this process. The anode is then ready for replacement in the electrolytic cell for use again. The MnO_2 product which is obtained in this manner consists of a variety of broken lumps of many sizes. It is then washed and milled to meet the specifications required for the particular battery application for which it is intended. Since the product contains some residual acid at this stage it is then neutralized by the addition of NH_4OH solution or solutions of Na_2CO_3, KOH or NaOH. For Leclanché cells, neutralization with NH_4OH is adequate, but if the product is intended for use in alkaline cells, use of KOH or NaOH solutions is preferred.

Table 1.27 shows the composition of several samples of electrolytic MnO_2. Three of the samples are from the Japanese manufacture of recent dates while the fourth shows the U. S. Army Signal Corps minimum specification used in the mid 1960s. It is clear that the purity of the product has increased since that time.

Fleischmann, Thirsk, and Tordesillas(133) have studied the kinetics of the electrodeposition of gamma MnO_2 on platinum electrodes. They conclude that the slow stage of the reaction is a rate-determining dehydration to form gamma MnO_2:

$$Mn(H_2O)_{n-4}(OH)_{ads} \rightarrow MnO_2 + (n-2)H_2O \tag{42}$$

Electrolytic MnO_2 usually possesses the gamma crystal structure although some workers have found that the beta variety may also be formed. Storey et al.(132) pointed out that if the temperature of electrolysis were low the beta type would be produced. A detailed investigation(134) has shown that the degree of crystallinity of the gamma type varies with electrolyte and type of anode used. In general the peak at 28° 2θ, which is characteristic of the gamma structure, tends to become higher and sharper at lower current densities and at higher electrolyte temperatures.

The discovery by Ramsdell(135) many years ago that a natural gamma MnO_2, later designated ramsdellite, by Fleischer and Richmond(136), could be converted to the more stable beta crystal structure by heating has prompted a considerable amount of study of the effect of heating on electrolytic MnO_2. It is well known that heating electrolytic MnO_2 at

TABLE 1.27. REPRESENTATIVE COMPOSITIONS OF ELECTROLYTIC MnO_2 SAMPLES

	Sudo Tekkosha Co.	Honda et al. Japan Metals and Chem. Co.	Takahashi et al. Tekkosha MnO_2	U.S. Army Signal Corps Specification
Total MnO_2	92.00%	92.8%	91.65%	85%
Total Mn	—	—	—	50
MnO	—	1.7	—	—
Comb. H_2O	1.49	4.4	1.77	—
Free H_2O	—	1.3	—	—
Total H_2O	—	—	—	3.0
SO_4	—	1.06	0.9	
Fe	0.008%	190 ppm	0.008%	0.25%
Cu	0.0003	5 ppm	0.0002	
Pb	0.0007	840 ppm	0.0005	0.25
Other heavy metals	—			0.05
Ni	0.0005	10 ppm	0.005	
Co	0.0008	9 ppm	0.002	
Sb	0.0001	3 ppm		
As	0.0003	1 ppm		
Mo	0.0002	1 ppm		
Cr	—	—	0.0003	
X-ray structure	Gamma			
BET surface area m^2/g	43.2			
pH		5.26		4.0–7.0
Particle size	92.4% below 24μ	50.2% thru 325 mesh	(A) 80% below 74μ (B) 95% below 44μ	
Real density		4.4		

temperatures below its decomposition point, where Mn_2O_3 is formed, releases some of the combined water present. This has been confirmed by studies with the thermobalance(137). They also showed that heating electrolytic MnO_2 at 200°C for 3 hours removed only a part of the combined water. However, samples prepared in this way showed a larger polarization when tested in a 3N NH_4Cl electrolyte at pH 7.

Fukuda(138) and Ninagi(139) have followed the change in x-ray diffraction when electrolytic MnO_2 is heated at temperatures up to 450°C. These show steps in the conversion of the original gamma pattern toward the beta pattern. Sasaki and Kozawa(140), Matsune(141), Kozawa and Vosburgh(82), Ninagi(139) and Muraki(142) have all investigated various phases of the change when electrolytic MnO_2 is heated. In general, these studies showed that the application of heat up to 250°C did not decrease the potential of the MnO_2 greatly but that higher temperatures were more ffective in this regard. Both Muraki(142) and a French Patent, 1,509,122(143) claim that heat treated electrolytic MnO_2 does provide increased battery service under certain conditions. Fox(144) describes the increase in electronic conductivity of gamma MnO_2 obtained by heating at 450° to 650°F under pressures of 10 to 50 tons/in.^{2}.

General Considerations on Manganese Dioxide

The Manganese-Oxygen Ratio   Neither natural nor synthetic types of manganese dioxide possess the formula $MnO_{2.00}$ where the ratio of manganese to oxygen is exactly 1:2. Both types are somewhat deficient in oxygen with formulas generally in the range of $MnO_{1.92}$ to $MnO_{1.98}$. Two explanations for the oxygen deficiency have been offered. The first, suggested by chemists, is that all samples of the oxide contain small amounts of some lower oxide, for example, Mn_2O_3. The amount is too small to appear in an x-ray diffraction pattern and the existence of this lower oxide is very difficult to prove. Even repeated digestion of a relatively pure sample of the dioxide with a mineral acid, other than HCl, does not give a product of the formula of $MnO_{2.00}$. Although the peroxidation can be increased slightly, it does not reach the desired ratio.

Efforts to prepare a pure synthetic MnO_2 by the thermal decomposition of manganous nitrate have given a product closely approaching $MnO_{2.00}$ in composition. However, the difficulties in completely removing the residual nitrate and its possible complicating effect on the analytical methods and the results, that is, the available oxygen and the total manganese contents, cannot be overlooked. Wiley and Knight(145) prepared beta MnO_2 by thermal decomposition and found that samples prepared at temperatures from 150° to 350°C closely approached $MnO_{2.0}$ in composition. Covington et al.(146) found that stoichiometric MnO_2 could be prepared by heating in oxygen when beta MnO_2 was used but that other varieties did not attain the stoichiometric value. Malati has reviewed the properties of manganese dioxide thoroughly(147).

Table 1.18 shows the compositions of a representative group of synthetic and natural oxides. These were obtained from very diverse sources and represent four different crystal structures. Yet the value of x in MnO_x ranges from 1.916 to 1.997. The fact that battery grade MnO_2 does not possess a stoichiometric composition does not apparently detract from the dry cell value.

The second explanation, offered by the solid state physicists, maintains that the oxygen deficiency is due to lattice imperfections, as evidenced by the fact that manganese dioxide possesses semiconductor properties. From the sign of the thermal EMF and the Hall constant measurements, Das(148) and Bhide and Dani(149) conclude that manganese dioxide is an

n-type semiconductor. On the other hand, manganese dioxide with p-type semiconductor properties has been reported by Elovich and Margolis(150). Malati suggests that these conflicting data could be due to different impurities in the samples studied.

The Composition of Manganese Oxides   Although the composition of both natural and synthetic manganese dioxides nominally approaches $MnO_{2.00}$, the deficiency of oxygen normally present in most samples invites the problem of calculating the true content of $MnO_{2.00}$. Many battery technologists assume the sample consists of a mixture of $MnO_{2.00}$ and Mn_2O_3. Using the analytical data for the total manganese content and the total MnO_2 content, as calculated from the available oxygen determination, the various important chemical characteristics of the sample are calculated.

The value of x in MnO_x is determined by

$$x = 1 + 0.632 \frac{\% \text{ total MnO}_2}{\% \text{ total Mn}} \tag{43}$$

The peroxidation, expressed in percent, is given by the last term in the above equation.

Part of the significance of the value of x in MnO_x has been described in the previous section. However, it furnishes a number which indicates the average degree of oxidation of all the manganese present in the sample. The term "peroxidation" represents the percentage of the total manganese present in the total MnO_2 as analytically determined. The term "free MnO_2" arises because the total MnO_2, as analytically determined, includes the MnO_2 combined in the form of Mn_2O_3. Thus the free MnO_2 is that portion of the total MnO_2 which is uncombined and therefore available for direct use in the cathode reaction. An example of the calculation follows. To make this calculation, it is assumed that the difference between the total Mn, as analytically determined, and the Mn content of the total MnO_2 is present as MnO which is combined with a part of the MnO_2 in the form of Mn_2O_3.

The composition of the electrolytic MnO_2 given in Table 1.18 has been selected for these calculations. The total Mn content is 58.07 percent and the total MnO_2 content is 87.43 percent.

$$\text{The Mn content of the total MnO}_2 = \frac{54.94 \times 87.43}{86.93} = 55.25\% \tag{44}$$

$$\text{The peroxidation} = \frac{55.25 \times 100}{58.07} = 95.10\% \tag{45}$$

The x value = 1.951

$$\text{Manganese combined as MnO} = 58.07 - 55.25 = 2.82\% \qquad (46)$$

$$\text{Equivalent amount of MnO} = \frac{70.93 \times 2.82}{54.94} = 3.64\% \qquad (47)$$

$$\text{Equivalent amount of combined MnO}_2 = \frac{86.93 \times 2.82}{54.94} = 4.46\% \qquad (48)$$

$$\text{The amount of Mn}_2\text{O}_3 \text{ present} = 3.64 + 4.46 = 8.10\% \qquad (49)$$

$$\text{The free MnO}_2 = \text{total MnO}_2 - \text{combined MnO}_2 = 87.43 - 4.46 = 82.97\% \qquad (50)$$

$$\text{The amount of total oxides} = \text{free MnO}_2 + \text{Mn}_2\text{O}_3 = 82.97 + 8.10 = 91.07\% \qquad (51)$$

The reader may be tempted to conclude that the battery quality of a particular sample of MnO_2 would be dependent on its content of free MnO_2. Unfortunately, so many other factors influence the battery value that such a conclusion cannot be made.

Crystal Structure of MnO_2 The relationship between the crystal structure and the depolarizer quality of MnO_2 samples has long tantalized battery technologists. St John[151] did pioneering work in this field and pointed out that the least crystallized samples were probably the best depolarizers. Fleisher and Richmond[136] summarized the geological and mineralogical properties of MnO_2 minerals and greatly helped the battery technologists become acquainted with this field. McMurdie[38] applied x-ray diffraction and electron microscope techniques to the types of MnO_2 that were then available in an effort to clarify the requirements of a battery depolarizer. The field has been reviewed by Ettel and Veprek-Siska[152] and also by Malati[147] in efforts to correlate and rationalize the many observations that have been published. This section summarizes the subject with the hope that the interested reader will consult the literature available for more details.

Table 1.28 presents the names of the recognized allotropic modifications of MnO_2 together with the Greek letter designations that have become shorthand symbols for them. A number of additional types, also designated by other Greek letters have been included. For example, Cole et al.[153] has designated certain samples as crystal type subgroups γ, γ' and γ'' and these have been referred to by Gattow[154] as η, η' and η''. Even the designation ϵ has been used to describe different products by different authors, for example, Glemser and Meisick[155]; Okada[156]. The ϵ type described by Pons and Brenet[157] was formed in an H_2O vapor atmosphere by the decomposition of $Mn(NO_3)_2$.

TABLE 1.28. STRUCTURES OF ALLOTROPIC FORMS OF MnO_2

Compound	Crystal System	Unit Cell Dimensions			Important Lines in X-ray Diffraction Pattern				
		a	b	c					
Pyrolusite	βMnO_2 Tetragonal	4.40	4.40	2.87	3.14	2.41	1.63	2.13	1.42
Psilomelane	Monoclinic	9.56	13.85	2.88	2.19	3.46	2.88	2.42	
Cryptomelane	αMnO_2 Tetragonal	9.84	9.04	2.86	3.11	2.40	1.54	6.9	4.9
Ramsdellite	Orthorhombic	9.27	4.53	2.87	4.07	1.61	1.36	2.54	
γMnO_2	Orthorhombic	9.35	4.44	2.85	3.98	2.44	1.40	2.14	1.61
ρMnO_2					4.0	1.65	1.61	2.43	
δMnO_2	Semi-Amorphous				7.4	1.43	4.2		

Above numbers represent angströms.

Table 1.28 shows the unit cell dimensions and positions of the important diffraction lines. Nye et al.(158) also summarized these data and
analyzed the structures of the various forms. It should be emphasized that
many samples of MnO_2 are poorly crystalline while others are highly
crystalline. Normally the beta type is characterized by sharp lines or
peaks denoting a highly crystalline structure. On the other hand the
gamma type(249) is often poorly crystallized and gives diffuse peaks.
Bystrom studied the gamma type(245). The delta type is also often poorly
crystallized.

Figure 1.40 shows a graphical presentation of the x-ray diffraction
patterns of the important types of MnO_2 available to the battery
industry(159). In this figure, a narrow vertical line represents a sharp
peak, the height of the line denoting the relative height of the peak. A
broad line denotes a diffuse peak.

Table 1.28 shows the diffraction lines expressed as the "d" values
whereas Fig. 1.40(159) shows the diffraction angle in degrees 2θ. These
two modes of presentation are related by the well-known equation

$$d = \frac{\lambda}{2 \sin \theta} \qquad (52$$

where λ is the wavelength of the radiation used. In obtaining the data for
Fig. 1.40 iron radiation was used. It has a weighted average wavelength of
1.93597 Å. Tables are available to convert the angles of diffraction
measured from the diffraction films or charts, into the corresponding "d"
values for the various types of radiation commonly used in x-ray
diffraction studies.

Miyake(160) reports that Fukuda(138)(161) has developed a com

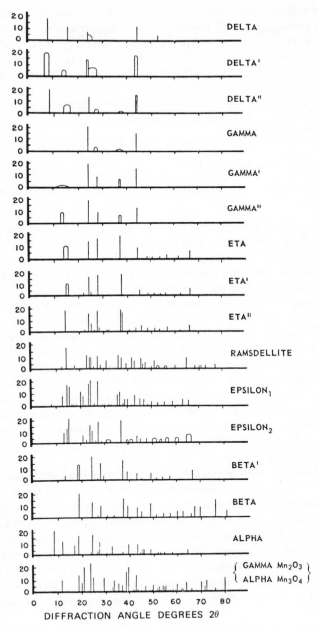

Figure 1.40 X-ray patterns of the MnO₂ minerals.

prehensive relationship between the crystal structures which differ somewhat from those presented above. Fukuda proposes two main systems.
The first, called the gamma series, contains seven phases, that is, three
synthetic hydrous ramsdellites, (s.h.r.) phases I, II, and III, three ramsdellite phases, I, II, and III, and an intermediate phase, gamma-beta, $(\gamma-\beta)$
between the two groups. The second system, called the beta series,
contains beta I, beta II, and beta III phases. On heat treatment, electrolytic MnO_2 changes in the following manner:

$$\text{s.h.r. I} \to \text{II} \to \text{III} \to (\gamma-\beta) \to \text{beta III} \to \text{beta II} \to \text{beta I} \qquad (53)$$

When natural ores are heated the following changes occur:

$$\text{ramsdellite I} \to \text{ram. II} \to \text{ram. III} \to (\gamma-\beta) \to \text{beta III} \to \text{beta II} \to \text{beta I.}$$
$$(54)$$

This concept is based on a careful study of the x-ray diffraction patterns
and has much to recommend it. It should be studied in the original text as
it is too lengthy for inclusion here.

At one time it was thought that the crystal structure of a particular
sample of MnO_2 could be used to indicate its battery value. With a better
understanding of the mechanism of the cathode reaction in the cell the
emphasis on the crystal structure has changed, leaving it today, primarily
as one important physical property but not the dominant one.

Ramsdell(135) recognized that a natural gamma form of MnO_2[4] could be
converted to the beta form by heating at temperatures of a few hundred
degrees Centigrade, that is, well below the temperature at which oxygen
loss occurs. This phenomenon has been examined by a number of
workers many of whom have applied thermogravimetric methods in their
efforts to further differentiate samples. These studies have shown that
beta MnO_2 is the most stable form and that other types can be converted
to it by suitable treatments.

Miyake(160) has summarized the thermal and hydrothermal methods
by which the crystal structures of the various allotropic forms of MnO_2
may be transformed. Milati(147) has also presented similar data in a
slightly different manner. A part of the data presented by Miyake(160) is
offered in Table 1.29. To serious workers in battery technology, much of
these data will be well known since they may have prepared several of the
conversions described. However, the tabulated data emphasize that the
crystal structure of a particular product does depend largely on the type
of process by which it was prepared. Thus, an understanding of the

---

[4]Fleischer and Richmond(136) were the first to use the name "Ramsdellite" to designate the
discovery by Ramsdell.

TABLE 1.29. CONVERSIONS OF THE CRYSTAL STRUCTURE OF MnO_2

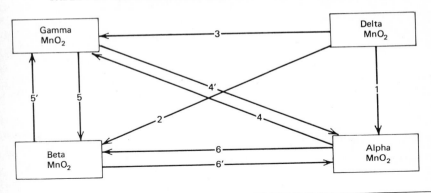

Change	Conversion		Process or Method	Reference
Delta to Alpha	1.	a.	Heat treatment at 350–500°C in air or O_2	Gattow & G 1961(154)
		b.	Autoclave in H_2O or in salt solution	McMurdie & G 1948(75) McMurdie 1944(38) Otto et al. 1944(204)
		c.	Heating with dilute acid solution	Maxwell & T. 1955(21)
Delta to Beta	2.		Heat treatment at 350–400°C in air	Gattow & G 1961(154)
Delta to Gamma	3.	a.	Heating at 120°C in air	Gattow & G 1961(154)
		b.	Heating with H_2O or dilute HNO_3	Feitknecht & M 45(76)
Alpha to Gamma	4.	a.	Boiling in dilute HNO_3	Glemser 1939(249)
		b.	Thermal decomp. to Mn_2O_3, then acid treat.	Gattow & G 1961(154)
Gamma to Alpha	4'.	a.	Heat treatment at 350–450°C in air	McMurdie & G 1948(75) Cole et al. 1947(153) Huber & S 1960(94)
		b.	Autoclave treat. in H_2O or salt sol'n.	Gattow & G 1961(154) Otto et al. 1944(204) Dubois 1935(106) McMurdie 1944(38)
Gamma to Beta	5.	a.	Heat treatment at 350–450°C in air	Gattow & G 1961(154) McMurdie & G 1948(75) Fukuda 1958(138) Huber & S 1960(94)
		b.	Autoclave with salt sol'n with oxidizer	Bystrom 1949(245)

TABLE 1.29 (*Contd.*)

Change	Conversion		Process or Method	Reference
Beta to Gamma	5'.		Thermal decomp. to Mn_2O_3, then acid treat.	Gattow & G 1961(154) Cole *et al.* 1947(153)
Alpha to Beta	6.		Heat treatment at 420–500°C in air	Buser & G 1955(53) Dubois 1935(106)
Beta to Alpha	6'.	a.	Autoclave in various salt solutions	Ninagi 1968(139)
		b.	Thermal decomp. to Mn_2O_3, then acid treat.	Cole *et al.* 1947(153)

conversion processes and their reversals goes far to remove much of the mystery originally associated with the allotropic forms of MnO_2.

The Evaluation of MnO_2 *in the Laboratory*   One of the long-time objectives in the battery industry has been the search for some laboratory method or group of determinations which would indicate the battery value of a sample of MnO_2. The hope was that such a test would eliminate the empirical approach of assembling actual cells and the time required for their tests. A great many workers have studied the problem from different viewpoints.

In the early days, most emphasis was placed on the chemical composition and purity of the material. It was well recognized that, even for materials of adequate purity, the composition of the oxide appeared to have little relation to its battery performance. Nevertheless, chemical analyses of MnO_2 samples are still a primary requirement. Improved methods such as atomic absorption spectrometry for impurity determinations, have been devised by Vink(162). Examples of representative compositions and impurity levels have been presented in earlier parts of this chapter.

Once the analysis of a sample shows it to be principally MnO_2 and the impurity levels are found to be within normally accepted limits, the technologist is faced with the question of how effectively will the sample operate in a dry cell. Many workers have devised laboratory tests in their attempts to measure the "depolarizing activity," an ill-defined term that is intended to show how successfully the material will operate in a cell. The search for such a test took two separate directions which are described under the headings: electrochemical methods, and gas evolution or chemical methods.

ELECTROCHEMICAL METHODS. It appears impossible to establish who made the first laboratory cell in an attempt to evaluate the depolarizing quality of a sample of MnO_2. Certainly the technique goes far back into the early days of dry cell technology, particularly since the sources of MnO_2 were many and the samples varied greatly in characteristics. In general, such cells were discharged on short time tests and the results bore little direct relation to the performance of cells made with the same depolarizer in cells used on the lighter drains common at that time. In retrospect, this lack of correlation may not have been as much a failure of the laboratory cell concept as it was a lack of understanding of the cathode reactions under various drain conditions. In any case the situation prompted a search for improved methods.

The very fact that the value of an MnO_2 sample in any cell lies in its property to furnish electrochemical energy kept technologists working with the laboratory bench cells as a direct approach to this intriguing problem. In addition to the hoped-for direct evaluation of MnO_2, such cells offered a means of studying other aspects of cathode operation, such as the relative effects of different types of carbon, various mix ratios, and electrolyte compositions. One versatile test apparatus, the utilization test unit(14) described earlier, was based on earlier work by French and MacKenzie(47). It had two unique features. The first was the provision of a continuous supply of fresh electrolyte to the cathode throughout the test. The second was a manually operated method of reading the open circuit potential of the cathode immediately after the circuit was opened. This scheme eliminated the problem of the IR drop in the unit. A considerable degree of success was obtained with this device in predicting the performance of many types of MnO_2 on a limited battery test. The failure of the test to predict the quality of a small number of MnO_2 samples definitely reduced its use as described earlier.

Morehouse and Glicksman(163) and Glicksman(164) developed a bench type cell which did not use flowing electrolyte. They employed it to test MnO_2 and other oxides as well as in their extensive studies on organic depolarizers. They too recognized the need to eliminate the IR drop and used an oscilloscope technique to obtain the open circuit potential values of their experimental cathodes.

Appelt and Purol(165) described their method, using a flowing electrolyte, for discharging a sample of MnO_2. They showed good correlation between the results obtained on their bench cell and the actual battery performance of 20 different samples of MnO_2.

Kornfeil(166) described still another type of a laboratory bench cell in which a nonflowing electrolyte was used. He utilized the interrupter developed by Marko and Kordesch(167) to eliminate the IR drop from the readings. In this manner he made the measurement of the open circuit potential of the cathode much easier. He too showed good correlation between the discharge time to 1.0 V, obtained on the bench cell, and the service of "A" size dry cells made with the same depolarizers discharged on 16 2/3 ohm continuous test.

Huber and Kandler(168) devised a new technique for preparing their experimental cathode. Particles of MnO_2, of about 40 microns diameter, were spread in a single layer on a conductive graphite-plastic foil. Suspensions of the MnO_2 particles in both an organic solvent solution of collodium and in aqueous media were used to prepare this thin layer. The electrodes were discharged in an aqueous ammonium chloride solution under both stirred and unstirred conditions. The discharge curves of both beta and gamma types of MnO_2 were in agreement with previous results and concepts.

Balewski and Brenet(169) described a bench cell in which a single layer of MnO_2 particles, selected between 90 and 110 microns in diameter, were embedded into the surface of a foil conductor under pressure while heated to 70°C. This method made the use of an organic solvent unnecessary (Huber and Kandler(168)) but it limited the amount of MnO_2 to about 18 mg/cm^2 of foil contacted. It too provided a cathode consisting of a single layer of MnO_2 particles, every one in contact with the conductive foil. A deoxygenated electrolyte was supplied to the cell. These authors showed that the discharge curves for gamma MnO_2 have two plateaus while those for beta MnO_2 have a single plateau. They presented the usual discharge curves along with charts showing the relation between the change in coulombs per change in volts potential plotted against the potential for several current densities. These experiments were directed toward a better understanding of the complex nature of MnO_2 during its electrochemical reduction and not toward the prediction of MnO_2 capacity in actual cells. However, the techniques presented by both Huber and Kandler and Balewski and Brenet offer unique methods of contacting the MnO_2 sample being tested.

Schweigart(170) and Olivier(171) have published on the subject of testing MnO_2 for electrochemical properties. Based on their work the Chemical Engineering Group of the South African Council for Scientific and Industrial Research has designed and marketed an instrument, designated the "Pulse Galvanostatic Analyzer,"(172) for evaluation of MnO_2 samples. The test cell is subjected to a constant discharge current which is interrupted periodically for short periods for measurements.

The results of such tests and the degree of satisfaction with the results that are obtained will necessarily vary with the objectives and the sophistication of the technologist. From the examples cited above and many others there quite obviously is a more or less direct relation between the performance of a sample of MnO_2 in a bench cell and its operation in an actual battery on selected tests. Whether or not this relation is an adequate basis for preliminary sorting of MnO_2 samples or for more extensive predictions of battery quality is a question. Thus the technologist is cautioned to use such procedures with care and to make comparisons with the performance of his sample in actual cells.

GAS EVOLUTION OR CHEMICAL METHODS.   Many chemists have believed that there should be some relation between the chemical reduction of a particular sample of MnO_2 and its electrochemical reduction in a dry cell. To that end, tests based on the chemical reduction have been devised. The first of these to receive attention was an oxalic acid reduction described by D'Agostino(173). In this test the reaction between MnO_2 and oxalic acid, in the presence of sulphuric acid, released carbon dioxide. The amount evolved and the rate of evolution were followed. The test was applied to natural ores, electrolytic and chemically prepared MnO_2 samples. The natural ores showed the lowest activity while the chemically prepared samples were most active with the electrolytic samples at an intermediate level. There was an objection that this test required an acid environment, quite different from the neutral to alkaline pH found in a battery cathode, and so the significance of the results were subject to some question.

A second chemical test was described by Drotschmann(244) and several other workers, for example, Fukuda(161), in which the MnO_2 sample is exposed to an alkaline solution of hydrazine sulfate under specific conditions. Nitrogen gas is released and the evolution is followed for about one half hour. Drotschmann(174) provided gas evolution data versus time for electrolytic MnO_2 as well as Ghana and Caucasian ores. He showed that the gas evolution rate is high at the beginning of the test but slows appreciably after 10 minutes. He pointed out that the quantity of gas evolved is greatest for the electrolytic sample, intermediate for the Caucasian and least for the Ghana material—roughly in the order of their BET surface areas of: electrolytic = 40, Caucasian = 10–13 and Ghana = 8 m^2/g. However, he recognized that Ghana ore was superior to Caucasian in actual cell performance and pointed out that the hydrazine method of testing lacked complete agreement with cell data in this comparison.

Fukuda(161) has also applied the hydrazine method for measuring the depolarizing activity of samples of MnO_2 from various sources. From this study he concludes that the chemical reduction of MnO_2 by hydrazine is similar enough to the electrochemical reduction in dry cells that the former may be used to characterize the samples. In his studies the activity was measured by adding a known amount of dried powdered MnO_2 to a special glass flask. A nitrogen atmosphere was maintained while measured volumes of 0.5 M NaOH and 0.1 M $N_2H_4 \cdot H_2SO_4$ solutions were added. The mixture was shaken for a specified time for the reaction between MnO_2 and N_2H_4 to occur. Then the concentration of N_2H_4 in the solution in the flask was determined with 0.1 M Iodine solution.

Table 1.30 presents a summary of some of the data offered by Miyake(160). It appears that the activity, determined by the above method, is generally related to the BET surface area of the MnO_2, although there are some exceptions in the table.

As has been pointed out by many authors the operation of an MnO_2 cathode is quite complex and it may be too much to assume that any single test will provide an evaluation of the many characteristics necessary for its operation in a dry cell. Nevertheless, as the above paragraphs indicate, the two groups of tests do provide some strong indications of the relative battery quality of an untried sample of MnO_2. As the technologist studies such comparisons he may reach the point where he wonders which route is best for the evaluation of a new sample of MnO_2, (i) a schedule of bench tests, or (ii) actual cell manufacture and testing. In all probability the serious worker will use both methods.

CARBON COMPONENTS

There are three types of carbon components normally used in dry cell manufacture. Two of these, acetylene black and graphite, are finely powdered materials used in the cathode mix while the third are the formed carbon or graphite parts, for example, electrodes used for electrically contacting the cathode mix. They are discussed separately below.

Powdered Carbon and Graphite

Leclanché recognized the advantages of adding powdered carbon to the cathode of his early cells because of the relatively low electrical conductivity of the particles of MnO_2.

The complex technology of producing uniform carbon powders had not developed at that time and the battery technologist of today can only imagine the frustrations that could have developed in making early batteries. Fortunately the technology of carbon and graphite developed in the United States at the turn of the century and provided timely help that the struggling dry cell business needed. Powdered coke was an improvement over powdered charcoal used in the manufacture of cathodes for early dry cells. Natural graphites of higher electrical conductivity than coke were also used but these often suffered from the presence of heavy metal impurities which shortened cell life considerably. The discovery of a method of making artificial graphite by Acheson(175) gave a further

TABLE 1.30. PHYSICAL AND CHEMICAL PROPERTIES OF MnO_2 FROM VARIOUS SOURCES

MnO_2 Source	Depolarizing Activity	Crystal Phases	Surface Area BET	Composition	
				% MnO_2	% Mn
Electrolytic	>100	s.h.r., I	28.2	90.7	53.7
Electrolytic	>100	near s.h.r., I, β III	30.0	90.0	53.8
Electrolytic	>100	s.h.r., II, β III	53.2	89.2	56.0
Electrolytic	>100	s.h.r., III, β III	52.0	90.9	—
Mn_2O_3 treated with H_2SO_4	>100	s.h.r., II, α	—	80.2	—
Precip. from $MnSO_4$ by $(NH_4)_2S_2O_8$	>100	near s.h.r., I, α	34.7	91.9	—
Precp. from $MnSO_4$ by KNO_3 and $KMnO_4$	96.7	near s.h.r., I, α	75.3	81.0	—
From $KMnO_4$ Soln + HCl	91.2	δ	44.9	80.3	40.5
Foreign ore	68.7	δ	31.6	67.9	53.3
Autoclaved MnO_2 in KCl Soln	65.1	α	58.2	89.5	55.0
Chemically prepared	59.5	s.h.r. III, β III	23.9	84.7	46.2
Japanese Ore A	52.0	Ram. IV, β II, β III	26.1	81.2	49.5
Japanese Ore B	41.9	Ram. IV, β III	16.1	73.6	50.4
Japanese Ore C	40.5	Ram. III, β II, β III	13.6	77.0	50.8
Japanese Ore D	38.6	Ram. IV, β I, β III	14.0	77.0	50.7
Japanese Ore E	36.6	α	27.7	77.7	42.9
Japanese Ore F	34.8	Ram. IV, β III	26.1	75.3	52.1
Japanese Ore (Perika)	32.0	β III	5.1	82.3	52.1
African ore	22.3	Ram. II, β III	—	84.0	55.6
Caucasian ore	14.8	β I	7.8	87.9	57.0
Java ore	12.5	β I		90.4	
MnO_2 from thermal decomposition of $Mn(NO_3)_2$	10.3	β I	2.2	98.0	59.9

Notes.
s.h.r. = synthetic hydrous ramsdellite.
I, II, III = different phases.
α, β, γ = the usual crystallographic designations.
Surface areas expressed in m²/g.

means to dry cell improvement. Burgess patented impalpable graphite(176) as a cathode mix ingredient thus making the industry aware of the importance of the particle size of this ingredient. Heise's(177)(178) work on milling of manganese ore and graphite simultaneously added materially to the basic knowledge of this phase of cathode manufacture. For many years powdered coke and graphite were standard ingredients in dry cell cathode preparations.

Carbon Blacks

The discovery that acetylene could be thermally decomposed to give hydrogen and a carbon black of unique characteristices for dry cell by Burger(179) in 1915 was not immediately accepted by the industry. However by 1932(60), acetylene black was commercially introduced by Shawinigan Chemicals Ltd. of Canada and battery manufacturers were surprised to find it offered a new means of incorporating more electrolyte in the cathode mix than had ever been possible before. Furthermore, the resulting cells were markedly improved in output and keeping quality. In the course of the next decade acetylene black became the standard carbon component for cathodes in most high-quality cells. When World War II made acetylene black a scarce commodity, a considerable technical effort(180) to find equivalent carbon blacks accomplished two important objectives. First, cooperative work with various carbon black producers did provide some special carbon blacks that were used in the emergency, and second it furnished the impetus to characterize acetylene black in a scientific manner(181). The then new electron microscope pictures of these samples emphasized the chainlike structure of acetylene black(182) and differentiated it from other similar materials. This property, together with its unique liquid absorbing properties explained why it had become attractive for use in cells. Measurements showed that the acetylene black particles range from 40 to 60 mμ in diameter and that the surface area is about 62 m^2/g as measured by the BET method. Figure 1.41 shows a representative electron microphotograph showing the typical structure. Bregazzi(246) presented an interesting comparison of acetylene black with a number of carbon blacks made by other processes. Battery performance data on "A" size cells is given to show the effects of mixing variables in the manufacture of cathode mix.

Molded Carbon Parts

The early discovery by Bunsen(183) that carbon could be used as a cathode current collector in a primary battery, instead of the more expensive platinum and other noble metals foreshadowed the important role it was to play later in the dry cell. In the form of carbon rods it has

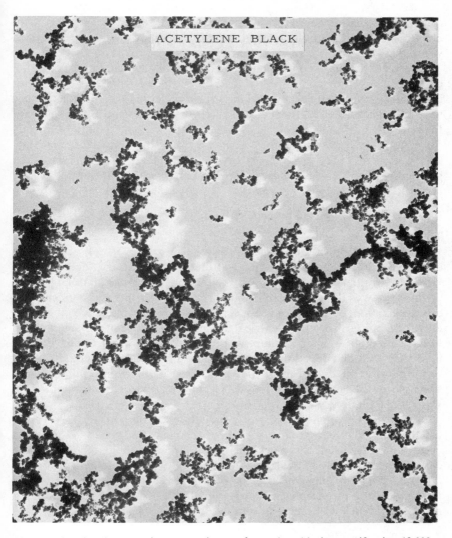

ACETYLENE BLACK

Figure 1.41 An electron microscope picture of acetylene black, magnification 12,000.

been used widely as the noncorrosive current collector in all types of cylindrical cells. The rods are prepared by conventional carbon technology, extruded, baked and then cut and ground to the precise dimensions required for today's automated handling machinery. For many cells the carbon rods are especially treated to make them electrolyte resistant. In other types of cells the carbon rod serves as the vent for any internally formed gas. Electrodes for other types are treated to render them electrolyte proof without closing the pores through which gas can be vented(184).

An alternative molded type current collector is that used in the external cathode cell. Here a mixture of carbon particles with wax and other ingredients is molded to form a thin current collector layer inside a paper tube which is the cell container(4). This layer must also be highly resistant to the transfer of moisture so as to maintain the keeping quality of the cell for as long a period as possible. The success of this unusual cell design, used for premium quality cells attests to the successful solution to such problems. Figure 1.3 shows a section through a representative cell of this type.

Conductive Carbon-Plastic Layers

The simple construction used in the Volta pile(5) where cell components were piled on top of one another to form a group of cells connected electrically in series had a unique attraction for battery technologists. However it was not until about 1920 that a successful commercial battery of this type was developed(185). The breakthrough was accomplished by the application of a carbonaceous plastic layer to one face of a zinc electrode to furnish the electrical connection between adjacent cells. Much later(186) thinner conductive coatings led to major improvements in the flat cell construction. Modifications such as a tray type(187) was representative of the effort at the time. Most of these special constructions were utilized in the "B" and "C" multicell batteries for radio and similar application.

In the above applications the conductive carbon plastic layer was usually applied directly to one surface of the zinc anode sheet. However a considerable amount of work has been done to develop a separate carbonaceous plastic film which is flexible and which can be incorporated into many battery designs. One of the most recent is that identified as "Condulon" a proprietary product which has found application in a number of types of battery products.

ZINC

Zinc has been used as the anode of the dry cells since the time of Leclanché. It has many properties which recommend it, particularly the ease with which it can be formed or shaped. The electrochemical properties are equally important and were discussed at length under the heading of "The Anode." However, the availability of zinc in a sufficiently pure state over the past hundred years has done much to make the dry cell a high quality commercial entity. Furthermore the relatively low price of zinc has greatly reduced if not eliminated the need for technological efforts to find substitutes. The production of dry cells in the world today is very large in terms of units manufactured (247). However, in spite of this scale of operations the use of zinc in dry cells represents only a few percent of the total zinc mined and used.

In the United States, all large zinc suppliers have worked closely with the battery industry to furnish the metal in rolled sheet or in other forms for direct use in batteries. Within the last decade, considerable interest in can manufacture by impact extrusion of zinc has developed especially in Europe. In other locations where small to medium quantities of special sizes of zinc cans are needed this method has also offered favourable economics. Aufenast (61) discusses the effect of zinc alloy composition on can manufacture by soldering, drawing and impact extrusion. He found that pure zinc was unsatisfactory and battery keeping quality was definitely improved by additions of alloying metals. A part of his data is summarized in Table 1.31.

THE SEPARATOR

The early wet batteries, including the original Leclanché (1)(2) unit, all had a finite layer of electrolyte between the anode and the cathode. This layer provided an ionically conductive path through the electrolyte but at the same time it furnished an electronic separation between the electrodes. Thus, two major functions of a battery separator were established. Somewhat later two-solution cells were introduced using a porous ceramic pot saturated with electrolyte to separate the two portions of the electrolyte. Obviously, the porous pot served some of the same purposes as the modern separator but its main function was to prevent mixing of the two electrolyte solutions, which were usually of quite different

TABLE 1.31. ZINC ALLOY COMPOSITIONS FOR BATTERY CANS

Alloying Metal	% Made by Drawing	% Made by Impact Extrusion
Lead	0.25	1
Cadmium	0.25	0.05
Iron	—	0.005
Copper	—	0.005
Tin	—	0.001

This study also included the structure and grain size of the alloy and its effect on amalgamation of the zinc can when it was used in a cell.

compositions. When Gassner(7) applied a layer of plaster of Paris to the inside of a zinc can he probably took one of the first steps toward the separator that we know today.

The plaster of Paris layer must have introduced a considerable amount of internal resistance into the cell system, thus lowering the working voltage. Moreover this layer must have presented many problems in cell manufacture that were difficult to overcome. It was followed by the bag-type construction in which a molded bobbin of depolarizing mix was wrapped in cloth and tied with several turns of twine(188). This unit was centered in a zinc can and the intervening space filled with a gelled electrolyte (Hellesen(189)). Often the gelled electrolyte was formulated from flour and starch added to the usual salt solutions. In the No. 6 cell, which was made commercially many years before the smaller sized units, a paper or pulpboard liner was used on the inside of the can instead of the cereal paste [H. T. Johnson(190)]. This change was apparently adopted more from manufacturing than from quality reasons since it enabled the cathode mix to be packed into the lined can. Improved battery performance was obtained when the cereal paste was combined with the paper layer(191) but this change was not generally practiced until about 1915. Improvements in the type of paper resulted in the use of thin kraft paper separators in such types of cells(192).

As smaller sizes of cells were developed many of the above separator types were applied. However, the flashlight cell utilized the bag type construction for many years. Here the molded bobbins were protected by wrappings of cloth or paper(193) or even by coatings of cereal paste itself(194). Mechanically stronger bobbins, developed some time later, largely eliminated the need for such wrappings(195). However, some types of small cells were manufactured with the so-called wrapped construction as late as World War II.

The cereal paste that found general use as the electrolyte thickener medium was the subject of considerable work. For over a decade the suspension of flour and/or cornstarch in dry cell electrolyte served admirably as a separator medium for commercial cells. It could be formulated so that it was rapidly gelled by the application of gentle heat during cell assembly, that is, cooked paste. On the other hand, other formulations would be converted into a stiff gel by the proper choice of electrolyte salt concentrations(196).

In the 1930s the industry started to use synthetic MnO_2 from many different sources as described elsewhere in this chapter. This material was a much more active oxidizing agent than natural ore and it was inevitable that some of the accepted materials of construction would need to be changed. Most important of this group was the cereal in the paste which was found to be the source of compounds capable of chemically reducing the active MnO_2 with resultant loss in cell voltage and capacity. In an effort to solve this problem many potential electrolyte thickeners were investigated. It was found that most cereals(197)(198) hydrolyze in dry cell electrolyte and the resulting hydrolysis products were active reducing agents as far as the synthetic MnO_2 was concerned. Obviously, a separator which would not hydrolyze in battery electrolyte was needed.

A fairly broad group of synthetic plastics including the alkyl ethers and hydroxy ethers of cellulose as well as certain polyvinyl compounds and the water soluble salts of cellulose glycolic acid were uncovered(197). These materials possessed the needed chemical stability, bibulous characteristics, and provided the required inhibition of the wasteful corrosion of the zinc anode. Moreover as a group they were film-forming materials and offered the unique opportunity to provide thin sheetlike layers either alone or on paper supporting sheets suitable for automated cell assembly methods.

The work on the film-type separators focused the attention on the requirements of a battery separator. Many functions must be served. The separator should provide an elastic, adhesive contact to the surface of the zinc anode and this contact should be maintained throughout the useful life of the cell. Furthermore, the separator should provide an ionic path between the two electrodes capable of easy diffusion of soluble electrolyte and reaction products. As pointed out earlier, the separator must provide an electronic barrier between the cell electrodes and one that is stable throughout the life of the cell. The separator often furnishes the means of adding the mercury salt or other inhibitor of corrosion to the anode part of the system. Above all the separator materials must be compatible with both the oxidizing conditions presented by the cathode as well as the reducing conditions at the anode.

It was found that these functions were best fulfilled by a double film type. The anode film softens in the electrolyte to give the adhesive contact to the zinc anode while the cathode film portion is highly bibulous but insoluble and acts as a barrier to the large molecules from which the anode film is prepared. Both layers are ionically permeable and only a few thousandths of an inch in thickness. Thus they offer an effective path for the diffusion of zinc chloride solution, produced at the anode surface during cell operation, to the cathode mix. One type of separator that was widely used consisted of a layer of methyl cellulose bonded to one side of a paper sheet. This combination provided the strength and uniformity needed to enable efficient cell assembly on a production basis. The methyl cellulose surface was placed adjacent to the zinc anode while the paper side contacted the cathode mix.

A variety of separators embodying the above principle have been developed. In the earliest work the cathode film was formulated from water-soluble methyl cellulose insolubilized to a limited degree by treatment with a tribasic organic acid such as citric acid(197). Later work showed that the cathode film could also be formulated successfully by control of the degree of substitution of the substituted group, for example, methoxy, ethoxy, carboxy methyl, and so forth on the cellulose unit(199). Further studies have shown that the anode layer need not be a film but that a layer of particles of the chosen electrolyte soluble resin can be bonded to a suitable cathode film(200) to make for easier cell assembly. When such a combined separator layer is exposed to electrolyte in the assembled cell the layer of film particles soften to become the gel to contact the anode surface. A special formulation of this type was devised for a separator for an aluminum cell(200).

Bruins et al. at the Brooklyn Polytechnic Institute(201)(202) examined an extensive series of possible separator materials for use in batteries with acid, neutral and alkaline electrolytes. Some of the products developed in this study were sheeted materials loaded with ion-exchange resins of various kinds. Unfortunately, none of these separator formulations have found a commercial market in the dry cell field.

In spite of the development of the plastic film separator, a group of workers were trying to find the reasons for the success in the use of flour and starch that had been the standby of the battery industry for so many years. An important part of this effort was done at the Bureau of Standards and it has been summarized by Hamer(49)(203). They developed a device to measure the gel strength of a sample of paste such as might be used in battery construction. Storing these samples with battery ingredients and periodically measuring the gel strength showed that MnO_2 caused a serious decrease, thus indicating some reaction between the depolarizer and the cereal components, thus confirming earlier data(198)(204).

One important finding from the studies of Morehouse, Hamer & Vinal(203) was the discovery that the gluten fraction of the flour, used in making the cereal paste for batteries, was a corrosion inhibitor. This led to an extensive study of other possible corrosion inhibitors which hopefully could replace mercury and chromates. Of the materials examined several, for example, compounds with carbonyl groups such as furfural and heterocyclic nitrogen-containing compounds like quinaldine, as well as several commercial products did inhibit zinc corrosion in laboratory tests. Unfortunately they were not successfully applied to dry cells. Some time later Zimmerman and Powers described(72) a series of organic compounds that were successfully used in production cells.

Improvements in separator technology have continued with the objective of providing more stable materials. In the field of starches the use of etherated starches patented by Morehouse and Welsh(205) opened the way for studies of the "Vulca" starches that were commercially available at that time. More recently a product developed in Japan is believed to be somewhat similar in nature. With proper formulation, these materials may be used in the paste lined construction for premium quality cells.

PACKAGING AND SEALING MATERIALS

A great deal of technical effort has been spent on the evaluation of packaging materials for batteries because of two important needs. First, there is the need for prevention of loss of moisture from the cell components during storage so as to maintain long shelf life. Second, many efforts to achieve a leakproof cell were based on containment of the electrolyte and special packaging methods were applied. In view of this complex situation many materials normally not considered for battery packaging have been evaluated. For example, lithographed steel was used by the Ray-O-Vac Company very early as a container for a unit cell(206). Alternatively, external cell wrappings of plastic films(207), often interleaved with paper sheets to provide moistureproof and leak resistant barriers to the electrolyte spew that develops during abuse testing, have been commercially applied.

Various waxes, asphalts, gilsonite, and other plastic materials have long been used to close the top of the unit cell before the external finish is applied. Most of these materials are thermoplastic so that a measured amount of a molten wax, or equivalent material, could be added to the top of each cell as it progressed along a production line. Special efforts are made to be sure the finished sealing layer is free from holes, often caused by air bubbles, and that the molten sealing material adequately bonds to the surfaces of the carbon electrode and the zinc can to effect a tight seal. Figure 1.2 shows a layer of asphalt while Fig. 1.3 shows a wax layer on the inside of the bottom closure where it serves a similar function.

The quality of the flat cell shown in Fig. 1.4 is dependent on the special packaging of both the individual cell and the stack of assembled cells. The most important item is the plastic envelope which contains the unit cell. This plastic layer must be highly impervious to moisture transfer so as to maintain the initial supply of moisture as long as possible. Many constructions of this type take advantage of the extra protection against water loss provided by a wax coating applied to the assembled stack of cells before they are placed in their final container.

Electrical connections to individual cells are usually contacts at the ends of the cell as shown in Figs. 1.2, 1.3, and 1.5. In use spring pressure is applied to metal strips contacting the cell terminals in the appliance in which the cell is placed. The terminals are usually made of sheet steel, tinplated to provide a contact of low electrical resistance. Multicell batteries have a variety of terminals, often metal snap fasteners or screw terminals of various designs, some of which are shown in Fig. 1.5.

BATTERY PERFORMANCE

In this section, representative discharge curves and associated data are presented as examples of how battery performance is measured in the laboratory. Of course, the consumer determines the useful life in a variety of ways, usually the time interval during which the battery lasts in his particular application.

A number of authors have attempted to develop a simple scheme for estimating the output of various types of batteries and cells and including dry cells. Such methods may treat the dry cell on a general basis and include the inherent assumption that all dry cells are substantially made from the same formulation. Thus, such calculation methods overlook the formulation art developed to a high degree by battery manufacturers. As a result the serious battery user should consult a qualified battery engineer to help him select the most satisfactory power source for any specific application (208)(209)(210)(211)(212)(213)(214).

The battery technologist is interested in relating the electrical output of a specific cell to the amounts of active materials incorporated during manufacture. He must therefore convert discharge curve data into ampere-minutes or ampere-hours so that these quantities can be compared with expectations derived from the chemical and electrochemical equivalents of active material present in the cell. As the art of making Leclanché cells continues to approach an exact science it is expected that the gap between theoretical and practical estimations of battery performance will be greatly reduced. When that time arrives, perhaps the extensive battery test programs now necessary can be drastically reduced.

The various aspects of battery performance are discussed under the headings: electrical output, keeping qualities, and leakage.

ELECTRICAL OUTPUT

It is unfortunate that the electrical capacity of a Leclanché cell cannot be predicted accurately from a knowledge of the amounts of materials used in its manufacture. The reasons for this are complex and perhaps not entirely unexpected in view of the fact that the cell output is the result of many factors working simultaneously. Thus the electrical output to be obtained from a Leclanché cell is dependent on such factors as its size, its formulation, construction, and the conditions under which it is discharged. For any given formulation, cells of large size produce more hours of service than small cells due to the greater amount of chemical fuel available in the former. The art of formulation is important since it is usually this factor which the battery manufacturer varies to obtain the specific service characteristics desired for particular applications. Included in the formulation are such variables as the type and amounts of depolarizer and carbon used as well as the proportion of $ZnCl_2$ and NH_4Cl incorporated. A group of factors may be classed as constructional items, which nevertheless influence the cell output because of their effect on the current density at one or both electrodes. The separator system incorporated in the cell plays an important part because its function is partly chemical and partly constructional. The conditions under which a cell is tested include not only the current drain, but also the period for which current is drawn as well as the rest period between such discharges and the temperature range of the cell during the test.

The above group of prime variables and many additional factors have made it necessary to standardize tests and testing conditions so that battery performance can be measured accurately. This is such an important phase of battery evaluation that a separate section entitled "Nomenclature and Testing Procedures for Primary Batteries" by W. J. Hamer appears as Chapter 8.

A large number of cells and batteries are available commercially and the performance characteristics of all such units would be so voluminous that space does not permit them to be included in this book. Instead only some representative examples will be offered in this section to show the important factors governing battery performance. The interested reader is referred to handbooks, for example, (211) for the detailed information on the performance of specific battery types and sizes of Leclanché cells.

The two types of tests selected for examples are the flashlight series and one group of radio tests simulating the use of batteries on a 4 hour schedule of service per day, characteristic of use in portable radios.

The series of flashlight tests are designed to cover the use of a flashlight on a range of operating schedules from continuous service to that equivalent to a use of only five minutes per day. The schedules are given in Chapter 8. Two types of flashlight lamps are now commonly used, the $\frac{1}{2}$ ampere and $\frac{1}{3}$ ampere types. These are simulated in testing by 2.25 ohm and 4 ohm resistances respectively. The cutoff voltages on these two series of tests are generally taken as 0.65 V for the 2.25 ohm test and 0.75 V for the 4 ohm series.

Figures 1.42 and 1.43 show respectively the family of four curves

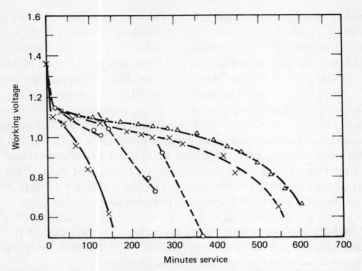

Figure 1.42 Discharge curves of "D" size flashlight cells on 2.25 ohm flashlight tests. Tests; continuous ×, heavy industrial flashlight ○, light industrial flashlight +, household intermittent △.

obtained on the 2.25 ohm and 4 ohm flashlight test series on "D" size flashlight cells (211). In each series of tests the load resistance is constant and the difference between individual tests is the total number of minutes discharge per day ranging from a continuous operation until the cell reaches the cutoff voltage, to a minimum of 5 minutes per day. Thus the shorter the actual discharge time in minutes per day, the longer is the actual total service obtained.

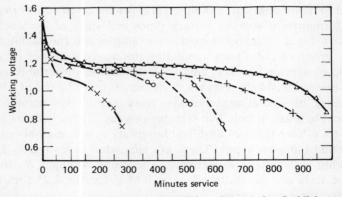

Figure 1.43 Discharge curves of "D" size flashlight cells on 4 ohm flashlight tests. Tests; continuous ×, heavy industrial flashlight ○, light industrial flashlight +, household intermittent △.

It is clear that the length of test is favored by the longer rest periods between discharges and vice versa. This situation emphasizes the importance of the cathode recuperation reactions discussed earlier in this chapter. Since these reactions are chemical and operate at relative low specific rates, the shorter tests do not allow time for these reactions to keep up with the primary cathode reaction which is electrochemical and operates at a rate dependent on the discharge current. As the length of the test is increased the two sets of cathode reactions in the cell tend to approach equality and the overall efficiency of the cathode is increased to give higher service. An additional result is a flatter discharge curve.

The discharge curve of any Leclanché cell on a constant resistance test resembles in general shape those given in Figs. 1.42 and 1.43. However, it is often convenient to compare performance under a wider range of test conditions. Figure 1.44 shows the performance of a "D" size cell on a variety of constant resistance loads, all imposed for a four-hour period per day(211). A logarithmic scale is used here to enable the direct comparison of services covering three orders of magnitude resulting from starting drains of 10 to 300 ma. This type of presentation emphasizes the lower voltage part of the discharge curve which is the important part for many applications.

Figure 1.45 shows a comparison of the discharge curves of a range of different sizes of cylindrical cells all on the same starting current of 10 ma and on a discharge schedule of four hours per day. Table 1.32 presents the

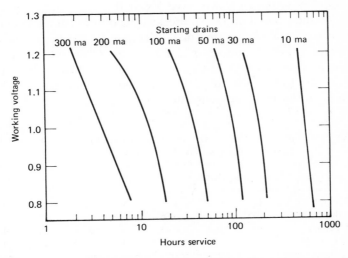

Figure 1.44 The service obtained from a "D" size flashlight cell at various starting drains on a 4-hour-per-day discharge schedule(211).

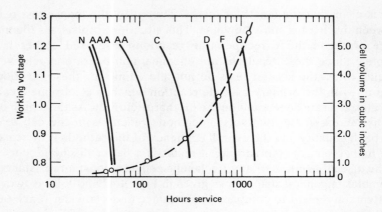

Figure 1.45   A comparison of the discharge curves of a range of cell sizes, all discharged at a starting drain of 10 ma on a 4-hour-per-day test schedule. ○ = The output in hours to 0.90 V plotted against the cell volume (211).

sizes and volumes of each cell in the group. It is clear that the larger cells produce considerably more hours of service than the smaller units. The dashed line on the chart shows the relation between the hours of service to 0.9 V and the cell volume expressed in cubic inches. In Table 1.33 the output expressed in hours per cubic inch is compared for a few types of cells on starting drains of 10 to 200 ma. On the lightest drain of 10 ma, all seven cells, regardless of size, give about the same output per unit of volume. As the current drain is increased the output per unit volume falls but it falls more rapidly for the small cells than for the larger ones. Figure 1.46 illustrates this situation graphically for the C, D and G sizes of cells. It is evident from Table 1.33 that as the current drain is increased, the service of any given size of cell falls somewhat more than in direct proportion to the increased current.

TABLE 1.32. THE ELECTRICAL OUTPUTS OF UNIT CELLS OF VARIOUS SIZES
DISCHARGED ON A TRANSISTOR TYPE TEST (211)

Cell Size	N	AAA	AA	C	D	F	G
Cell length	1.180"	1.75"	1.93"	1.87"	2.25"	2.74"	4.19"
Cell diameter	0.445"	0.254	0.365	1.073	1.22	1.22	1.22
Cell volume	0.16 in.³	0.20	0.48	1.25	2.76	4.22	4.92
Hours to 0.90 V	40	45	110	280	620	890	1400

The test schedule was a four-hour discharge per day on a constant resistance load such that the starting drain was 10 ma.

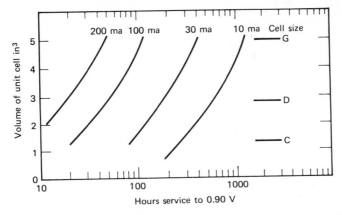

Figure 1.46 A comparison of the services to 0.90 V given by unit cells of three sizes, "C," "D," and "G," on a 4-hour-per-day schedule and at four levels of starting drains(211).

The temperature at which batteries are tested plays an important role. Generally speaking, the higher the testing temperature, the greater will be the output of a given battery. The reverse is also true and special formulations are required for batteries to be used at lower than normal temperatures. A special section of this book describes these special batteries and their operation (see Chapter 5, "Low Temperature Batteries").

TABLE 1.33. THE OUTPUT PER UNIT VOLUME FOR UNIT CELLS OF VARIOUS SIZES ON A RANGE OF STARTING LOADS

		Starting Drain							
		10 Ma[a]		30 Ma		100 Ma		200 Ma	
Cell Size	Cell Volume	Hours to 0.9 V	Hours p in.³	Hours to 0.9 V	Hours per in.³	Hours to 0.9 V	Hours per in.³	Hours to 0.9 V	Hours per in.³
N	0.16 in³	40	249						
AAA	0.20	46	230						
AA	0.48	110	230						
C	1.25	280	224	86	68.9	20	16.0		
D	2.76	620	225	185	67.0	50	18.1	18	6.5
F	4.22	890	2.10						
G	4.92	1300	2.64	440	89.6	115	23.4	47	9.53

[a]The test schedule was a four-hour per day discharge on a constant resistance load such that the starting drains were the levels given above.

Normal testing of batteries is done in controlled atmosphere rooms maintained at 70°F ±1°F and 50 percent relative humidity. Testing is also performed at other temperatures, for example, 113°F. Specific battery applications often dictate the temperature to be used. Thus, batteries for electric watches are commonly tested at 90°F since they are used where the temperature approximates that of the human body.

The tables and charts presented above show some of the complex relationships existing between the performance of different sizes of cells on different testing schedules. The consumer is usually concerned with the number of minutes or hours of operating life to be expected from a given unit cell or battery and charts of this type are very useful to him.

Figure 1.47 presents an example of a three-dimensional figure showing the relation between hours per day on test, starting drain on a constant resistance test, and the ampere-hours output obtained from a "D" size cell. It is clear that, at the lightest drains, the greatest outputs are achieved. However, the figure clearly shows that if the drain is small, for example, 10 ma, and the test schedule is two hours per day, less total

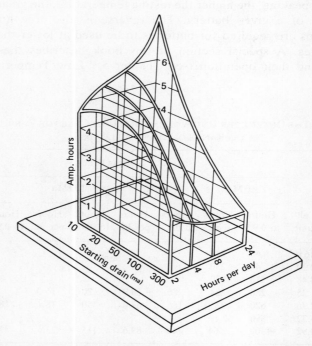

Figure 1.47 Battery service life as a function of initial current drain and duty cycle ("D" size carbon zinc cell)(211).

output is obtained than at a higher drain of 25 ma on the same schedule. The reason for this apparent discrepancy is that, at the lowest drain, the length of the test is extended to such an extent that shelf factors reduce the total output.

The short circuit amperage of a cell is often thought to be an indicator of its service capacity. Unfortunately this is not true as the amperage mainly reflects the internal resistance of the unit. Examples of the relation between amperage and internal resistance of fresh cells are given in Table 1.34. Further data on this aspect of dry cell behavior are given in Chapters 8 and 10. The short circuit amperage may be adjusted over a wide range of values by changes in cell formulation. For example, the "D" size photoflash cell has a much higher amperage than the general purpose cell of the same size. On a flashlight test the latter is superior to the photoflash type.

The problem of determining the remaining life in a partly discharged cell is difficult, since readings of open circuit voltage provide no indication. The simplest and most reliable method is to use a voltmeter with a resistance across its terminals to impose a load on the battery amounting to approximately half the maximum drain recommended for that particular type and size of cell. With some experience, readings of this type can be a useful guide to estimating the residual capacity. Some commercial instruments are designed to utilize this principle.

The Leclanché cell occupies a unique place in the group of primary cell systems that have been commercialized to date. As pointed out in the preceding discussion, it has a number of limitations in current capacity and in output particularly on heavy drain tests. On the other hand it is

TABLE 1.34. GENERAL LEVELS OF INITIAL
AMPERAGE AND INTERNAL RESISTANCE
FOR FRESH UNIT CELLS

USASI Cell Size	Average Flash Current (amperes)	Approximate Internal Resistance (ohms)
N	2.4	0.69
AAA	3.5	0.44
AA	5.4	0.29
C	3.3	0.47
D	5.8	0.27
F	9.0	0.17
G	11.0	0.11
6	38.0	0.03

manufactured from low cost raw materials on a scale which has not been approached by any other cell system. A survey made a few years ago revealed that the manufacture of Leclanché cells was the second largest electrochemical industry in dollar volume, surpassed only by the chlor-alkali industry. It is indeed difficult to visualize serious competition for the Leclanché system in view of its unique combination of low cost, ready portability, ease of replacement, ready availability, good keeping qualities and a service level which is adequate for a great variety of applications.

KEEPING QUALITIES OF LECLANCHÉ CELLS

Every primary cell, other than reserve types, consists of a powerful reducing agent, the anode, separated from a powerful oxidizing agent, the cathode, by a layer of electrolyte usually absorbed in a separator sheet. It is inevitable that reactions between either electrode and the electrolyte will occur and these with time will reduce the original battery capacity to some extent. Battery manufacturers have gone to great effort to study the many phases of this problem so that the customer will obtain as near the original battery capacity as is possible.

In Leclanché cells, there are a number of factors that may be responsible for a slow deterioration of battery quality during storage. These factors have been previously discussed under the heading "Shelf Reactions." The military forces are important users of dry batteries and the U. S. Navy sponsored extensive studies to establish the rates at which the deterioration in service capacity would occur under various storage temperatures. These tests were begun in 1951 and Hellfritsch(215) and Warburton(216) have reported on this study as it progressed. They examined about 24,000 military batteries, storing portions of the group at temperatures of −34°C, −12°C, 4°C, and 21°C. One of the objectives of this study was to determine the optimum storage temperature at which dry batteries should be maintained awaiting use.

Figure 1.48, taken from Warburton's paper, shows the effect of the four storage temperatures and the time of storage on the percent of the initial capacity retained. It is obvious from these data that the lower storage temperatures favor the better keeping quality of the batteries, a factor that had been well recognized by the industry. Hellfritsch(215) points out that batteries have been successfully stored for 12 years at −34°C with substantially no loss of average capacity. Since the freezing point of battery electrolyte is usually between −20°C and −26°C, it appears that the originally operable Leclanché cell was converted to the equivalent of a reserve unit by freezing it at −34°C. When the temperature was again raised to the normal range the battery returned to operating status again.

It should be emphasized that the batteries used in the above study were

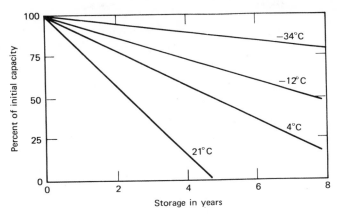

Figure 1.48 Average capacity retention of Leclanché batteries stored at four temperature levels.

military types manufactured about 1951. Thus the data reported on these units accurately reflect the state of the art at that time. The battery art and science has developed considerably since then and improved designs and formulations now available should show improved keeping quality as well as improved service capabilities.

Davis(217) examined the storage quality of "D" size cells which had been assembled with a cathode mix containing electrolytic MnO_2 and a separator of cereal coated paper. This design represented an up-to-date construction of a high quality cell intended for good performance on the HIF test. His program covered 7000 cells which were examined periodically over 12 months. Most of the cells were contained in hermetically sealed tin cans which initially were supplied with atmospheres of air, nitrogen, oxygen, and a partial vacuum. Part of the group was stored at 70°F and part at 113°F. Analyses of the gases generated in representative containers were made during storage.

Davis' findings confirmed those of earlier workers(74) in that both hydrogen and carbon dioxide are generated during storage. Hydrogen arises from the wasteful corrosion of the zinc anode while the carbon dioxide is presumed to develop as a result of the reaction between cereal paste and MnO_2. He found that the storage in sealed cans greatly improved the retention of HIF battery capacity. After 12 months at 70°F, the canned cells retained 80 to 90 percent of their initial capacity whereas the control cells possessed 40 percent, as judged by their performance to the 1.1 V cutoff. At 113°F, there was a greater loss in capacity than at 70°F as would be expected. The hydrogen and carbon dioxide generated during storage developed a maximum pressure of 50 to 60 inches of mercury during the 12-month period. It should be noted that the particular storage can used in this work provided a free space of 42 ml for each cell contained in the can, a factor that was probably important in limiting the pressure developed.

LEAKAGE OF LECLANCHÉ CELLS

When a Leclanché cell is subjected to a continued short circuit or to a heavy load until the voltage approaches zero it is not unusual for some of the electrolyte to find its way to the exterior of the cell and to corrode the cell contacts, the flashlight, or other metal cell container. This peculiarity of dry cells, termed leakage, was taken for granted during the first thirty years of this century. In the mid 1940s, one U. S. battery manufacturer(206) introduced a leakproof dry cell to the consumer market. Although this unit was not leakproof under all conditions of discharge, it did represent a decided improvement in packaging that appealed to the consumer. Competition in this field developed rapidly as shown by the large number of patents issued to the inventors of a great variety of new methods of constructing and packaging unit dry cells. Within ten years after the introduction of the first so-called leakproof unit cell, substantially all U. S. battery manufacturers had their own leakproof cells on the market.

Progress in attaining a cell that is completely leakproof under all conditions has been slow although the majority of these special cells do show a greatly reduced tendency to leak under most of the discharge test conditions met in every day use. In 1964 the Federal Trade Commission(218) ordered all dry cell makers to refrain from using the term "leakproof" on their labels. This ruling was apparently based on the conclusion, made by the F.T.C., that most cells of this type were leak resistant rather than wholly leakproof. In spite of the fact that the objective of a truly leakproof cell had not then become a commercial reality a great deal of technical and engineering effort has been spent in this direction.

Few publications have appeared on this interesting area of dry cell performance but the patents offer excellent examples of the many methods that have been developed. Not all of these have found commercial use but as a group they cover the field.

In view of the large number of patents pertaining to the design of leakproof cells, it is impossible to review them all in detail or even to list the group. Instead, representative patents will be given below as examples of the various techniques studied. The interested reader is referred to class 136 "Batteries"(219) of the U. S. Patent files for more complete specifications.

It should be recognized that, in spite of the optimistic claims and examples given in many of the patents cited, no satisfactory solution to the leakage problem by a containment or packaging method alone has yet been put into commercial production. Thus the examples offered are primarily of historical interest. A truly leakproof Leclanché cell, a new patented approach that is highly successful, is described in a later section. As a result, a Leclanché leakproof cell is now available on the commercial market.

Furthermore, the zinc chloride cell described at the end of the chapter, is being actively studied in many battery laboratories. It too is substantially leakproof and thus these two new solutions to the leakproof problem are at long last available.

The subject of dry cell leakage is covered in the following sections: a. causes of cell leakage, b. testing cells for leakage, c. attempts to alleviate leakage (packaging techniques, chemical additions, electromechanical methods), and a truly leakproof Leclanché cell.

Causes of Cell Leakage

The anode and cathode reactions were presented in an earlier part of this chapter. At low and medium current drains, these reactions operate properly. However, their success depends on the diffusion of the anode product, $ZnCl_2$, in solution, from the anolyte to the cathode. Under heavy drain conditions, especially during continuous drain tests, the formation of the electrode products is more rapid than their diffusion into other parts of the cell. As a result, these products accumulate where they are formed.

Thus the initial cell, which contained a single composition of electrolyte distributed through both anode and cathode, has been changed into a two-electrolyte cell. On heavy continuous drain the following reactions can be visualized:

Electrolyte dissociation

$$NH_4Cl \rightarrow NH_4^+ + Cl^- \qquad (55)$$

Anode reaction

$$Zn^\circ + 2Cl^- \rightarrow ZnCl_2 + 2e^- \qquad (56)$$

Cathode reaction

$$2MnO_2 + 2NH_4^+ + 2e^- \rightarrow 2MnOOH + 2NH_3 \qquad (57)$$

The products of these two reactions again react in one or two ways depending on the concentration of NH_4Cl at the reaction site. If sufficient NH_4Cl is present, that is, above about 5 percent, the following reaction occurs:

$$ZnCl_2 + 2NH_3 \rightarrow ZnCl_2 \cdot 2NH_3 \qquad (58)$$

If the NH_4Cl level is below the 5 percent value, the probable reaction is

$$5ZnCl_2 + 8NH_3 + 8H_2O \rightarrow ZnCl_2 \cdot 4Zn(OH)_2 + 8NH_4Cl \qquad (59)$$

A cell that has been on a continuous discharge for a matter of days or weeks is certainly very thoroughly exhausted. In such a cell the products formed in reactions (56) and or (57) build up at the anode-cathode interface. The products of these reactions are soluble in their own electrolytes, that is, the NH_3 is soluble in the catholyte and the $ZnCl_2$ is soluble in the anolyte. When these two soluble phases meet at the cathode-anode interface, reactions (58) or (59) occur and a solid product is formed. Thus the formation of this layer of products is a self-healing process in that any diffusion that does occur produces more product. The diffusion through this precipitated layer is very slow and may be infinitesimal. In a "D" size cell of the pasted type the concentration of $ZnCl_2$ in the anolyte layer may reach about 55 percent and have a pH of about 2.5. In such an environment the cereal contents of the paste layer become hydrolyzed to form a sticky, highly viscous, electrolyte which entraps gas bubbles easily. Moreover this acid electrolyte attacks the zinc surface and may release hydrogen. If a cell is discharged and left connected to the load resistor, additional hydrogen can be electrochemically generated at the cathode as long as any zinc metal remains in contact with the negative terminal.

The combination of the viscous electrolyte with the hydrogen, which can develop some pressure presents a difficult problem of containment, a situation well exemplified by the many patented designs which attempt it. The viscous electrolyte often exudes from openings at the ends of the cell package and the cell is then said to have "leaked."

Testing Cells for Leakage

Once leak-resistant cells became a factor in the competitive battery industry, it became necessary to devise methods of testing the many experimental types evaluated. A large number of test conditions were initially considered and many of them were used for comparative purposes. Even today there is no widely accepted test that the battery industry uses as a standard. This is surprising in view of the fact that all that is being measured is the tendency for a cell to develop exudate when the user leaves it in his appliance, flashlight, radio, recorder, and so forth, and forgets to turn the switch off. As a result the battery is discharged well beyond the normal cutoff, probably to nearly zero volts. Under these conditions, a number of deteriorating processes operate and often consume the anodically soluble parts of the cell.

Since leakage of cells is a phenomenon that is usually associated[5] with heavy current discharge most of the recent emphasis on developing a leakage test has concentrated on using the $2\frac{1}{4}$ ohm continuous flashlight test as an experimental medium. Cells are placed on the load resistance and, in many instances, are not removed until the test is completed in a matter of months. In other instances, the resistance is detached from the cell after selected periods. Both heavier and lighter loads are also used. Thus a direct short circuit is often used as is a 1 ohm resistor to simulate cell behavior on loads characteristic of toy operation. Lighter loads representing radio drains are often added to the list although cells are not as prone to show leakage on these lighter loads as on the heavier type. In most cases the cells placed on the leakage tests are examined periodically for signs of leakage. The results are reported as the number of days before leakage develops on a specific test. Thus, cells that show no leakage at 100 days are not considered as successful as those from a duplicate test, which show no leakage at 200 days.

Attempts to Alleviate Leakage

Packaging Techniques   When the leakage problem first was given serious consideration, it was thought that the solution lay in the application of a tight container for the cell. A great deal of effort in many laboratories has been spent on this approach and many patents have been issued. Each of these has described a method of cell construction and closure which has offered a step in the progress toward a more leak-resistant cell. It must be admitted that an effective containment for the cell does help make the cell more leak resistant. However, continued improvement of the cell enclosure has now progressed for over 20 years and as yet no real solution to the problem appears to be attained by this approach. A series of representative patents are offered to show the types of construction that have been seriously considered by the industry. To conserve space in presenting these examples, we have included a limited number of illustrations and that part of the text of the patent which deals with the description of the invention. The interested reader is referred to the original patents for more details.

One of the first patents on a leakproof construction was issued to Anthony (U.S. Pat. No. 2,198,423)(206) in 1940. This patent introduced the concept of a nonabsorbent layer of paper or other insulating material to separate the outside of the zinc can from the inside of a steel sleeve or

---

[5]There are instances where cells discharged on light drain tests also show leakage.

sheath which became the external container for the unit. The ends of the steel sleeve were mechanically spun over to provide tightly sealed joints at both ends of the unit cell. Figure 1.49 shows a cross section of the patented construction with a legend to identify the many parts used.

A second Anthony patent (U.S. Pat. No. 2,243,938)(206) shows an alternative construction in which a zinc can with an asphalt washer protecting the inside bottom surface is used. An insulating sheath is applied to the cell and then a metal sheath is used to cover the side walls. In both of these patents the bottom of the zinc can serves as the negative terminal of the cell.

In the embodiment illustrated in Fig. [1] inclusive. A designates a zinc cup which serves as a negative electrode for the cell; B, a carbon electrode provided with a depolarizing-mix B'; and C, a metal sheath which encloses the cell.

The sheath C preferably is made of sheet iron and may be of tubular form with its lower end turned inwardly to form a flange 22. A rubber sealing washer 23 may be placed on the flange 22 before the cell is inserted. In most cases it is desirable to insulate the casing A from the jacket C, and this may be done by wrapping a sheet 24 of insulating material such as rubber, treated paper or the like, around the cell and tucking the lower ends 25 inwardly. The cell is then forced into the jacket and pressed firmly against the washer 23. The top closure 20 is then forced over the cap 19 and the top edges of the steel jacket are spun inwardly so that the jacket tightly embraces both ends of the cell. Preferably, a spun flange 26 bites into a groove provided in the closure 20 so that a very tight joint is provided.

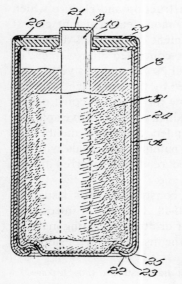

Figure 1.49 Leakproof dry cell, H. R. C. Anthony, U.S. Pat. 2,198,423, April 23, 1940.

It was quickly recognized by the battery technologists in the industry that there were some disadvantages to the constructions patented by Anthony. One handicap was that the volume available for the collection of cell exudate was limited to that within the cell container. Glover (U.S. Pat. No. 2,396,693)(220) patented a construction in 1946 which provided for an increased volume in which the exudate could be collected. It was achieved by arranging a porous mesh structure or layer between the external jacket and the outside of the zinc can in which the cell was built. Figure 1.50 shows a typical construction described in this patent. One novel feature was a hollow carbon electrode to help vent the hydrogen gas evolved within the cell during abuse tests. The need for such a venting electrode emphasizes the important role that venting plays in achieving leakage resistance in commercial cells.

The use of the hollow carbon electrode was also patented by Glover (U.S. Pat. No. 2,552,091)(221) in a group of cell designs that combined improved space for exudate collection within the cell with the improved venting.

Referring to the drawing, and more particularly to Figs. 1 and 2 thereof, the improved dry cell of the present invention may comprise a cylindrical zinc can or container electrode Z, a cylindrical carbon electrode C disposed centrally within and spaced from the outer electrode Z; a cylindrical mix M, consisting of depolarizing material and conductive carbon particles wet with electrolyte, in

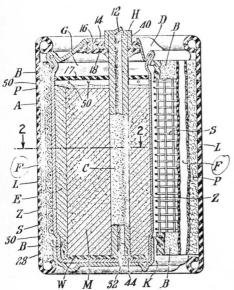

Figure 1.50 Dry cell and process for manufacturing the same, R. L. Glover, U.S. Pat. 2,396,693, March 19, 1946.

which the inner end of the electrode member C is embedded; and a bibulous electrolyte-immobilizing material E between the mix M and the electrode Z.

The space A is bounded on the inside by the container electrode Z, on the outside by an electrolyte-proof, waterproof surface or film F, and at the top and bottom of the cell by an electrolyte-proof and waterproof material B, which latter preferably is bonded both to the film and to the container electrode. The film F may be carried by a layer of supporting material, such as a thin paper backing P as illustrated in Figs. 1 and 2, or by direct application to the inside surface of the outer label L as illustrated in Fig. 3. To provide an effectively distributed expansion space, the container electrode is wrapped with a layer of uniformly thick and uniformly distributed spacing material S, so constituted and arranged as to provide a large number of voids adapted to collect and retain cell exudate.

In 1952, Teas(4) patented an external cathode construction (U.S. Pat. No. 2,605,299) as shown in Fig. 1.51. This construction represented a

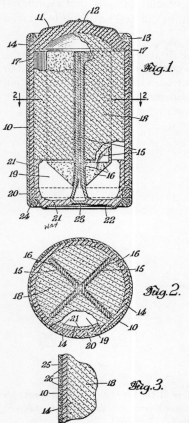

Figure 1.51 Primary galvanic cell, J. P. Teas, U.S. Pat. 2,605,299, July 29, 1952.

radical departure from the conventional cell and offered the opportunity to enclose the cell components more completely than previous constructions. The elimination of the zinc can as the container for the cell was a significant step toward fabricating the unit in an entirely new manner.

In the accompanying drawing:

Fig. 1 is a vertical section of a primary galvanic cell according to the invention;

Fig. 2 is a horizontal section taken on the line 2-2 of Fig. 1 looking in the direction of the arrows; and

Fig. 3 is a fragmentary sectional view of a modification of the cell of the invention.

The dry cell of this invention comprises a non-corrodible tube 10 preferably composed of fibrous material such as ordinary wrapping paper. One end of the tube 10 is closed by a metal cap 11 preferably dished outwardly and provided with a central boss 12, the latter serving as a terminal for the cell. The cap 11 is tightly secured to the tube 10, the edges of the tube and the cap being rolled or crimped together as shown for example at 13.

Adjacent to the inner surfaces of the tube 10 and the cap 11 and in juxtaposition therewith is the cathode 14 of the cell. The cathode 14 is an electrically conductive carbon composition which preferably is molded in situ, for instance by injection molding, impact-extrusion molding or other conventional molding operation, although it may be produced in any convenient manner.

The anode 15 of the cell, preferably composed of zinc; is placed centrally in the cell and is provided with radial vanes or fins. In cross section the anode 15 is preferably X-shaped as shown in Fig. 2. It is provided by wrapping, spraying or dipping with a conventional separator 16 of bibulous material such as paper or gel. A fibrous washer 17 may be used at one end of the cell to separate the anode 15 from the cathode 14. The vanes of the anode 15 are short enough that their edges do not come into contact with the cathode 14 at any point.

The intervening space in the cell between the anode 15 and cathode 14 is substantially filled with depolarizer mix 18 of any desired composition leaving an air space 19 through which the stem of the anode 15 extends in the direction of the open end of the tube 10. A fibrous support ring 20 is provided at the bottom of the tube 10 and the cell is sealed internally by a conventional seal 21 of wax or other sealing material. A metal bottom plate 22 to which the stem of the anode 15 is electrically connected at 23, as by soldering, serves as an external closure-terminal member for the cell. The bottom plate 22 is tightly secured to the tube 10, its edges and the edges of the tube being rolled or crimped together as shown at 24.

The tube 10 for the cell of the invention as already stated is preferably composed of a fibrous material such as wrapping paper.

It is important that the electrical conductivity of the cathode be high. Unfortunately, the presence of a binder in the cathode composition detracts from the normally excellent conductivity of the carbon. To increase the conductivity of the cathode it may be desirable in fabricating the tube 10 of the cell of the invention to use a metal foil, for example lead, tin, or aluminum, for the inner ply 25 (Fig. 3) of the tube. In such case, to maintain gas permeability, the metallic inner ply 25 should be provided with openings, for example small perforations 26.

In 1953 Reinhardt and Stapleton (222) obtained a patent (U.S. Pat. No. 2,642,471) on a very different approach to cell packaging than had been used up to that time. It is characterized by a one-piece molded plastic top closure and sleeve which is applied to the partly finished cell to enclose all but the positive and negative terminals in plastic.

Also in 1953, Lang (223) patented (U.S. Pat. No. 2,649,491) an improved dry cell construction. He poured the soft seal directly on the top of the cathode mix in the cell airspace as contrasted to other constructions where a spacing washer was used. He used a top washer to divide the airspace into two parts and closed the top of the cell with a two-piece metal cover such as was in common use at that time. The bottom closure was fabricated of copper, silver, or of some base metal plated with either of these. He claimed that this coating made the bottom closure electropositive to the cell cathode, when it is deeply discharged on abuse tests, and so prevents anodic attack and resultant perforation.

In 1956, Glover (207) patented (U.S. Pat. No. 2,773,926) the plastic-bonded, spiral-wound paper jacket that has been used extensively to package unit cells. This container has many of the advantages of the metal jacket described previously and is considerably cheaper. Figure 1.52 shows a typical construction of this type.

A multi-ply jacket 34 is provided about the outside of the cupped electrode 10. The innermost ply 36 (see in enlarged detail in Fig. 8) of the jacket 34 adjacent to the electrode 10 is composed of bibulous material such as kraft or other paper. An intermediate barrier ply 38 is composed of electrolyte-impervious organic resin for instance a polymer or copolymer of a vinyl derivative. Adjacent to the barrier ply 38 is another ply 40 to which is applied a similar ply 41, both the plies 40 and 41 being composed of electrically non-conductive material such as kraft or other paper, which serves to support the barrier ply and to provide adequate strength for the jacket 34 when the innermost ply 36 becomes wet with cell exudate. Additional plies may of course be provided if desired. The bottom of the jacket 34 is provided with a metallic (suitably tin-plated steel) closure 42, the edges of the closure 42 and the bottom edges of the jacket 34 being curled in liquid-tight fashion as indicated at 44. The configuration of the bottom closure 42 is such that the closure is in contact with the outwardly deformed portions 12, 14 of the cupped electrode 10.

A great deal of attention has been given to the problem of protecting the negative terminal cover of the cell from electrolytic corrosion. This cover or end plate is usually made of tin-coated steel and as long as the zinc can

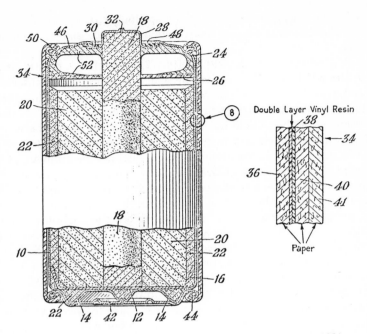

Figure 1.52 Dry cell, R. L. Glover, U.S. Pat. 2,773,926, December 11, 1956.

remains intact it anodically protects the negative terminal cover. However, when the metallic zinc is electrochemically exhausted, the negative cover may become the anode in the cell and both chemical and electrochemical attack can occur. Such attack can result in serious corrosion, rusting, and perforation of the negative cover and result in cell leakage.

Urry recognized that one of the weak points of the many leakproof constructions that had been tried was the electrolytic attack on the negative cover which contacts the negative terminal of the cell. In 1963, he patented (224) a construction in which the negative terminal was appropriately protected by plastic, wax, and the like, and yet was electrically connected to the zinc can of the contained cell by a metal foil. Examples of this construction are given which emphasize how the edge of the metal foil layer is effectively pressed against the negative terminal cover within the curl where the end of the jacket is locked in place. Improved versions were patented in 1965 (U.S. Pat. No. 3,185,694)(225).

In 1963, Reilly et al. (226) also approached the problem of protecting the negative cover by using a suitable layer of plastic, wax, and the like. The electrical contact between the negative cover and the bottom of the zinc can was made by a button, sphere, or spring, or other suitably shaped piece of carbon, graphite, or metal such as lead, silver, titanium, or chromium.

Urry (U.S. Pat. No. 3,202,549), in 1965(227), applied a plastic shielding layer to the whole exterior of the zinc can and to the upper interior side walls and also the inside of the bottom. This distribution of the shielding layer left unprotected only the active inside side-wall surface of the zinc can. Along with this shielding he used an annular space between the jacket and the shielded can to provide space for the exudate.

The viscous nature of the exudate that collects in the airspace of a cell, discharged under abuse conditions, presents problems, since it coats all available surfaces of electrode, cathode mix, seals, etc., and this coating effectively prevents the normal venting of hydrogen gas through the porous media provided for this purpose. Often the carbon electrode is waterproofed but remains permeable to hydrogen and furnishes a vent for this gas. In 1967, Urry patented(228) a partition seal (U.S. Pat. No. 3,338,750) which hung from the positive cover assembly with its lower rim pressed into the upper surface of the cathode mix. This plastic annular ring permitted the accumulation exudate on the space outside the ring but prevented exudate from entering the space inside the ring. Thus the normal venting paths permitted hydrogen to vent from the space inside the annular ring.

A patent, (U.S. Pat. No. 3,214,298) granted in 1965(229) shows the use of the partition seal in conjunction with a sleeve type zinc anode. Figure 1.53 shows a cross section through such a cell.

Below the top closure a partition seal of the type disclosed in my above-referred copending application is provided comprising a tube 30 having an inwardly flanged upper end 32 which is fitted over the flanged cap 24 and on which rests the inner peripheral edges of the top cover 26, thereby securing the tube 30 within the top closure and insulating the metal cap 24 from the top cover 26. The tube 30 surrounds but is spaced from the carbon electrode 12 and is embedded at its lower end within the depolarizer mix 14 at the junction 34, defining a free space 36 between the tube 30 and the carbon electrode 12 and an outer exudate chamber 38. The tube 30 preferably is molded together with the flanged end 32 from a plastic material, for example, polyethylene. The top cover 26 preferably is coated on its underneath side with a thin layer 40 of a liquid-repellent material, for example, vinyl resin or lacquer, for preventing against its corrosion. A layer 42 of a moisture-proof material is also preferably applied over the top surface of the depolarizer mix 14 for prohibiting the ingress of atmospheric oxygen into the cell, but at the same time permitting gas to be vented from within the depolarizer mix 14, asphalt, vinyl resin or grease having been used successfully for this purpose.

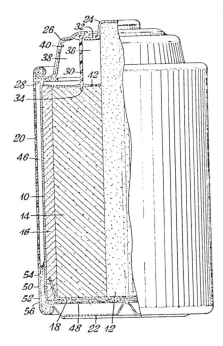

Figure 1.53 Leak-resistant dry cell, L. F. Urry, U.S. Pat. 3,214,298, October 26, 1965.

In 1967 Yamamoto et al. (230) was granted a patent (U.S. Pat. No. 3,320,094) covering an improved construction for a leak-resistant dry cell. Several unique features are described.

A molded plastic closure, for example, polyethylene with a thin section permeable to hydrogen, was combined with a plastic sleeve of thermo-contractile (shrink) vinyl tubing to enclose the side wall. A metal jacket was used to enclose the unit and to clamp the vinyl tubing at both ends of the cell to complete the package. The patent includes cell formulations and gives discharge curves comparing the conventional cells with those described in the patent. The formulation indicates that the patented cells contained "treated electrolytic manganese dioxide" and indeed the discharge curves confirm this statement. Table 1.35, comparing the leakage of the patented and conventional cells on seven selected leakage tests, shows that the patented cells leaked to a limited extent on only one test while the conventional cells leaked on six of the tests listed. Without trying to detract from this invention, the battery technologist might

TABLE 1.35. ELECTROLYTE LEAKAGE
The results of tests with regard to electrolyte leakage are shown below. The conventional cells under test show wide ranges of variations in number of leakage default.

Test Procedure	Number of Cells Leaked	
	Present Invention	Conventional
Continuously discharged through 4-ohm resistance until the time when terminal voltage drops to 0.7 V	0	0
Continuously discharged through 4-ohm resistance for 24 hours, and then left open-circuited at 20°C for 15 days	0	0–75
Continuously discharged through 4-ohm resistance for 24 hours, and then left open-circuited at 20°C for 50 days	0	15–100
Continuously discharged through 4-ohm resistance for 24 hours, and then left open-circuited at 30°C for 30 days	0	0–100
Continuously discharged through 1-ohm resistance at 0°C for 30 days	0	20–100
Continuously discharged through 4-ohm resistance at 30°C for 30 days	0	0–100
Continuously discharged through 125-ohm resistance at 30°C for 60 days	2	6–100

It is then clear that there is little fear of electrolyte leakage in dry cells of the present invention.

wonder to what extent the "treated electrolytic manganese dioxide," used in these cells, contributed to their resistance to leakage. No further description of the depolarizer is given in the patent.

In 1967, Jammet patented(231) a different approach to a leak resistant cell (U.S. Pat. No. 3,342,644). This patent represents an improvement over his 1960 patent (U.S. Pat. No. 3,168,420)(232). Here, a metal cup is pressed over the bottom of a zinc can used in the construction of a cell. A tubular plastic jacket with a serrated lower edge is applied followed by a second metal cup with an open bottom. The assembly is then subjected to a drawing operation to reduce the diameter of this part of the container and thus mechanically seal the bottom of the cell against leakage of exudate.

In 1970 Reilly and Bemer(233) patented a cell construction that incorporated several novel features. The airspace at the top of the cell was divided into two portions by a top collar which was permeable to gas

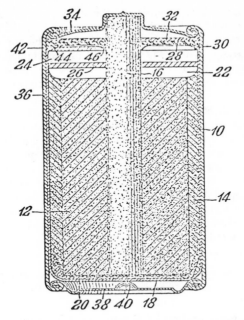

Figure 1.54 Primary dry cell, T. A. Reilly and W. Bemer, U.S. Pat. 3,506,495, April 14, 1970.

but not to liquid. This improvement helped in the separation of these two exudate components. Figure 1.54 shows a cross section through the completed cell. Not shown is the detail of the cell container jacket. This container consisted of several layers.

1. One layer of high strength liquid-impermeable plastic, for example, poly(ethylene terephthalate).
2. Two layers of a thermoplastic material, for example, polyethylene.
3. Three layers of a fibrous cellulosic material, for example, kraft paper.

Referring now to Fig. 1 of the drawing, a primary dry cell embodying the invention may comprise a cupped electrode 10 of a consumable metal (e.g., zinc) having therein a depolarizer mix 12, an immobilized electrolyte 14 and a porous carbon electrode 16 embedded within the depolarizer mix 12. Both the depolarizer mix 12 and the carbon electrode 16 may be suitably provided in the form of a conventional bobbin. Separating the bobbin from the bottom of the cupped electrode 10 is a conventional bottom insulator washer 18, suitably of cardboard or paper. Atop the washer 18 is a fibrous or paper cup 20 which fits around the bottom edges of the depolarizer mix 12.

Positioned within the upper open end of the cupped electrode 10 and so placed as to define a lower free space 22 above the depolarizer mix 12 and an upper free space 24 is an inner seal composed of a liquid impermeable top collar 26. The top collar 26 is fitted tightly within the upper end of the cupped electrode 10 and around the carbon electrode 16. The top collar 26 is permeable to gas but impermeable to liquid and serves as a liquid barrier, and may be made of paper or other fibrous material which is coated with a liquid-repellent material, for example, a plastic such as polyethylene. Spaced above the top collar 26 and within the upper free space 24 is a liquid and gas impermeable seal washer 28. The seal washer 28 fits tightly around the carbon electrode 16 and rests on the upper peripheral edges of the cupped electrode 10, which peripheral edges are turned slightly inwardly as indicated at 30. The seal washer 28 is impermeable to both liquid and gas and is electrically nonconductive and may be composed of a suitable plastic material such as polystyrene. Mounted on top of the seal washer 28 is a layer 32 of a highly absorbent or bibulous material such as conventional blotting paper, for example. The seal washer 28 and absorbent layer 32 may be formed from a laminated sheet of polystyrene having adhered to one side a layer of blotting paper or other suitable bibulous material.

In 1970, Wilke patented(234) a construction that utilizes a somewhat different way of closing the top of a conventional cell. He recognized that the usual plastic poured seal was placed within the top of the zinc can, thus using space that could be better utilized for cathode mix. Therefore he turned inward the top edge of the can, placed a washer on the top of the cell, and poured the plastic seal over the washer and around the outside of the zinc can. The completed cell has a metal exterior jacket.

In any series of structure designs there is always the temptation to combine parts where possible to simplify the assembly operation. Ohki has patented (U.S. Pat. No. 3,556,859)(235) a design in which the top closure and cell container, usually two separate parts, are molded as a single unit. The conventional cell is placed in this combined top closure and container and a metal bottom closure is added. The final mechanical squeeze on the edges of the bottom closure locks the plastic container and bottom together to provide a completely enclosed cell.

Although most of the patents cited in this section show constructions that are designed to prevent leakage of the cell after an abuse test, there are other conditions that may cause leakage. Thus, a cell on shelf may develop an increased volume of electrolyte, as when exposed to a high temperature and solid NH_4Cl dissolves. If this electrolyte is greater in amount than that which can be held by the bibulous materials within the cell it will appear as liquid electrolyte and some of it may leak out of the container. To help control this situation Heise and Cahoon patented (U.S. Pat. No. 2,849,521)(236) a multiply washer to be used at the top of the bobbin. This special washer consisted of several layers of bibulous paper separated by layers of a suitable colloid adhesive such as carboxy methyl cellulose. Data included in the text of the patent show that it was helpful in preventing leakage under the above conditions.

Chemical Additions One of the attractive approaches to the leakage problem was the incorporation, in the airspace of the cell, of some reagent that might react chemically with the exudate and render it less likely to leave that area. There are many chemicals that will react with zinc chloride and form insoluble compounds and some of these have been tried. Materials like zinc oxide, calcium hydroxide, and other similar compounds react to form cementlike hard substances, for example, "sorrel cement." Unfortunately the high viscosity of the exudate in the airspace of cells effectively interferes with this reaction and renders such additions almost useless.

A very similar approach to preventing leakage is the use of materials other than the customary wheat flour and corn starch in the manufacture of the electrolyte paste layer. Morehouse and Welsh(205) obtained a patent (U.S. Pat. No. 2,748,183) on the use of modified starches in the manufacture of dry cells. The starches were modified by treatment with ether-forming bifunctional reagents such as epichlorohydrin, propylene dichloride, glycerol dichlorohydrin or dichlorobutane. These materials were commercially available at that time (1952) as the "Vulca" starches. The chemical reaction of the bifunctional ether-forming reagent with starch consists principally in the formation of an ether linkage with two hydroxyl groups of the starch per molecule of the reagent, a substantial proportion of which is believed to result in cross-linking of the starch molecules. Their patent states:

By means of our invention a dry cell is produced showing excellent performance characteristics combined with an outstanding degree of resistance to leakage or swelling of the cell. . . . Probably the most severe such test is to short circuit a cell for a period of 21 days, whereby commercial leakproof cells display serious leakage in 40–100 percent of the cells tested. Cells of the above construction have shown no serious leakage under this test and only a small percentage, 10 to 30 percent show slight leaks.

Electromechanical Methods The methods of alleviating leakage described in a previous section have been mainly constructional changes applied to the conventional dry cell. These might be classed as strictly mechanical approaches to the problem. A somewhat different method is described by Cahoon(237) in a patent (U.S. Pat. No. 2,876,271) shown in Fig. 1.55. This patent describes an electrolytic fuse placed in the airspace of the cell in such a position that it will be in contact with the exudate as it is driven into the airspace. When the exudate and fuse contact the latter

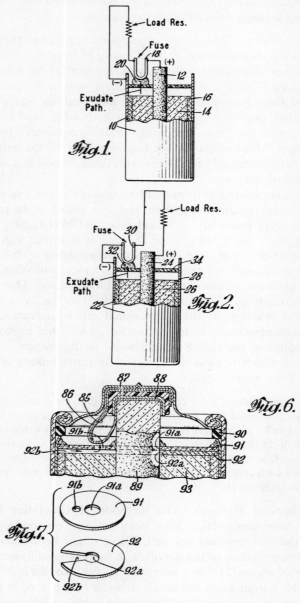

Figure 1.55   Electrolytic fuse, N. C. Cahoon, U.S. Pat. 2,876,271, March 3, 1959.

becomes the anode of a shorted cell and dissolves anodically. The experimental fuses were made of thin sheet zinc or aluminum which are anodically active in the acid exudate. When the fuse is broken in this way the electrical discharge circuit is opened and the cell returns to a resting state. Since the fuse parts may be designed to operate at any point in the cell discharge after exudate develops the cell does not go into the deep continuous discharges characteristic of the usual abuse test. Thus the cell does not continue to develop exudate after the fuse opens the circuit. Table 1.36 shows the important improvement in leakage resistance obtained with the electrolytic fuse. On every abuse test in vogue at that time (1956), fused cells showed no leakage. It is unfortunate that production costs prohibited the commercial manufacture of this successful cell design.

TABLE 1.36. THE RESISTANCE TO LEAKAGE OBTAINED WITH THE ELECTROLYTIC FUSE

Short Circuit at 70°F				
Days	1	15	35	55
Leakproof cell "A"	0	100	100	100
Fused cell	0	0	0	0
0.6 Amp Lamp Continuous at 45°C (113°F—50% R.H.)				
Leakproof cell "B"	14	78	90	92
Leakproof cell "C"	0	94	100	100
Leakproof cell "A"	0	33	33	50
Fused cell	0	0	0	0
0.6 Amp Lamp Continuous at 35°C (95°F—50% R.H.)				
Leakproof cell "B"	8	50	56	56
Leakproof cell "C"	0	22	26	42
Leakproof cell "A"	0	84	84	84
Fused cell	0	0	0	0
0.6 Amp Lamp Continuous at 70°F				
Leakproof cell "B"	8	22	44	52
Leakproof cell "C"	0	42	80	84
Leakproof cell "A"	0	67	100	100
Fused cell	0	0	0	0

TABLE 1.36 (*Contd.*)

Days	1	15	35	55
5 Ohm Continuous at 70°F				
Leakproof cell "A"	0	0	0	0
Fused cell	0	0	0	0
2.25 Ohm Heavy Intermittent Flashlight to 0.65 V, Then Started Through 0.6 Amp Lamp				
Leakproof cell "A"	0	0	67	67
Fused cell	0	0	0	0
2.25 Ohm Light Intermittent Flashlight to 0.65 V, Then on 0.6 Amp Lamp Continuously at 70°F				
Leakproof cell "B"	0	6	10	14
Leakproof cell "C"	0	0	2	4
Leakproof cell "A"	0	0	33	33
Fused cell	0	0	0	0
150 ma Constant Current for 20 Hours, Then on Open Circuit Shelf				
Leakproof cell "B"	0	0	2	10
Leakproof cell "C"	0	30	46	52
Leakproof cell "A"	0	0	0	0
Fused cell	0	0	0	0

Note. The numbers in the table show the percent of the experimental cell that showed leakage after the time period given.

More specifically, Fig. 1 illustrates a typical dry cell having a zinc can 10, serving as its anode, a carbon electrode 12, a conventional electrolyte-wet cathode mix indicated by reference character 14, separated from the cathode by paste or other separator means 16. Interposed between the external terminal or brass cap (not shown) and the cathode 12 is a strip of zinc metal constituting the dry cell fuse 18. This fuse in effect is the anodic component of a separate cell system which may be represented as follows:

$$MnO_2/ZnCl_2\text{-}NH_4Cl/fuse$$

Under the conditions provided by the outlined circuit, the fuse is electro-chemically attacked by exudate, which may rise above normal mix level, as indicated in exaggerated form by reference character 20.

Where the dry cell fuse operates by electro-chemical action, it need not be composed of a metallic material having a greater electro-negativity than the dry cell anode. This conclusion obtains inasmuch as the voltage between the cathode and the dry cell fuse is higher than that between the cathode and the anode. Consequently, the dry cell fuse strip will be attacked by exudate in preference to the anode.

Fig. 2 illustrates a variant of the invention actuated by chemical attack. Here a conventional dry cell having a container zinc anode 22, a carbon electrode 24, and an electrolyte-wet depolarizing mix composed of manganese dioxide (as in Fig. 1) 26, separated from the cathode by paste or other separator means 28, is fitted with an aluminum fuse strip 30 in series with the cell anode, and, eventually, with the load resistance (i.e., a bulb, in the case of flashlight cells).

From Fig. 2 and reference character 32 it is obvious that when exudate rises above normal mix level and bypasses washer 34, it then is able to attack, corrode and destroy the cell fuse, thereby breaking the circuit and stopping further cell discharge. The fuse-equipped cell then returns to equilibrium, leakage having been forestalled.

Fig. 6 shows a modification of the invention in which mechanical means in the form of a metal spring accelerate circuit breaking. As indicated, a metal spring 85 presses against the dry cell fuse 86, which in this modification is borne on a combined plastic cell closure and fuse support 87. To insure electrical contact, a brass cap 88 is fitted over the top of carbon electrode 89. As shown, sufficient space is provided in the top washer 90 for the indicated spring action. A direct low resistance path to guide the exudate to the fuse is provided by top collars 91 and 92.

The version of the invention represented on Fig. 6 is particularly attractive for small hearing aid cells and the like, owing to its combined plastic cell closure and fuse support, which may be fabricated to small dimensions, and similarly conserving space.

A Truly Leakproof Leclanché Cell

Schumm has described(238) the reduction in leakage obtained when a pasted type cell is assembled with a somewhat smaller-than-normal bobbin diameter and a corresponding thicker paste wall. Table 1.37 presents his data obtained on "D" size cells. It is clear that cells having paste wall thicknesses of 0.176 and 0.276 in. (0.44 and 0.70 cm) respec-

TABLE 1.37. THE EFFECT OF PASTE WALL THICKNESS ON EXCESS ELECTROLYTE MOVEMENT

Group	Paste Wall Thickness		Paste Weight Grams per Cell	Excess Electrolyte Weight
I	0.071"	0.18 cm	7.5	4.4 g
II	0.126	0.32	13.8	2.2
III	0.176	0.44	18.1	0.0
IV	0.276	0.70	25.3	0.0

tively, show no development of excess electrolyte, that is, leakage, on the "short circuit for 24 hours" test which he used. This is a very severe test and the results are significant. Two additional unexpected improvements resulted from the increase in paste wall thickness.

1. A porous, fluffy precipitate layer formed on the surface of the cathode in contrast to the tight, highly adherent layer normally found in thin paste wall cells.
2. If a small amount of electrolyte movement occurs, the excess liquid is weakly alkaline rather than acidic as in the thin paste wall cells.

On a standard 2.25 ohm LIF test the cells with the thickest separator layer gave 13 percent more output than those with the thinnest layer shown in Table 1.37. Schumm explains the increased output by the reduction in anode polarization due to the greater volume of anolyte available in the cells with the thicker paste walls.

After over 30 years of continued efforts in many laboratories around the world the results shown above represent an important breakthrough and an amazing achievement in Leclanché cell technology.

THE ZINC CHLORIDE BATTERY SYSTEM

The battery technologist may find it hard to believe that the removal of the ammonium chloride from the Leclanché cell formulation actually results in a new battery system. This situation can well be understood when it is recognized that a great part of many Leclanché cell experimental programs involved the study of the effects of varying the amount of ammonium chloride used in its manufacture. Nevertheless, work in recent years has shown that a cell made with an electrolyte consisting of an aqueous solution of $ZnCl_2$ does have some distinct advantages over the conventional Leclanché cell insofar as leakage resistance and heavy service are concerned.

Cells are being produced commercially using the $ZnCl_2$ system, but since there has been little published it must be concluded that the information is considered proprietary. On the basis of the publications to date the discussion here is limited to the probable reaction mechanism in the $ZnCl_2$ cell, and one development in separator technology.

THE PROBABLE REACTION MECHANISM IN THE $ZnCl_2$ CELL

Huber has described (239) the $ZnCl_2$ cell as one of the two ways that a leakproof cell can be made, the other being the use of the alkaline MnO_2-Zn system (cf. Chap. 7, Vol. I). He gives the following overall reaction for this cell:

$$8MnO_2 + 4Zn + ZnCl_2 + 9H_2O \rightarrow 8MnOOH + ZnCl_2 \cdot 4Zn(OH)_2 \cdot 5H_2O$$
(60)

He points out that the cell continues to dry out with continued discharge, presumably due to the large amount of water bound up in the reaction product $ZnCl_2 \cdot 4Zn(OH)_2 \cdot 5H_2O$. He does not give any supporting evidence for the degree of hydration that he has reported, and the reader is left to wonder if this is the same salt as the usual hydroxy chloride $ZnCl_2 \cdot 4Zn(OH)_2$ referred to in the literature. Although there are a number of oxychlorides and hydroxychlorides of zinc described in the literature, only three products have been identified by x-ray diffraction.

Two German patents by Krey(240) (2,201,285 and 6) and a U.S. patent(241) have been granted which deal with the zinc chloride system. These two patents are very similar and both describe the use of an aqueous electrolyte containing 15 to 40 percent by weight of $ZnCl_2$. This electrolyte is used in the manufacture of a cathode mix which contains 23 to 39 percent by weight of water, a very much higher level than usually found in Leclanché mixes. The patents point out that as the water content of the cathode is increased to about 39 percent no leakage is observed from the cells made from this cathode mix.

ONE DEVELOPMENT IN SEPARATOR TECHNOLOGY

Battery technologists who have worked with the Leclanché system will at once recognize that cereals, usually employed as paste thickeners, will probably hydrolyze in the $ZnCl_2$ cell electrolyte. Even in electrolytes containing ammonium chloride the cereal breaks down as reported previously(197). Perhaps in an effort to avoid this difficulty Indian Pat. No. 112,711(242) describes the use of ion-exchange resins in the separator layer for a $ZnCl_2$ cell. In a very similar, if not identical patent(242) (U.S. Pat. No. 3,558,364), examples of discharge curves are given, as shown in Fig. 1.56. The two cells employed for this test were entirely identical except that in one cell an anion exchange compound was used and in the other cell a cation exchange compound was employed. In one example the anion exchange compound was a polystyrene cross-linked with less than 3 mol percent of divinylbenzene and having active groups constituted by quaternary ammonium groups. The cation exchange compound was of the same type except that the active groups were sulfonated groups; that is, the compound was cross-linked with Na-polystyrene sulfonate. The ion exchange resins were used as powders with particle diameters between 20 and 70 microns. They were dispersed in an acetone solution of a butadiene acrylonitrile copolymerisate and applied to the inside of the

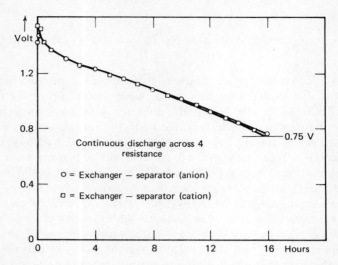

Figure 1.56 Leakproof galvanic cell employing granular anion exchange compound separator, W. Krey, U.S. Pat. 3,558,364, January 26, 1971.

zinc can as a suspension, and allowed to dry. The electrolyte was a solution containing 30 percent of $ZnCl_2$ as the sole electrolyte salt.

The patent states that the cell made in this way was completely leakproof even under conditions of great stress such as under short-circuit conditions. It further states that the capacity on heavy discharge is 15 to 25 percent higher than that obtained from conventional cells.

REFERENCES

1. Leclanché, G., *Les Mondes*, **16**, 532 (1868); *Compt. Rend.*, **83**, 54 (1876); *Compt. Rend.*, **83**, 1236 (1876); Fr. Pat. 71,865 (1866).

2. Leclanché, G. and Barbier, Fr. Pat. 124,108(1878); Davis, J., Electrochem Tech. **5**, 487 (1967).

3. Heise, G. W. and Cahoon, N. C., *J. Electrochem. Soc.*, **99**, 174C (1952); Heise, G. W., *The Primary Battery*, Vol. I, J. Wiley, New York (1971).

4. Teas, J. P., U.S. Pat. 2,605,299 (July 29, 1952).

5. Volta, A., Letters XIII and XIV to van Marum (Aug. 30 and Oct. 11, 1792); *Phil. Trans. Roy. Soc. London*, **83**, 10 (1793); **90**, 403 (1800).

6. Crocker, F. B., *Trans. Electrochem. Soc.*, **10**, 107 (1906).

7. Gassner, C., Ger. Pat. 37,758 (April 8, 1886); U.S. Pat. 373,064 (Nov. 15, 1887).

8. Meerburg, *Z. Anorg. Chem.*, **37**, 206 (1903).

9. Cahoon, N. C., *Trans. Electrochem. Soc.*, **92**, 159 (1947).

10. Takahashi, T., Nakauchi, H., and Sasaki, K., *Electrochem. Soc. Japan J.*, I, **22**, 589 (1954); II, **22**, 593 (1954); III, **24**, 171 (1956); IV **24**, 221 (1956); V, **24**, 414 (1956); VI, **24**, 471 (1956); VII, **24**, 516 (1956); VIII, **25**, 58 (1957); IX, **25**, 118 (1957); X, **25**, 277 (1957).

11. Holler, H. D. and Ritchie, L. M., *Trans. Electrochem. Soc.*, **37**, 607 (1920).
12. Martin, T. J. and Helfrecht, A. J., *Trans. Electrochem. Soc.*, **53**, 83 (1928).
13. Cahoon, N. C., *Trans. Electrochem. Soc.*, **68**, 177 (1935).
14. Cahoon, N. C., *J. Electrochem. Soc.*, **99**, 343 (1952).
15. Burgess, C. F., U.S. Pat. 1,305,250 (June 3, 1919).
16. McMurdie, H. F., Craig, D. N., and Vinal, G. W., *Trans. Electrochem. Soc.*, **90**, 509 (1946).
17. Sasaki, K., *Mem. Fac. Eng. Nagoya Univ.*, **3**, 81 (1951).
18. Johnson, R. S. and Vosburgh, W. C., *J. Electrochem. Soc.*, **99**, 317 (1952).
19. Kozawa, A., *J. Electrochem. Soc.*, **106**, 79 (1959).
20. Kozawa, A. and Sasaki, K., *J. Electrochem. Japan*, **22**, 569 (1954).
21. Maxwell, K. H. and Thirsk, H. R., *J. Chem. Soc.* **1955**, 4057.
22. Muller, J., Tye, F. L., and Wood, L. L., *Batteries 2*, Ed. D. H. Collins, p. 201 Pergamon Press Ltd., London (1965).
23. Benson, P., Price, W. B., and Tye, F. L., *Electrochem. Tech.* **5**, 517 (1967).
24. Caudle, J., Summer, K. G., and Tye, F. L., *J. Chem. Soc. Far. Trans.* I (1973).
25. Korver, M. P., Johnson, R. S., and Cahoon, N. C., *J. Electrochem. Soc.*, **107**, 587 (1960).
26. Cahoon, N. C., *Extended Abstracts, No. 19, Battery Division, Electrochem. Soc.*, Vol. 8, 57 (1963).
27. Getman, F. H., *J. Phys. Chem.*, **35**, 2749 (1931).
28. Cahoon, N. C. *Extended Abstracts, No. 39, Battery Division, Electrochem. Soc.* Vol. 10, 106 (1965).
29. Drucker, C. and Huettner, R., *Z. Physik. Chem.*, **131**, 237 (1928); Drucker, C. and Finkelstein, A., *Galvanische Elemente und Akkumulatoren*, Akad. Verlags, G.m.b.H., Leipzig p. 43 ff. (1934).
30. Thompson, M. deKay, *Trans. Electrochem. Soc.*, **68**, 167 (1935).
31. Kaneko, S., *J. Soc. Chem. Ind. (Japan)*, **32**, 120 (1929).
32. Drotschmann, C., *Moderne Primärbatterien*, Nikolous Branz, Berlin Schoenberg, p. 11 ff. (1951).
33. Latimer, W. M., *The Oxidation States of the Elements and their Potentials in Aqueous Solutions*, 2nd Ed., Prentice-Hall, Englewood Cliffs, N.J. c. (1952).
34. Tomassi, D., *Traité des Piles Électriques*, Georges Carré, Paris (1889).
35. Priwoznik, E., *Pogg. Ann.* **142**, 467 (1871).
36. Daniels, F. *Trans. Electrochem. Soc.*, **53**, 45 (1928).
37. Friess, R., *J. Am. Chem. Soc.*, **52**, 3085 (1930).
38. McMurdie, H. F., *Trans. Electrochem. Soc.*, **86**, 191 (1944).
39. Takahashi, T., Sakurai, M., and Sasaki, K., *Electrochem. Soc. Japan J.*, II, **22**, 593 (1954).
40. Takahashi, T. and Sasaki, K., *Electrochem. Soc. Japan J.*, IX, **25**, 118–122 (1957).
41. Isambert, F., *Compt. Rend.*, **66**, 1259 (1868); *Int. Crit. Tables*, VII, 252, 253 (1930) McGraw-Hill Book Co., New York.
42. Bottone, S. R., *Galvanic Batteries*, Whittaker, New York and London (1902).
43. LeBlanc, M., *A Text-Book of Electrochemistry*, Macmillan, New York (1907).
44. Davidson, A. W., *J. Chem. Education*, **25**, 536 (1948); Thompson, M. deKay, and Crocker, E. C., *Trans. Electrochem. Soc.*, **27**, 155 (1915).
45. Divers, E., *Chem. News and J. Phys. Sci.*, **46**, 259 (1882).
46. Cahoon, N. C. and Heise, G. W., *J. (and Trans) Electrochem. Soc.*, **94**, 214, 385 (1948); Drotschmann, C., *Chem. Ztg.*, **64**, 244 (1940).
47. French, H. F. and MacKenzie, A. A., Unpublished; cf. Cahoon, N. C., *J. Electrochem. Soc.*, **99**, 343 (1952).

48. Cahoon, N. C., *Extended Abstract, No. 22, Battery Division, Electrochem. Soc.* (October, 1956).
49. Cowley, J. M. and Walkley, A., *Nature*, **161**, 173 (1948); Ferrell, D. T., Jr. and Vosburgh, W. C., *J. Electrochem. Soc.*, **98**, 334 (1951).
50. Cahoon, N. C., Johnson, R. S., and Korver, M. P., *J. Electrochem. Soc.*, **105**, 296 (1958).
51. Korver, M. P., Johnson, R. S., and Cahoon, N. C., *J. Electrochem. Soc.*, **107**, 587 (1960).
52. Cahoon, N. C. and Korver, M. P., *J. Electrochem. Soc.*, **109**, 1 (1962).
53. Buser, W. and Graf, P., *Helv. Chim. Acta.*, **38**, 810 (1955).
54. Kozawa, A., *J. Electrochem. Soc.*, **106**, 79 (1959).
55. Vosburgh, W. C., *J. Electrochem. Soc.*, **106**, 839 (1959).
56. Euler, J., *Electrochemica Acta*, **4**, 27 (1961).
57. Brunnauer, S., Emmett, P. H., and Teller, E., *J. Am. Chem. Soc.*, **60**, 309 (1938).
58. Moore, W. G., U.S. Pat. 3,414,440 (Dec. 3, 1968).
59. Copeland, L. C. and Griffith, F. S., *Trans. Electrochem. Soc.*, **89**, 495 (1946).
60. McQueen, A., *A Romance in Research*, Instruments Publishing Co., Pittsburgh (1951); Technical Appendix, Storey, O. W., p. 364.
61. Aufenast, F., *Proc. Int. Symposium on Batteries*, paper (b), (October, 1958) Christchurch, Hants, England.
62. Aufenast, F., Muller, J., in "Batteries," Collins, D. H. *Proc. 3rd Int. Symposium* Pergamon Press, New York, p. 333 (1962).
63. Chaney, N. K., *Trans. Am. Electrochem. Soc.*, **29**, 183 (1916).
64. Friend, J. N., *Engineering*, **127**, 354 (1929), from *C.A.*, **23**, 2372 (1929).
65. Coleman, J. J., U.S. Pat. 2,598,226 (May 27, 1952).
66. Pushin, N. A., *Zeit. Anorg. Chem.*, **36**, 201 (1903); *Circ. Bur. Stds.*, No. 395, U.S. Govt Printing Office (1931).
67. Hellesen, W. F. L., U.S. Pat. 439,151 (October 28, 1890).
68. Hambuechen, C., U.S. Pat. 1,292,764 (Jan. 28, 1919); Schulte, W. B., U.S. Pat 1,370,056 (Mar. 1, 1921).
69. Hamer, W. J., *J. Res. Nat. Bur. Stds.*, **40**, 251 (1948).
70. Cahoon, N. C., U.S. Pat. 2,534,336 (Dec. 19, 1950).
71. Benner, R. C., U.S. Pat. 1,276,714 (Aug. 27, 1918); Ruhoff, O. E., U.S. Pat. 1,331,877 (Feb. 24, 1920); cf. ref. to McQueen, A.(60).
72. Zimmerman, H. M. and Powers, R. A., U.S. Pat. 2,900,434 (Aug. 18, 1959); U.S. Pat. 2,971,044 (Feb. 7, 1961).
73. Aufenast, F. and Davis, J., *Proc. 2nd Int. Symposium on Batteries*, paper 9/1 (October, 1960), Bournemouth, England.
74. Hamer W. J., *Trans. Electrochem. Soc.*, **90**, 449 (1946).
75. McMurdie, H. F. and Golvato, E., *J. Res. Natl. Bur. Stds.*, **41**, 589 (1948).
76. Feitknecht, W., and Marti, W., *Chim. Acta.*, **28**, 149 (1945).
77. Coleman, J. J., *Trans. Electrochem. Soc.*, **90**, 545 (1946).
78. Scott, A. B., *J. Electrochem. Soc.*, **107**, 941 (1960).
79. Daniel-Bek, V. S., *Trudy Soveshchaniya Elektrokhim. Akad. Nauk S.S.S.R., Otdel. Khim. Nauk*, **1950**, 513 (1953).
80. Kornfeil, F., *J. Electrochem. Soc.*, **109**, 349 (1962).
81. Brouilet, Ph., Gund, A., Jolas, F., and Mellet, R., Int. *Symp. Batteries*, Brighton (1964).
82. Kozawa, A. and Vosburgh, W. C., *J. Electrochem. Soc.*, **105**, 59 (1958).
83. Ferrell, D. T., Jr. and Vosburgh, W. C., *J. Electrochem. Soc.*, **98**, 334 (1951).
84. Chreitzberg, A. M., Jr. and Vosburgh, W. C., *J. Electrochem. Soc.*, **104**, 1 (1957).

85. Chreitzberg, A. M., Jr., Allenson, D. R., and Vosburgh, W. C., *J. Electrochem. Soc.*, **102**, 557 (1955).
86. Johnson, R. S. and Vosburgh, W. C., *J. Electrochem. Soc.*, **100**, 471 (1953).
87. Nauman, K. and Fink, W., *Z. Electrochem.*, **62**, 114 (1958).
88. Era, A., Takehara, Z., and Yoshizawa, S., 17th Meeting CITCE, Tokyo (1966).
89. Brenet, J. *Proc. Int. Comm. Electrochem. Thermodynam. and Kinetics*, 8th Meeting, Madrid (1956), Butterworths Scientific Pubs. 7.4, 394 (1956).
90. Brenet, J. in "Batteries," Collins, D. N. *Proc. 3rd. Int. Symp.* (1962).
91. Gaband, J. P., Morignat, B., and Laurent, J. F., 14th Meeting, CITCE, Moscow (1963).
92. Brenet, J., Private communication.
93. Laurent, J. F. and Morignat, B., *Batteries, Res. and Dev. in Nonmechanical Elec. Pwr. Sources*, Collins, D. H., Pergamon, New York (1963).
94. Huber, R. and Schmier, A., *Electrochim. Acta.*, **3**, 127 (1960).
95. Gosh, S. and Brenet, J., *Electrochimica Acta*, **7**, 449 (1962).
96. Feitknecht, W., Oswald, H. R., and Feitknecht-Steinmann, U., *Helv. Chem. Acta*, **43**, 1947 (1960).
97. Bode, H., Schmier, A., "Batteries," Collins, D. H. *Proc. 3rd Int. Symposium* (1962).
98. Kozawa, A. and Powers, R. A., *Electrochem. Tech.*, **5**, 535 (1967).
99. Miyazaki, K., *J. Electrochem. Soc.*, **116**, 1469 (1969).
100. Miyazaki, K., *The 7th International Power Sources Symposium*, pp. 607–624, Brighton, England (1970).
101. Zwicker, W. K., Groeneveld Meijer, W. O. J., and Jaffe, H. W., *Am. Mineralogist*, **47**, 246 (1962).
102. Burgess, C. F., U.S. Pat. 1,305,251 (June 3, 1919).
103. Tamura, H., *Electrochemistry of Manganese Dioxide and Manganese Dioxide Batteries in Japan*, Vol. II, p. 189 ff., Takahashi, K., Yoshizawa, S., and Kozawa, A., U.S. Branch Office of the Electrochemical Soc. of Japan, Cleveland (1971).
104. Hanshin Yosetsu Kizai Co., *Electrochemistry of MnO₂ and Manganese Oxide Batteries in Japan*, Vol. II, p. 159 ff., U.S. Branch Office, Electrochem. Soc. Japan, Cleveland (1971).
105. Drucker, C. and Finklestein, A., *Galvanische Elemente und Alkumulatoren*, Akad. Verlags G.m.b.H. Leipzig (1932).
106. Dubois, P., *Ann. de Chim.*, **5**, 411 (1936); *Compte. Rend.*, **700**, 1107 (1935).
107. Holmes, M. E., U.S. Pat. 1,148,230 (July 27, 1915).
108. Heil, A., German Pat. 259,999 (1910); U.S. Pat. 1,053,390; 1,053,505 (Feb. 18, 1913).
109. Kaplan, M. L., U.S. Pat. 1,178,927 (Apr. 11, 1916).
110. Kaplan, M. L., U.S. Pat. 1,287,041 (Dec. 10, 1918).
111. Dean, R. S., *Mining. Eng.*, **4**, 55 (1952).
112. Welsh, J. Y. and Peterson, D. W., *J. Metals*, **9**, 762 (1967).
113. Bradley, W., U.S. Pat. 1,937,508 (Dec. 5, 1933).
114. Henn, J. J., Peters, F. A., Johnson, P. W., and Kirby, R. C., *U.S. Bur. Mines Rep. Invest.*, 1968, No. 7156.
115. Nossen, E. E., *Ind. and Eng. Chem.*, **43** (7) 1695 (1951).
116. Welsh, J. Y., Private communication.
117. Welsh, J. Y., U.S. Pat. 2,956,860 (Oct. 18, 1960).
118. Honda, T., Mizumaki, K., and Tanabe, I., *Electrochemistry of Manganese Dioxide and Manganese Dioxide Batteries in Japan*, Vol. I, p. 203 ff., Takahashi, K., Yoshizawa, S., and Kozawa, A., U.S. Branch Office of the Electrochemical Soc. of Japan, Cleveland (1971).
119. Van Arsdale, G. D. and Maier, C. B., *Trans. Electrochem. Soc.* **33**, 109 (1918).

144   LECLANCHÉ AND ZINC CHLORIDE DRY CELLS

120. Nichols, C. W., Ibid, **57**, 393 (1932).
121. Johnson, N. J., Private communication.
122. Lee, J. A., *J. (and Trans.) Electrochem. Soc.* **95**, 2P–13P (1949); McQueen, A., *A Romance in Research*, Instruments Publishing Co., Pittsburgh (1951), p. 366, ff.
123. Kissin, G. H., "Electrolytic Synthesis of Battery Active Manganese Dioxide," Final Report, Contract W-36-039-SC-32239, Georgia Inst. of Technology (July 31, 1949).
124. Kozawa, A., *Electrochemistry of Manganese Dioxide and Manganese Dioxide Batteries in Japan*, Vol. I, 57 (1971), U.S. Branch Office of Electrochem. Soc. Japan, Cleveland.
125. Miyazaki, K. and Suemoto, T., Ibid, Vol. I, 159 (1971).
126. Kato, T., Ibid, Vol. I, 219 (1971).
127. Takahashi, K. and Kozawa, A., Ibid, Vol. II, 23 (1971).
128. Koshiba, J. and Nishizawa, S., Ibid, Vol. II, 85 (1971).
129. Sudo, A., Ibid, Vol. II, 101 (1971).
130. Araki, I., Ibid, Vol. II, 115 (1971).
131. Welsh, J. Y., *Electrochem. Tech.*, **5**, 504 (1967).
132. Storey, O. W., Steinhoff, E., and Hoff, E. R., *Trans. Electrochem. Soc.*, **86**, 337 (1944).
133. Fleischmann, M., Thirsk, H. R., and Tordesillas, I. M., *Trans. Faraday Soc.*, **58**, 1865 (1962).
134. Era, A., Takehara, Z. and Yoshizawa, S., *Denki Kagaku*, **35**, 339 (1967).
135. Ramsdell, L. S., *Am. Mineral*, **17**, 143 (1932), C. A. 4722 (1932).
136. Fleischer, M. and Richmond, W. E., *Econ. Geol.* **38**, 269 (1943).
137. Era, A., Takehara, Z., and Yoshizawa, S., 17th Meeting of CITCE, Tokyo (Sept., 1966).
138. Fukuda, M., *National Technical Report*, **4**, 321 (1958); **5**, 1 (1959).
139. Ninagi, S., Doctoral Thesis, Kyoto Univ. (Feb., 1968).
140. Sasaki, K. and Kozawa, A., *Denki Kagaku*, **25**, 273 (1957).
141. Matsuno, S., *J. Chem. Soc. Japan, Industrial Chem. Sect.*, **44**, 621, 909 (1941); Ibid, **46**, 605 (1943).
142. Muraki, I., Japanese Pat. Application No. 14512 (published June 20, 1968).
143. Société des Accumulateurs Fixes et de Traction, French Pat. 1,509,122 (Dec. 4, 1967).
144. Fox, A. L., U.S. Pat. 2,591,532 (April 1, 1952).
145. Wiley, J. S. and Knight, H. T., *J. Electrochem. Soc.*, **111**, 656 (1964).
146. Covington, A. K., Cressy, T., Lever, B. G., and Thirsk, H. R., *Trans. Faraday Soc.*, **58**, 1975 (1962).
147. Malati, M. A., *Chem. and Ind.*, 446 (1971).
148. Das, J. N., *Z. Phys.*, **151**, 345 (1958).
149. Bhide, V. G. and Dani, R. H., *Physica*, **27**, 821 (1961).
150. Elovich, S. Yu. and Margolis, L. Ya., *Dokl. Akad. Nauk SSSR*, **107**, 112 (1956).
151. St. John, A., *Phys. Rev.*, **21**, 389 (1923).
152. Ettel, V. and Veprěk-Siška, J., *Chemické Listy*, **57**, 785 (1963).
153. Cole, W. F., Wadsley, A. D., and Walkley, A., *Trans. Electrochem. Soc.*, **92**, 133 (1947).
154. Gattow, G. and Glemser, O., *Z. Anorg. Allg. Chem.*, **121**, 309 (1961).
155. Glemser, G. and Meisick, H., *Naturwissenschaften*, **44**, 614 (1957).
156. Okada, S., Uei, I., and Chin, H., *Denki Kagaku*, **15**, 79 (1947).
157. Pons, L. and Brenet, J., *Cr. Hebd. Seanc. Acad. Sci.*, Paris, **259**, 2825 (1964); **260**, 2483 (1965).

158. Nye, W. F., Levin, S. B., and Kedesdy, H. H., *Proc. Annual Power Sources Conf.*, **13**, 125 (1959).

159. Tauber, J. A., *X-Ray Diffraction Key to Manganese Oxide Minerals*, E. J. Lavino, Philadelphia (1964).

160. Miyake, Y., *Electrochemistry of Manganese Dioxide and Manganese Dioxide Batteries in Japan*, Vol. I, S. Yoshizawa, K. Takahashi, and A. Kozawa, Electroch. Soc. Japan, U.S. Branch Office, Cleveland (1971) pp. 115, ff.

161. Fukuda, M., National Technical Report, **4**, 321 (1958); Ibid, **5**, 1 (1959); *Denki Kagaku*, **28**, 67 (1960); *National Technical Report*, **3**, 1 (1957).

162. Vink, J. J., *Analyst*, **95**, 399 (1970).

163. Morehouse, C. K. and Glicksman, R., *J. Electrochem. Soc.*, **103**, 149 (1956).

164. Glicksman, R., *J. Electrochem. Soc.*, **108**, 1 (1961).

165. Appelt, K. and Purol, H., *Electrochimica Acta*, **1**, 326 (1959).

166. Kornfeil, F., *J. Electrochem. Soc.*, **106**, 1062 (1959).

167. Marco, A. and Kordesch, K., U.S. Pat. 2,662,211 Dec. 8, (1953).

168. Huber, R. and Kandler, J., *Electrochimica Acta*, **8**, 265 (1963).

169. Balewski, L. and Brenet, J. P., *Electrochem. Tech.*, **5**, 528 (1967).

170. Schweigert, H., South African Council for Scientific and Industrial Research, *Special Reports Chem.*, **74** (revised 1969); *Chem.*, **142** (1971).

171. Olivier, I., Ibid, *Chem.*, **173** (1971).

172. Zulch, B. J., *Tech. Info. for Industry*, **9**, 4, 1 (1971), Pretoria, South Africa.

173. D'Agostino, O., *Ricerca Sci.*, **9**, I 195 (1938).

174. Drotschmann, C., *Batterien*, **20** 2, 887 (1966).

175. Acheson, E. G., U.S. Pat. 568,323 (May 19, 1896); Symanowitz, R., *Edward Gooderich Acheson*, Vantage Press, New York (1971).

176. Burgess, C. F., U.S. Pat. 1,162,449 (Nov. 30, 1915); 1,211,363 (Jan. 2, 1917).

177. Clymer, W. R., U.S. Pat. 1,480,533 (Jan. 8, 1924).

178. Heise, G. W., U.S. Pat. 1,553,530 (Sept. 15, 1925).

179. Burger, P., German Pat. 280,098 (1912); U.S. Pat. 1,123,843 (Jan. 5, 1915).

180. Payne, V. F., *Trans. Electrochem. Soc.*, **86**, 345 (1944).

181. Mrguditch, J. N. and Clock, R. C., *Trans. Electrochem. Soc.*, **86**, 351 (1944).

182. Watson, J. H. L., *Trans. Electrochem. Soc.*, **92**, 77 (1947).

183. Bunsen, R., *Pogg. Ann.*, **54**, 417 (1841); **55**, 265 (1842); **57**, 110 (1843); *Compte. Rend.*, **16** (1843).

184. Chaney, N. K., U.S. Pat. 1,809,471 (June 9, 1931).

185. Huntley, A. K., *Trans. Electrochem. Soc.*, **68**, 219 (1935); Read, C. J. and Morrill, M. T., U.S. Pat. 690,772 (Jan. 7, 1902).

186. French, H. F., *Proc. Inst. Radio Engrs.*, **29**, 299 (1941); U.S. Pat. 2,272,969 (Feb. 10, 1942).

187. Franz, A. O., Martinez, J. M., and Koppelman, M. D., U.S. Pat. 2,416,576 (Feb. 25, 1947).

188. Bottone, S. R., *Galvanic Batteries*, p. 221 ff., Whittaker, New York and London (1902).

189. Hellesen, W. L. F., U.S. Pat. 439,151 (Oct. 28, 1890).

190. Johnson, H. T., U.S. Pat. 487,839 (Dec. 13, 1892).

191. Coleman, C. J., U.S. Pat. 495,306 (April 11, 1893).

192. Heise, G. W. and Schumacher, E. A., U.S. Pat. 1,808,410 (June 2, 1931); 1,890,178 (Dec. 6, 1932).

193. Schulte, W. B., U.S. Pat. 1,370,054 and 1,370,055 (March 1, 1921); Heise, G. W., U.S. Pat. 1,788,870 (Jan. 13, 1931).
194. Schorger, A. W., U.S. Pat. 1,316,597 (Sept. 23,1919); 1,370,052 (March 1, 1921); Staley, W. D., U.S. Pat. 1,760,090 (May 27, 1930); Heise, G. W., U.S. Pat, 1,842,871 (Jan. 26, 1932).
195. Hambuechen, C., U.S. Pat. 1,292,764 (Jan. 28, 1919).
196. Hambuechen, C., Ibid; Schulte, W. B., U.S. Pat. 1,370,056 (March 1, 1921).
197. Cahoon, N. C., U.S. Pat. 2,534,336 (Dec. 19, 1950); Cahoon, N. C. and Korver, M. P., *J. Electrochem. Soc.*, **105**, 293 (1958).
198. Cahoon, N. C., *Trans. Electrochem. Soc.*, **86**, 327 (1944) discussion.
199. Cahoon, N. C. and Korver, M. P., U.S. Pat. 2,900,433 (Aug. 18, 1959).
200. Korver, M. P. and Cahoon, N. C., U.S. Pat. 3,048,647 (Aug. 7, 1962).
201. Bruins, P. F., *Proc. 11th Annual Res. and Devel. Conf.*, U.S. Army Signal Corps, Ft. Monmouth, N.J., p. 50, 62 (1957).
202. Higgins, T. W., U.S. Pat. 3,018,316 (Jan. 23, 1962).
203. Morehouse, C. K., Hamer, W. J., and Vinal, G. W., *U.S. Bur. of Stds. Res. Paper* RP1863, **40** (Feb., 1948).
204. Otto, E., Vinal, G. W., and Ostrander, E. H., *Trans. Electrochem. Soc.*, **86**, 327 (1944).
205. Morehouse, C. K. and Welsh, J., U.S. Pat. 2,748,183 (May 29, 1956).
206. Anthony, H. R. C., U.S. Pat. 2,198,423 (April 23, 1940); U.S. Pat. 2,243,938 (June 3, 1941).
207. Glover, R. L., U.S. Pat. 2,773,926 (Dec. 11, 1956).
208. Yeaple, F. D., *Product Engineering*, 760 (March, 1965).
209. Stephens, C., *Electronics Illustrated*, 103 (1963).
210. Kaye, G. E., *Radio Electronics*, **46** (June, 1963).
211. *Eveready Battery Applications and Engineering Data*, Battery Division, Union Carbide Corp., New York (1971).
212. Cameron, D. B., *Electronics World*, 33 (1961).
213. Grimes, C. G. and Herbert, W. S., No. 269D, SAE., *Trans. Soc. Automotive Engrs.* (March 1, 1961).
214. Pipal, F. B., *Electronics World*, 42 (Oct., 1965).
215. Hellfritsch, A. G., *Proc. 11th Power Sources Conf.* Ft. Monmouth, N.J. (May, 1957); *Proc. 6th Int. Power Sources Symposium*, 251 (Sept., 1968).
216. Warburton, D. L., *Proc. 17th Power Sources Conf.* p. 138, Ft. Monmouth, N.J. (1963).
217. Davis, J., *4th Int. Symposium*, "Batteries" 2, Ed. D. H. Collins, Pergamon Press, New York, 233 (1965).
218. *Wall Street Journal* (May 18, 1964).
219. U.S. Patent Classification System.
220. Glover, R. L., U.S. Pat. 2,396,693 (March 19, 1946).
221. Glover, R. L., U.S. Pat. 2,552,091 (May 8, 1951).
222. Reinhardt, O. K. and Stapleton, T. C., U.S. Pat. 2,642,471 (June 16, 1953).
223. Lang, M., U.S. Pat. 2,649,491 (Aug. 18, 1953).
224. Urry, L. F., U.S. Pat. 3,115,428 (Dec. 24, 1963).
225. Urry, L. F., U.S. Pat. 3,185,694 (May 25, 1965).
226. Reilly, T. A., Beckman, J. R., and Bishop, H. K., U.S. Pat. 3,115,429 (Dec. 24, 1963).
227. Urry, L. F., U.S. Pat. 3,202,549 (Aug. 24, 1965).
228. Urry, L. F., U.S. Pat. 3,338,750 (Aug. 29, 1967).
229. Urry, L. F., U.S. Pat. 3,214,298 (Oct. 26, 1965).
230. Yamamoto, S., Nakaiwa, K., Kawauchi, S., and Kanai, H., U.S. Pat. 3,320,094 (May 16, 1967).

231. Jamet, J. F., U.S. Pat. 3,342,644 (Sept. 19, 1967).
232. Jamet, J. F., U.S. Pat. 3,168,420 (Feb. 2, 1960).
233. Reilly, T. A. and Bemer, W., U.S. Pat. 3,506,495 (April 14, 1970).
234. Wilke, M. E., U.S. Pat. 3,524,770 (Aug. 18, 1970).
235. Ohki, S., U.S. Pat. 3,556,859 (Jan. 19, 1971).
236. Heise, G. W. and Cahoon, N. C., U.S. Pat. 2,849,521 (Aug. 26, 1958).
237. Cahoon, N. C., U.S. Pat. 2,876,271 (March 3, 1959).
238. Schumm, B., J. Electrochem. Soc., Extended Abstracts, 73-2, 70 (1973); U.S. Pat. 3,615,859 (Oct. 26, 1971).
239. Huber, R., Metalloberfläche, 24, 293 (1970).
240. Krey, W., Ger. Pat. 2,021,285 (Nov. 11, 1971); Ger. Pat. 2,021,286 (Nov. 18, 1971).
241. Krey, W., U.S. Pat. 3,358,364 (Jan. 26, 1971).
242. Varta-Pertrix-Union Ges., Indian Pat. 112,711 (Dec. 16, 1968); U.S. Pat. 3, 558, 364 (Jan. 26, 1971).
243. Kemp, K. T., New Edinburgh Phil. J., 70 (1828). From P. Benjamin, Ed. The Voltaic Cell, John Wiley, New York (1899).
244. Drotschmann, C. and Wyatt, R., Chem. Weekblad 56, 265 (1960).
245. Bystrom, A. M., Acta. Chim. Scandanavia, 3, 163 (1949).
246. Bregazzi, M., Electrochem. Tech., 5, 507 (1967).
247. Hamer, W. J., Electrochem. Tech., 5, 490 (1967).
248. Gouge, F. H. and Orban, R. L., Electrochem. Tech., 5, 501 (1967).
249. Glemser, O., Ber., 72B, 1878 (1939).
250. Gabano, J. P., Morignat, B., and Laurent, J. F., Electrochem. Tech., 5, 531 (1967).
251. Huber, R., "Batteries," Vol. I, Manganese Dioxide, Ed. K. V. Kordesch, Marcel Dekker, New York, 1–234 (1974).
252. Kozawa, A., "Batteries," Vol. I, Manganese Dioxide, Ed. K. V. Kordesch, Marcel Dekker, New York, 385–516 (1974).
253. Swartz, D. E. and Cahoon, N. C., Electrochem. Soc. Extended Abstracts, Battery Division, 9, 101 (1964).

ADDITIONAL REFERENCES

Bolen, M. and Weil, B. W., "Literature Search on Dry Cell Technology," Special Report No. 27, Engineering Experiment Station, Georgia Institute of Technology (1948).
Kordesch, K. V., Manganese Dioxide, Vol. I, Marcel Decker, New York 1974.
Vinal, G. W., Primary Batteries, John Wiley, New York (1950).
Martin, L. F., Dry Cell Batteries: Chemistry and Design, Noyes Data Corp., Park Ridge, N.J. (1973).
Hamer, W. J., "25 Years of Primary Batteries," 25th Annual Proc. Power Sources Conf., (May, 1972).
Hamer, W. J., "Major Revisions Made in New Dry Cell Standards," The Magazine of Standards, 306–308 (1965).
Kozawa, A. and Powers, R. A., J. Chem. Education, 49, No. 9, 587 (1972).

2.

Magnesium Cells

JOHN L. ROBINSON

GENERAL CONSIDERATIONS

The advantages to be derived from the use of magnesium as a soluble primary-battery anode have long been recognized. Though the theoretical potential ($E° = -2.37$ V) is not realized in practice, the observed value, usually of the order of -1.2 to -1.4 V, is still 0.5 to 0.6 V more anodic than zinc ($E° = -0.76$ V). In addition, magnesium has a very favorable electrochemical equivalent weight: only 0.45 g/amp-hr as compared with 1.2 g for zinc. In theory, therefore, a pound of magnesium might yield the watt-hour equivalent of about five pounds of zinc under battery conditions; however, wasteful corrosion and other factors lead to a lower but still favorable ratio (one pound of magnesium is the equivalent of about two and one-half pounds of zinc).

Corrosion has been the greatest single deterrent to the exploitation of magnesium as a battery anode. The metal owes its intrinsic corrosion-resistance to an oxidic film, the removal of which, by acid, electrolysis, or amalgamation, exposes the metal to accelerated attack in aqueous solutions. Inhibitors, for example, chromates, form films that can substantially prevent corrosion during periods of idleness, but these are disrupted by electrolysis and offer only limited protection during cell discharge.

It has been suggested (1) that the excessive corrosion rate might be due to the anodic dissolution of magnesium to form univalent ions, the latter unstable and reacting with water:

$$2Mg^+ + 2H_2O \rightarrow Mg(OH)_2 + Mg^{++} + H_2 \qquad (1)$$

Considered as favoring this explanation was the transient reducing power of the resultant electrolyte but, in view of Uhlig's (2) demonstration that this phenomenon can be caused by dissolved atomic hydrogen and not by a magnesium-ion species, the "univalent magnesium" theory lacks credibility.

There is, however, considerable evidence (3) for the "chunking" or sloughing off, under load, of colloidal or discrete magnesium particles, which might account, at least in part, for the rise in wasteful corrosion with increasing current density.

Under these circumstances, it is not surprising that anode efficiency is quite low and dependent in large measure on the length of time the magnesium electrode, deprived of its protective oxide coating while under load, is subjected to wasteful corrosion. Thus anode efficiency, generally of the order of 50 percent, may be 40 percent or less in low-drain intermittent service and reaches 60 to 70 percent only under the most favorable conditions of continuous operation.

Fortunately, the advantages of the magnesium anode far outweigh its disadvantages. Substantial progress has been made in cell development, and a bright future for the magnesium electrode seems indicated.

EARLY HISTORY

An early, and perhaps the first, reference to a magnesium anode may be credited to Bultinck(4) who, in 1865, noted the increased electromotive force obtained by substituting the metal for zinc in a couple with silver (cathode). Subsequently, and well into the 20th century, magnesium anodes were tried in many different systems, most of them of little importance and, because of excessive wasteful anode corrosion, without significant practical results. Mention is made, for example, of modified Bunsen cells(5), variously described(6)(7), as magnesium: carbon two-fluid elements, with magnesium sulfate or dilute sulfuric acid electrolyte and bichromates, permanganates or nitric acid as cathodic reactants. The EMF reported was of the order of 3.5 V(7).

A modified "Rowland" Daniell cell: $Mg:MgSO_4::CuSO_4:Cu$ has also been reported, in which magnesium in the form of a bar or plate coated with paraffin, "is fed slowly down as it dissolves"(8). Neither these nor other examples that might be cited seem to have contributed to the acceptance of the magnesium anode in modern battery technology. Indeed, as long as the metal was scarce and expensive, there can have been little incentive for battery development, and the experimental results obtained were of academic interest rather than of practical value.

In the 20th century, the availability of magnesium in abundant supply and at moderate cost (about 38 cents per pound in 1972 and 85 cents per pound in 1975), together with new areas of usefulness for primary batteries, encouraged a fresh attack on the problem of the magnesium anode. Most of the work was concentrated on a modified Leclanché-type dry cell, but a number of other potentially important systems have received serious consideration.

CELLS WITH INERT CATHODES

Where current demand is relatively low and wasteful corrosion can be tolerated, magnesium with its high anodic potential can be used in simple cells without cathodic reactants. When the quantity of electrolyte, for example, an ordinary salt solution, is limited, the cell reaction may be considered as

$$Mg° + 2NaCl + 2H_2O \rightarrow MgCl_2 + 2NaOH + H_2 \uparrow \qquad (2)$$

with hydrogen liberated at the cathode. As alkali is formed, the electrolyte pH rises until the point of $Mg(OH)_2$ precipitation, in the neighborhood of 9.5 to 10.5, is reached:

$$MgCl_2 + 2NaOH \rightarrow Mg(OH)_2 + 2NaCl \qquad (3)$$

The overall cell reaction may, therefore, be considered as

$$Mg° + 2H_2O \rightarrow Mg(OH)_2 + H_2 \uparrow \qquad (4)$$

Thus the magnesium anode has been combined with a steel wool cathode in a reserve-type saltwater(9) cell capable of delivering (at 0.5 V) about 100 Whr/lb of dry, not activated, weight.

CATHODIC PROTECTION

Cathodic protection, whereby a countercurrent is imposed to prevent the anodic dissolution of metals like iron or copper, is today a principal means of preventing corrosion. If not supplied from an external source, such current can be obtained by coupling a metal of high anodic potential (the "sacrificial anode"), with the work to be protected (the cathode) to form a simple voltaic cell. This method was first used as early as 1824 by Sir Humphry Davy(10) who showed that zinc could be used to protect the copper sheathing of ships from corrosive attack in sea water.

In the widespread use of cathodic protection, whether for underground structures, for example, pipelines and cables, for hot-water tanks or for chemical equipment, magnesium has become the favored sacrificial anode metal, its annual consumption for this purpose estimated at 15 million pounds per year. A typical cell installation, involving the protection of an underground pipeline(11), is shown in Fig. 2.1.

The open-circuit potential of the Mg-Fe cell is about 1 V and the current density required to protect steel in soils, (roughly 0.0001 to 0.005 amp/ft^2) or about 0.001 to 0.05 amp/m^2, is quite low. It follows that despite low anode efficiency, about 500 amp-hr/lb (1100 amp-hr/kg) or one-half the theoretical value, a single anode can protect a large area of steel. Specifically, since pipelines usually have an insulating coating and only

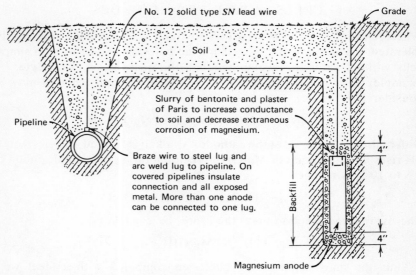

Figure 2.1 Cathodic protection of pipeline with magnesium sacrificial anode.

the fortuitously exposed surfaces need be considered, a single electrode may protect as much as 5 miles (8 km) of coated pipe as compared with only 100 feet (30 m) if the surface is bare(12).

SEAWATER BATTERIES

The magnesium:inert-cathode cell has also been suggested for a number of marine applications, especially those remotely situated which, like sonobuoys and other electronic devices, demand long-lasting, maintenance-free power sources.

A typical cell, designed for immersion in seawater is described by Marzolf(13). The assembly consisted of 12 magnesium plates connected in parallel, interspersed between 13 nickel plated iron cathodes (also in parallel) with $\frac{3}{32}$ in. (0.238 cm) spacing between electrodes. The active electrode area of the single cell was 8.7 ft^2 (0.8 m^2). For terminal voltages between 0.2 and 0.3, current densities ranged from about 0.7 to 1.7 amp/ft^2 (7.5 to 18 amp/m^2).

Numerous other cells, varying mainly in details of construction, have been suggested. The low-drain Kirk battery(14), designed to deliver 2 W of power at 6 V, for at least two years, operated at 0.44 V per cell with a current density, low because of wide (0.81 in.) (2.0 cm) interelectrode spacing, of about 0.053 amp/ft (0.57 amp/m^2); with reduced interelectrode spacing (e.g., 0.03 in.) (0.076 cm) and copper cathodes platinized to minimize hydrogen overvoltage, current densities of 9.4 to 9.6 amp/ft^2 (100 amp/m^2) at cell voltages of 0.37 to 0.52 have been obtained(15).

THE MAGNESIUM DRY CELL

The development of a magnesium dry cell proved to be much more than the substitution of magnesium for zinc in the conventional $Zn:MnO_2$ (Leclanché) system. Much of the early difficulty was associated with excessive wasteful corrosion, and considerable work was done to find a suitable electrolyte.[1] But this effort led to no practical results, and it was not until the mid 1940s(18) that a satisfactory dry cell began to emerge.

In the development of the magnesium cell, it was the anode itself and the electrolyte that required research study, the carbon:manganese-dioxide cathode mix of the conventional dry battery proving to be quite adequate for the work in hand. The cathodic behavior of MnO_2 is discussed in detail in Chap. 1 and need not be repeated here.

ANODE

Two major obstacles to the use of magnesium in the dry cell are the high corrosion rate during discharge and the so-called "delayed action," that is, the time lapse after the circuit is closed, before operating voltage is achieved.

Corrosion

Under static conditions, that is, in the absence of current flow, wasteful corrosion of magnesium in nonacidic electrolyte is usually slight. The oxidic film normally coating the metal gives considerable protection, and treatment with chromates or bichromates, as with zinc, further increases corrosion resistance. However, as previously noted, when the protective film is removed during cell discharge, for example, by electrolysis, the bare metal can decompose water, and rapid attack occurs during the time required to reestablish a protective film. This is serious not only because of the loss of metal, but also because the reaction abstracts water from the electrolyte [cf. Eq. (4)], an important consideration in dry cells. In addition, the hydrogen formed must be permitted to escape, complicating the problem of adequate sealing of batteries.

[1]R. T. Wood, in 1928(16), suggested an ammonium-chloride solution with additions of NH_4NO_3 and $Na_2Cr_2O_7$ to reduce corrosion. Ferrabino, in 1930(17), recommended electrolytes containing $CaSO_4$ and $MgCl_2$ with an inhibitor, for example, calcium ferrocyanide $(Ca_2Fe(CN)_6)$.

Delayed Action

When a circuit containing a magnesium cell is closed, there is a definite time lapse before the anode's protective film is disrupted and full operating voltage is achieved(19)(20)(21) as illustrated in Fig. 2.2(21a). Originally, this "delayed action" required several seconds, a serious handicap in many types of service. Considerable improvement has been accomplished in the course of cell development, chiefly through the development of special magnesium alloys. Electrolyte composition is also a factor, and the delay has been reduced to a second or less.

Anode Composition

The most useful alloying ingredients for the magnesium electrode were found to be aluminum and zinc, the former having beneficial effect on apparent current efficiency (cf. Fig. 2.3)[2], the latter reducing the time lapse

[2]The low anode efficiencies in Fig. 2.3 are to be attributed to the fact that cells were discharged intermittently—5 minutes per day— and subject to wasteful corrosion during the entire test period.

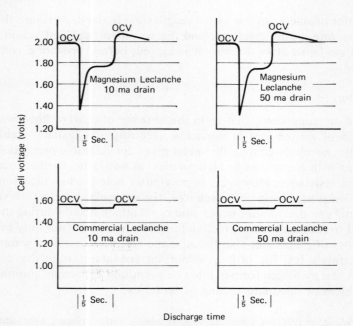

Figure 2.2 Oscillographic measurements of delay in magnesium versus none in Leclanché cells. Change in voltage of AA size Mg and Zn Leclanché type cells, subjected to 10 and 50 ma constant current drains(21a).

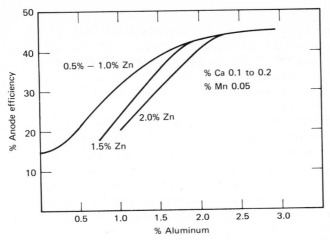

Figure 2.3 Effect of alloy composition on magnesium anode efficiency. These cells were discharged on the 4 ohm general-purpose intermittent test.

of delayed action (cf. Fig. 2.4). The optimum alloy, designated as AZ21A,[3] was thus determined as containing 1.5 to 2.0 percent aluminum and 1.0 to 1.5 percent zinc(22). Minor (0.1 to 0.3 percent) additions of calcium(23) and indium(24) have also been reported as reducing delayed-action time without affecting anode efficiency.

[3]Magnesium alloys are identified by a coding system issued by the American Standard Testing Materials Committee B-7 under their designation B 275-63.

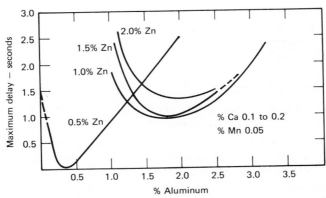

Figure 2.4 Effect of anode composition on delay. These cells were discharged on the 4 ohm general-purpose intermittent test.

ELECTROLYTE

The alkali and alkaline earth bromides show less corrosive attack on magnesium than the corresponding chlorides(25)(26), and the selection of $MgBr_2$ electrolyte, buffered by the addition of $Mg(OH)_2$, marked an important advance in technology of the magnesium cell(27). Maximum capacity is attained in the concentration range of 225–400 g of $MgBr_2$ per liter (ca. 2.5 to 4.5 normal).

An alternate electrolyte containing magnesium perchlorate, $Mg(ClO_4)_2$, has resulted from the work of Lozier(28). It is used in the same normality range as its $MgBr_2$ analogue and yields about equal service capacity, with some indications of less pronounced "delay action"(29). It appears to be somewhat less corrosive though it too should be buffered, and it has largely superseded $MgBr_2$ electrolyte in military applications.

With either electrolyte, wasteful corrosion "on shelf" can substantially be eliminated by the addition of soluble chromates, usually as the lithium salt, which, as previously noted, form protective films. If the additions are small, for example, of the order of 0.2 g/l, they do not increase the time lag of "delayed action." This amount would not be adequate under normal intermittent-service conditions, since the protective film is destroyed every time the circuit is closed; it is customary, therefore, to add a few percent of poorly soluble chromate, for example, $BaCrO_4$, to the cathode mix to serve as a reservoir to replace lost inhibitor(23).

CATHODE

The carbon:manganese dioxide cathode mix used in the magnesium dry cell is essentially the same used in the conventional zinc unit (cf. Chap. 1). A typical formulation might be:

Dry mix (parts by weight)
    Manganese dioxide   87
    Acetylene black   10
    Barium chromate   3
Electrolyte
    410 ml/kg of dry mix

Manganese dioxide:carbon ratios may range from about 8.5:1 to 15:1. Since water is removed by formation of $Mg(OH)_2$ in the cell reaction, the mixes with the higher carbon content, because of their greater water-absorbing capability, are favored for the lighter service applications.

With a synthetic MnO_2, more active than the natural ore, some bromine may be liberated on contact with $MgBr_2$ electrolyte, a reaction which can be suppressed by the addition of a small quantity, for example, one percent of magnesium hydroxide to the cathode mix.

CELL REACTION

With the addition of magnesium hydroxide, the cell electrolyte is alkaline to begin with and magnesium dissolved, either through battery discharge or by wasteful corrosion, is soon precipitated as $Mg(OH)_2$. The primary cell reaction may therefore be considered as simply

$$Mg° + 2H_2O + 2MnO_2 \rightarrow Mg(OH)_2 + 2MnOOH \tag{5}$$

Thus, but for the loss of water, the electrolyte is essentially nonvariant. However, since the anode efficiency is only of the order of 50 percent, the wasteful-corrosion reaction, equation (4), must also be taken into account. The overall cell reaction can then be considered as

$$2Mg° + 4H_2O + 2MnO_2 \rightarrow 2Mg(OH)_2 + 2MnOOH + H_2 \uparrow \tag{6}$$

Thus the wasteful corrosion further increases the loss of water and liberates substantial quantities of hydrogen which must be vented. These

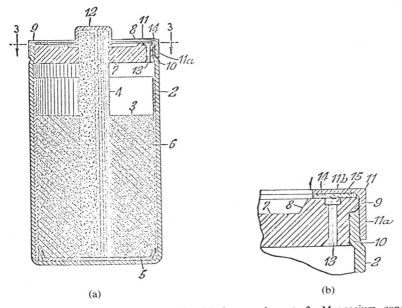

(a)   (b)

Figure 2.5 Cylindrical magnesium cell with improved vent. 2: Magnesium can; 3: Depolarizer mixture; 4: Carbon electrode; 5: Washer; 6: Bibulous paper liner; 7: Plastic cell closure; 8: Peripheral edge of cell closure; 9: Shoulder on edge of cell closure; 10: Magnesium can bent inward at this point; 11: Seal ring; 11a: Vertical leg of seal ring; 11b: Horizontal leg of seal ring; 12: Metallic positive terminal cap; 13: Venting aperature in plastic closure; 14: flat annular sealing gasket; 15: Opening at top of venting aperature. Taken from U.S. Patent 3,494,801 (30).

are some of the factors demanding consideration in cell design and composition.

CELL CONSTRUCTION

Round Cell

Cell design, in the main, has closely followed conventional dry battery practice (cf. Chap. 1), as indicated by the construction shown in Fig. 2.5. Mercury is omitted since amalgamation, by removing the protective oxidic film, would accelerate rather than inhibit the wasteful corrosion of the magnesium anode. Uncoated Kraft paper is generally used as separator-liner material. Adequate sealing has been made possible by the design of self-sealing vents (30) which allow the escape of hydrogen but prevent moisture loss or the access of atmospheric oxygen (which would accelerate wasteful corrosion).

Due largely to the formation of $Mg(OH)_2$, the cell reaction is accompanied by an actual increase in volume which may exert enough pressure

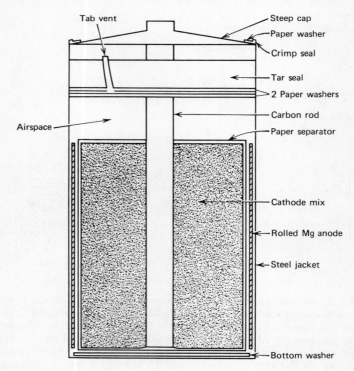

Figure 2.6 Round cell with steel jacketed construction.

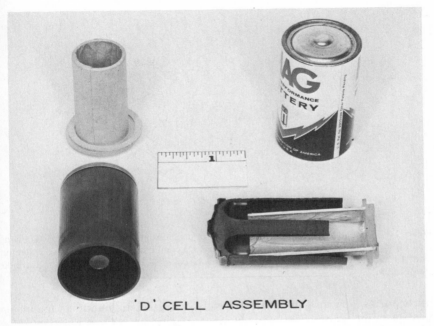

Figure 2.7 Balaguer construction of a magnesium cell. A: liner covered magnesium anode; B: molded carbon cathode collector; C: shows the manner in which parts A and B are assembled; D: the finished cell; E: rule 1½ inches long for comparison. Courtesy of the Battery Corporation of America, Hollywood, Florida.

to bulge or even to split the magnesium can. To avoid this difficulty, the steel-jacketed construction shown in Fig. 2.6 has been proposed(31). This has the added advantage of substituting a rolled magnesium cylinder for the more expensive impact-extruded magnesium cup.

In another modification (cf. Fig. 2.7) suggested by Balaguer(32), a carbon cup molded with a central rod serves as cathode collector and a magnesium cylinder, Kraft-paper-coated, is inserted into the depolarizing mix. This construction has the added advantage of utilizing both sides of the magnesium anode.

Flat Cell

As with conventional zinc units, flat cell construction (cf. Chap. 1) was found advantageous, particularly in multicell batteries. With constructions such as the "cathodic envelope" cell(33) (cf. Fig. 2.8) both sides of the anode can be used. Flat cell constructions result in a better balance between the active materials and minimize the amount of the relatively expensive magnesium anode material required. Expansion of the cell

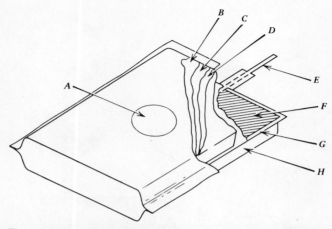

Figure 2.8 The cathode envelope construction for magnesium cells. A: positive contact with metal foil; B: plastic film wrapper; C: metal foil; D: conductive film; E: negative collector lead; F: anode; G: separator; H: depolarizer mix cakes(33). Courtesy of Union Carbide Corporation.

mass during service can be controlled by using elastic banding tape or by incorporating a compressible cushion, for example, a foam plastic, in the cell stack.

SERVICE CHARACTERISTICS

From the point of view of service, the outstanding characteristic of the magnesium dry cell (open circuit voltage, 1.9) is its high operating potential: 0.3 to 0.4 V above that of the conventional zinc unit under normal load. This is particularly advantageous in series connected multiple-cell batteries, because of the flatter discharge curve and smaller number of magnesium cells required. Indicated discharge curves for zinc and experimental magnesium cells are shown in Fig. 2.9(33a).

Since the magnesium electrode is protected by corrosion-inhibiting chromate additions and moisture losses can be kept small, magnesium cells have excellent shelflife(34), as much as 85 percent of their original service capacity being retained, even after two-year storage at 130°F.

Ampere-hour capacity is generally adequate for normal service applications, but may show considerable variation for different operating conditions. Service-limiting wasteful corrosion, for example, is enhanced whenever the protective film on magnesium is removed, as it is during cell operation, and the bare metal is exposed, its magnitude depending on such factors as duration of test, number and duration of discharge periods, and the like. Thus, anode efficiency, calculated on the basis of divalent

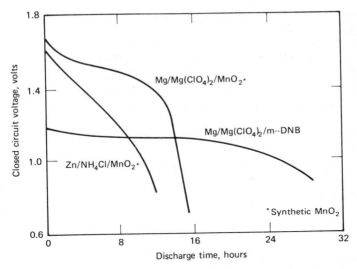

Figure 2.9 Discharge curves of magnesium and zinc "A" size cells on $16\frac{2}{3}$ ohm test. Discharge curves for dry cells tested at 70°F ("A" cells)—$16\frac{2}{3}$ ohm drain. m-DNB represents meta dinitrobenzene. Reprinted with permission of CHEMICAL TECHNOLOGY.

magnesium, may be as high as 70 percent on continuous discharge, or as low as 40 percent on long, drawn-out intermittent tests (e.g., the "Household Intermittent Test" simulating flashlight service at the rate of only five minutes use per day), with the latter showing a reduction in ampere-hour capacity of the order of 25 percent.

On the other hand, capacities two to three times those of conventional zinc batteries have been obtained under service conditions favoring the magnesium cell, namely continuous and high-rate discharge(18)(27) (33)(34). Military applications are of this type and much effort has been spent to tailor the magnesium cell to give optimum performance, for example, on the BA-4386 test.[4]

In military applications(33a) the magnesium battery has shown the capability of providing twice the energy density (watthours per pound), of the conventional Leclanché cell at medium discharge rates(50). The magnesium battery also possesses excellent storage properties particularly at elevated temperature. After 12 weeks at 160°F they give 75 percent of their initial capacity, a considerable improvement over the Leclanché system.

[4]The BA-4386 battery is a unit with a maximum voltage of 15, and is designed to furnish two outputs for a transceiver application.

A part of the advantage of the magnesium battery is that it requires fewer cells to provide a given voltage than is the case with zinc cells. A second advantage is the flatter discharge curve of the magnesium cell, as shown in Fig. 2.9, due, in part, to the more nearly constant cathode pH as contrasted to the steadily rising pH of the Leclanché cathode under heavy drain conditions. As a result it is possible to use three magnesium cells in series to replace four zinc cells in series in many applications.

As yet, substantially the entire output of magnesium dry cells is being channeled into military applications, where the system is favored for its high, well-sustained operating voltage and superior shelf quality at elevated temperatures. Its capability over a wide range of operating conditions has been established and its early entry into the civilian market may reasonably be expected.

MISCELLANEOUS SYSTEMS

Research and development has been centered, for the most part, on the conventional dry cell but much work has also been done on the utilization of magnesium anodes in other systems, none, as yet, of commercial importance. Special attention has been directed toward "one-shot" systems which, though intrinsically of poor shelf quality and capable of only limited service life, can be activated or assembled just prior to discharge. The $Mg:AgCl$ and $Mg:Cu_2Cl_2$ cells described in Chap. 7, are of this type and gas-depolarized batteries (e.g., $Mg:Cl_2$) generally fall into the same category.

SUBSTITUTES FOR MnO_2

In extension of his work with magnesium anodes and $Mg(ClO_4)_2$ electrolyte, Lozier(28) made experimental cells containing cathodic reactants other than MnO_2. Characteristic (approximate) operating voltages obtained for representative materials were as follows:

Material	Operating Voltage
NiO	2.0 to 1.6
AgO	1.7 to 1.5
HgO	1.6
MnO_2	1.6
m-dinitrobenzene	1.1
Bi_2O_3	1.1
CuO	1.0 to 0.9

Cells made with these materials showed flat, well-sustained voltage-discharge curves and several systems have been given more detailed study. According to Cupp(35), the feasibility of the wet-type HgO cell has been established but this, apparently, holds only for a reserve unit which must be used shortly after activation, since "wet shelf" is limited to one or two months. Further deterrents to commercial exploitation are the lack of weight or volume advantage over its MnO₂ counterpart, and a cost at least three times as great. Copper oxide has also been shown to be practicable both in conventional wet cells(36)(37) and in seawater batteries(38), but it, too, lacks attractiveness because of high cost and low operating voltage.

MAGNESIUM: m-DINITROBENZENE

Of particular interest is the possibility of using organic depolarizers (cf. Chap. 4), especially those which, with zinc anodes, would yield cells of undesirably low voltage. Research and development have largely been centered on m-dinitrobenzene(28)(39)(40) which, if reducible to m-phenylenediamine

$$C_6H_4(NO_2)_2 + 8H_2O + 12e^- \rightarrow C_6H_4(NH_2)_2 + 12OH^-$$

would mean an electrochemical equivalent of almost 2 amp-hr/g (0.52 g/amp-hr), about six times the MnO₂ equivalent (0.33 amp-hr/g). Test cells made with m-dinitrobenzene show flat, well-maintained discharge curves(41) as shown in Fig. 2.9. Their watt-hour capacity appears to be somewhat larger than that of MnO₂ cells, their greater ampere-hour capacity offsetting their lower operating voltage. There has been some concern regarding their temperature-range of usefulness, experimental cells having shown rather serious reductions in capacity when temperature was reduced much below 70°F(41).

MAGNESIUM: AIR CELLS

Air-depolarized cathodes are particularly attractive with magnesium anodes, since the operating voltage obtained, nominally 1.2 to 1.3, are 0.3 to 0.4 V above those registered with zinc. Efforts to substitute an oxygen electrode for manganese dioxide in the magnesium dry cell of conventional construction have been generally unsuccessful, since air access causes accelerated corrosion and moisture loss, with correspondingly poor shelf quality.

A flat-reserve-type dry cell has been described(25) in which the component parts (cf. Fig. 2.10), anode (A), separator (B) and a carbonaceous mix cake (C) wet with NH₄Br electrolyte were assembled in a re-usable frame immediately before discharge. Normal drain at 0.04 amp/in.² (0.006 amp/cm²) gave the discharge curve shown in Fig. 2.11, which emphasizes the higher working voltage of the magnesium unit.

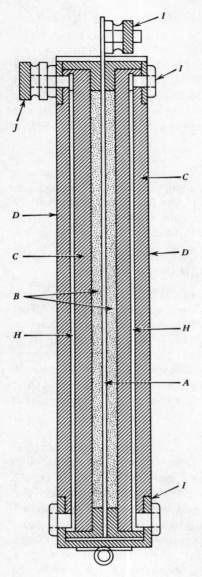

Figure 2.10 One-shot magnesium:air cell. *A*, magnesium anode; *B*, separator; *C*, cathode mix; *D*, air-pervious contact member.

Liquid-activated cells, too, have received attention. In its simplest form, designed for emergency or signal lighting, a unit of this type comprises merely a magnesium anode and a conventional active-carbon cathode, and is activated by the addition of a solution of common salt

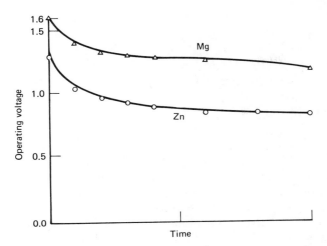

Figure 2.11   Discharge curves of magnesium versus zinc anodes in air-depolarized systems. One-shot air-depolarized cells on continuous discharge at 0.04 amp. per inch².

(e.g., 5 to 10 percent). The cathode, though "waterproofed," is not as resistant to salt water as it is to the caustic alkali electrolyte of conventional $Zn:O_2$ batteries; however, penetration of liquid is too slow to affect the performance over a period of days or even weeks. Indeed, for a $Mg:C$ assembly designed for seawater operation (off-shore signal lighting and other marine applications), capability for "unattended operation for periods of six months or longer" has been predicted[42].

A heavy-drain magnesium : air battery (test cycle: 9 minutes at 1.5 and 1 minute at 40 amp/ft² (0.16 to 4.3 amp/dm² respectively)) has been described by Kent[43]. This battery, designed for high output, comprised 22 or 23 cells in series in a package $11\frac{5}{8}$ in. $\times 4\frac{3}{16}$ in. $\times 9\frac{1}{2}$ in. (29.5 × 10.6 × 24.1 cms) and weighing not more than 8 lb without the electrolyte. It was expected to provide 30 cycles of operation each of 12 hours duration. The air cathode was a polymer bonded nickel-based electrode carrying 4 g of platinum per square foot. The separator was of glycerinated Cellophane and the anodes were AZ-61 alloy (a magnesium alloy containing 6 percent aluminum and 1 percent zinc). The electrolyte was an aqueous solution of sodium chloride for example, 1.2 N. The nominal battery working voltage was 24 with a minimum of 22 and it was tested at 105°F. The initial working voltage was 1.3 V per cell equivalent to about 30 V per battery. A new anode and a fresh charge of electrolyte were placed in the cell before each cycle. The experimental program confirmed the feasibility of a carried energy density of 50 Whr/lb for the magnesium-air battery.

MAGNESIUM: HALOGEN CELLS

The magnesium:chlorine system, though not of immediate commercial importance, has been studied because of its intrinsically high voltage, 2.8 to 2.9 as compared with 2.05 for its zinc counterpart, and its excellent heavy-drain characteristics(44). Both wet and dry cell constructions have been considered.

Dry Cells

A "one-shot" cell essentially similar in construction to the unit described in Fig. 2.10, when exposed to chlorine at atmospheric pressure, showed a flat discharge curve over a period of 10 hours at a drain of 6 amp/ft^2 (0.65 amp/dm^2). Operating voltage (2.4 to 2.5) was 0.5 to 0.6 V above that of the corresponding zinc cell, but ampere-hour capacity was only about half as great. The discrepancy can be explained on the basis of the (40 percent) excess corrosion rate of the magnesium anode and, loss of water due to the formation and precipitation of the end-product, $MgCl_2 \cdot 6H_2O$.

Extremely high current densities with only slight reduction in operating voltage can be obtained by utilizing chlorine at elevated pressure. Thus, magnesium:chlorine cells(44) designed for short (e.g., 10-minute) bursts of power, operating at the vapor pressure of chlorine (5 to 7 atmospheres at normal room temperatures) delivered current at 13 amp/dm^2 (about 120 amp/ft^2). Operating voltage was 2.4 to 2.25 as compared with 1.8 for its zinc counterpart. The complete cell, constructed as indicated in Fig. 2.12, was only 0.125 in. (0.318 cm) thick.

Wet Cells

Neipert and Carr(45) have described a system in which the cathode reactant, bromine, dissolved in $MgBr_2$ electrolyte, entered the cell through a porous carbon cathode. Open-circuit voltage was given as 2.29; operating voltage ranged from 2.20 at 29 amp/ft^2 (3.1 amp/dm^2) and 1.65–1.70 at 125 amp/ft^2 (13.4 amp/dm^2) to 1.18 at 200 amp/ft^2 (21.5 amp/dm^2).

In a similar cell, the electrolyte is a sodium bromide solution, in which the cathodic reactant, chlorine, is much more soluble than in the chloride electrolytes previously used(46). The resulting Mg:Cl_2 system shows improved heavy-drain characteristics: operating voltage 2.0 at a current density of 125 amp/ft^2 (13.4 amp/dm^2).

SYSTEMS WITH HIGH ANODIC EFFICIENCY

Parasitic corrosion or low anode efficiency is perhaps the greatest single adverse property of the magnesium cells under consideration. However,

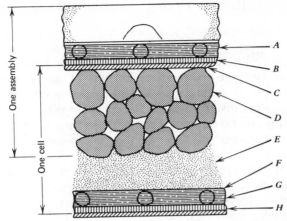

Figure 2.12 Magnesium:chlorine heavy drain battery. Construction of chlorine-depolarized battery. A—paste layer with netting reinforcement, B—anode, C—inert, non-porous impervious conducting coating, D—diffuser element of porous carbon particles, E—wet carbon cathode, F—cellophane, G—paste layer, H—anode(44).

systems where the anode corrosion is low or the anode efficiency high have been reported.

Chromic-Acid Electrolyte Cells

Barbian and McNulty(47) described an experimental cell in which as-semblies of magnesium and carbon plates were submerged in an electro-lyte of 35 to 40 percent chromic acid (the cathodic reactant), plus 5 to 25 percent phosphoric acid. The cell operating voltage at $3\,amp/ft^2$ $(0.32\,amp/dm^2)$ ranged from 0.8 to 1.4 and about 1.2 to 1.6, the higher voltages obtained with heat-treated magnesium anodes. Anode efficien-cies ranged from 85 to 95 percent, but the cathode efficiency (reduction of Cr^{+3}) was only of the order of 25 percent. The resulting low ampere-hour capacity per unit volume of the chromic acid electrolyte solution militated against commercial development of this system.

Organic Salt Electrolytes

Anode efficiencies approaching 100 percent have been reported with electrolytes of organic salts such as magnesium acetate(48), but the high resistivities (greater than 100 ohms-cm) of these electrolytes limited their use in practical cells. However, the work with chromic acid and the organic salt electrolytes indicates that the wasteful corrosion of the magnesium anode can be further reduced or eliminated and coulombic efficiency increased in operating systems.

SUMMARY AND PROSPECTUS

Magnesium, as an anode, offers about 0.5 V more driving potential than zinc and nearly double the ampere-hour capacity per unit weight in continuous heavy-drain service. However, with low-rate and highly intermittent service, the ampere-hour capacity of magnesium is about the equal of zinc, reflecting the high wasteful corrosion rate under discharge conditions. Another disadvantage is the transient voltage behavior which results in a time lag—about one second or less for a fresh cell, but longer in a partially discharged one—in reaching full operating voltage after the circuit is closed.

To date, the principal electrochemical use of magnesium has been in cathodic protection, where it is generally preferred over other potential anode materials, for example, zinc or aluminum, because of its high anodic potential.

The magnesium anode has also been used for several years in the "one shot" Mg: AgCl and Cu_2Cl_2 cells (Chap. 7), reserve systems for which, in general, call for a high energy level and good heavy-drain characteristics are demanded.

The most promising application of the magnesium anode is in the Mg:MnO_2 dry cell, a system now well established as a major power source for military communication equipment, for which well over 50 million unit cells [about 1 in. (2.54 cm) in diameter and 3 in. (7.6 cm) in height] were produced in 1971. General use should soon follow, though the fact that most commercial devices have been designed to operate at the voltage level of the zinc cell which may limit the rate of penetration of the commercial market.

REFERENCES

1. Petty, R., Davidson, A., and Kleinberg, J., J. Am. Chem. Soc., 76, 363 (1954); cf. Uhlig, H. H., Corrosion and Corrosion Control, John Wiley, New York (1963) p. 191.
2. Uhlig, H. H., and Krutenat, R. C., J. Electrochem. Soc., 111, 1303 (1964); cf. Krutenat, R. C. and Uhlig, H. H., Electrochim. Acta, 11, 469 (1966).
3. Hoey, G., and Cohen, M., J. Electrochem. Soc., 105, 245 (1958); Marsh, G., and Schashl, E., ibid, 107, 960 (1960); James, W. J., Straumanis, M. E., Bhatia, B. K., and Johnson, J. W., ibid, 110, 1117 (1963); Mueller, C. E. and Bowers, F. M., ibid, 118, 394 (1971).
4. Bultinck, Compt. rend., 61, 585, (1865); cf. Cazin, A. Piles Électriques Gauthier-Villars, Paris, 1881 (p. 131).
5. Heim, G., Elektrotech. Z., 8, 472, 518 (1887).
6. Morehouse, C. K., J. Electrochem. Soc., 99, p. 187C (1952).
7. Bottone, S. R., Galvanic Batteries, Whittaker, London and New York, 1902 (p. 241).
8. Id, ibid (p. 213).
9. Optiz, C. L., Steel (Oct. 6, 1969).
10. Davy, H., Phil. Trans. Roy. Soc., 114, 151, 242, 328 (1824–5).

11. Uhlig, H. H., *Corrosion and Corrosion Control*, John Wiley, New York (1963), p. 185.
12. Id., ibid, p. 188.
13. Marzolf, J. M., *U.S. Nav. Res. Lab. Rept.* 5961 (May 1963) (AD406782); cf., Goldenberg, L. and Fidelman, M., U. S. Pats. 3,036,141-2 (1962).
14. Kirk, R. C., U.S. Pat. 3,228,800 (1966).
15. Kirk, R. C. and Carr, R. E., U. S. Pats. 3,177,099 and 3,185,592 (1965).
16. Wood, R. T., U.S. Pat. 1,696,873 (1928).
17. Ferrabino, G., Swiss Pat. 137,015 (1930).
18. Kirk, R. C. and Fry, A. B., *Trans. Electrochem. Soc.*, **94**, 277 (1948); cf. Fry, et al. U. S. Pats. 2,597,907-8-9 (1951).
19. Robinson, H. A., *Trans. Electrochem. Soc.*, **90**, 485 (1946).
20. Robinson, J. L. and King, P. F., ibid, **108**, 36 (1961).
21. King, P. F., ibid, **110**, 1113 (1963).
21a. Glicksman, R., *J. Electrochem. Soc.*, **106**, 457 (1959); Glicksman, P. and Morehouse, C. K., ibid, **102**, 273 (1955).
22. Robinson, J. L., U.S. Pat. 3,038,019 (1962).
23. Kirk, R. C., Fry, A. B., and George, P. F., U.S. Pats. 2,621,220 (1952) and 2,712,564 (1955).
24. Stevens, J. A., U. S. Pat. 2,934,583 (1960).
25. Heise, G. W., U. S. Pat. 1,899,615 (1933).
26. Gordon, C. J., U. S. Pat. 2,050,172 (1936), Brit. Pat. 526,601 (1940).
27. Kirk, R. C., George, P. F., and Fry, A. B., *J. Electrochem Soc.*, **99**, 323 (1952).
28. Lozier, G. S., U. S. Pat. 2,993,946 (1961); *Proc. Power Sources Conf.*, **16**, 134 (1962).
29. Murphy, J. J. and Wood, D. B., *Proc. Power Sources Conf.*, **21**, 100 (1967).
30. Urry, L. F., U. S. Pat. 3,494,801 (1970).
31. Reid, R. W., and Kirk, R. C., U.S. Pat. 2,850,557 (1958).
32. Balaguer, R. R., and Schiro, F. P., *Proc. Power Sources Conf.*, **20**, 90 (1966); cf. Balaguer, R. R., U. S. Pat. 3,272,655 (1966).
33. Moffitt, W. E., *Proc. Power Sources Conf.*, **14**, 129 (1960).
33a. Murphy, J. J., *Chem. Tech.*, **1**, 487 (1971).
34. Hovendon, J. M., ibid, **10**, 8 (1956); id and Wood, D. B., ibid, **19**, 88 (1965).
35. Cupp, E. B., ibid, **19**, 92 (1965); **20**, 93, (1966).
36. Almerini, A. L., ibid, **20**, 95 (1966).
37. Knapp, H. R. and Almerini, A. L., ibid, **17**, 125 (1963).
38. Ehrlich, M. P. and Stoller, M., U. S. Pat. 3,441,445 (1969).
39. Lozier, G. S. *Proc. Power Sources Conf.*, **14**, 132 (1960).
40. Endrey, A. L. and Reilly, T. A., ibid, **22**, 91 (1968).
41. Doe, J. B. and Wood, D. B., ibid, **22**, 97 (1968).
42. Shotwell, J. J., Kirk, R. C., and Nelson, A., *Extended Abstracts of the Battery Division of the Electrochemical Society*, (Oct. 1957).
43. Kent, C. E., *Proc. Power Sources Conf.*, **21**, 106 (1967).
44. Heise, G. W., Schumacher, E. A., and Cahoon, N. C. *J. Electrochem. Soc.*, **94**, 94 (1948); Heise, G. W. and Schumacher, E. A., U. S. Pat. 2,612,532 (1952).
45. Neipert, M. P. and Carr, R. E., U. S. Pat. 3,134,698 (1964).
46. Blue, R. D., et al., U. S. Pat. 3,019,279 (1962).
47. Barbian, H. A., and McNulty, R. E., *Trans. Electrochem. Soc.*, **91**, 387 (1947).
48. Robinson, J. L., *Proc. Power Sources Conf.*, **17**, 142 (1963).
49. Wood, D. B., "Batteries", Vol. 1, *Manganese Dioxide*, Ed. K. V. Kordesch, Marcel Dekker, New York, 521 (1974).
50. Urry, L. F., *J. Electrochem. Soc.*, **122**, 715 (1975).

3.

Aluminum Cells

John J. Stokes, Jr. and David Belitskus

Aluminum is a very attractive anode material for primary batteries. Its relatively low atomic weight of 26.98 along with its trivalence result in a gram-equivalent weight of 9.00 and a corresponding electrochemical equivalent of 2.98 amp-hr/g (1352 amp-hr/lb) as compared with 2.20 (1000 amp-hr/lb) for magnesium and 0.82 (372 amp-hr/lb) for zinc. From a volume standpoint, aluminum should yield 8.04 amp-hr/cm^3, as compared with 5.85 for zinc and 3.83 for magnesium. Additionally, in common with zinc but not with magnesium, the hydroxide of aluminum is amphoteric in nature so that an aluminum cell should operate over a wide pH range. Finally, aluminum is both an abundant and relatively inexpensive metal.

Like magnesium (cf. Chap. 2), aluminum owes its stability to an oxidic film, and despite its high reversible electrode potential (Table 3.1), its actual potential in cells is about the same as or slightly lower than that of zinc. Amalgamation, which removes the oxide layer, raises potential, but only at the expense of reduced resistance to corrosion.

TABLE 3.1. STANDARD REDUCTION
POTENTIALS(1)

	E_0 Acid, volts	E_0 Base, volts
Mg	−2.37	−2.69
Al	−1.66	−2.35
Zn	−0.76	−1.25

HISTORICAL INTRODUCTION

Aluminum has been considered as a battery electrode, though originally as cathode rather than anode, since the 1850s when Hulot(2) described the cell zinc (mercury): aluminum with dilute H_2SO_4 electrolyte. According to Cazin(3), this cell, at the start, delivered current "at least comparable to that of the zinc:platinum element."

Aluminum as anode was used in the Buff cell of 1857(4), a two-fluid battery,

$$Al: \text{dilute acid in porous pot}: HNO_3: C,$$

with an EMF of 1.377 V. The Mennons patent of 1859(5) and the Eager and Milburn patent of 1891(6) cover essentially the same system. The latter also mentions a single-fluid cell:

$$Al: HCl: NH_4Cl: C,$$

and a subsequent patent(7) suggests sodium bichromate as cathodic reactant, claiming that (when used) with "electric glow lamps," the resulting cell gave better and steadier light than cells of their other patent.

Brown's patent of 1893(8) describes a cell with an anode of an amalgamated aluminum-zinc alloy and a carbon cathode, but does not specify the electrolyte. A 1902 patent granted to Anderson(9) claims a cell with an aluminum anode, carbon cathode, and an electrolyte of dilute nitric acid, potassium bichromate, and calcium fluoride, the latter purported to reduce wasteful corrosion. The addition of zinc chloride to the electrolyte was recommended for stronger currents in closed-circuit work. In 1904, Noble and Anderson(10) patented a cell with aluminum and carbon electrodes and an electrolyte of dilute nitric acid kept at a temperature of 150°F or greater. They reported an EMF of 1.5 V with a very low internal resistance. Peek's patent of 1911(11) describes a cell with one electrode of aluminum and an electrolyte of aqueous ammonium borate and ammonium tartrate, along with glycerin. This electrolyte reportedly kept the surface films on the aluminum to a minimum. In 1913, Paine(12) reported "poor results" with pure aluminum anodes but claimed improved properties and an increase in EMF of about $\frac{1}{2}$ V when the metal was amalgamated in the system:

$$Al(Hg): NaCl: \text{porous diaphragm}: FeCl_3: C \quad (E = 2 \text{ V}).$$

Vince's patent of 1932(13) reported a cell with concentric spirals of aluminum and copper separated by a spacer material, the latter also serving as an absorbent for the sodium hydroxide electrolyte. In 1935, Pennock and Lee(14) patented a cell with a liquid aluminum amalgam anode, a carbon cathode, and a sodium hydroxide electrolyte, the cell so constructed that the anode could be replenished with aluminum powder, shot, or filings during service. The inventors claimed a 1.3 V EMF and a steady and constant current for this cell.

The use of aluminum or amalgamated aluminum as an anode in heavy-duty chlorine-depolarized batteries was reported in 1948 by Heise, Schumacher, and Cahoon(15). With amalgamated aluminum, open circuit voltage was as high as 2.45 V (0.4 V above the value for zinc or for unamalgamated aluminum), but wasteful corrosion limited the use of these cells to short-time operation, well below that with zinc.

In 1956, Lozier, Glicksman, and Morehouse(16) patented aluminum anode cells with cathodes of positive halogen organic compounds, such as N,N-dichloro-p-toluenesulfonamide or N,N-dichlorodimethylhydantoin. These cells were shown to have high capacities on continuous drain (4-ohm), but other properties such as shelf life or capacity on intermittent drain were not disclosed.

The aluminum:oxygen system was studied by Zaromb(17) and by Bockstie, Trevethan, and Zaromb(18), who found that the addition of zinc oxide or certain organic inhibitors, such as alkyldimethylbenzyl-ammonium salts to the electrolyte, significantly decreased the corrosion of amalgamated aluminum anodes in 10M sodium or potassium hydroxide solutions. Replacement of the carbon-air cathode normally used in batteries of this kind with a porous, sintered nickel sheet impregnated with carbon, silver, and Teflon (electrolyte-resistant but permeable to oxygen) permitted a reduction in thickness and weight. Shortcomings of this system included frequent and unpredictable wasteful corrosion of aluminum and the increased anode polarization which accompanied the use of most of the inhibitors.

With the metal priced at $5.00 per pound as late as the mid 1880s, there was little incentive for systematic study or exploitation of the aluminum anode. It is hardly surprising, therefore, that early work was casual and empirical, contributing little or nothing to battery science and technology. Some of the systems described must have been quite unworkable, and such factors as shelf life and capacity seem frequently to have been ignored. It was not until serious efforts were made on the utilization of aluminum anodes in Leclanché-type dry cells that important progress toward a commercially acceptable battery was made.

An early reference to an aluminum analog of a Leclanché-type cell is in Sargent's patent of 1951(19), covering the system:

$$Al : aqueous\ NaOH + ZnO : porous\ membrane : MnO_2(C).$$

The inventor suggested that aluminum precipitated zinc from the caustic-alkali zincate solution, to form a protective coating on the anode, at least while the cell was not in operation. Ruben's patent of 1953(20) disclosed the use of manganous chloride tetrahydrate as the electrolyte salt in an aluminum:manganese dioxide cell. An EMF of 1.56 V, with relative freedom from wasteful anode corrosion and correspondingly low gas production, was reported. Further improvements by Stokes(21)(22)(23) furnish much of the basis for the work on dry cells discussed in this chapter.

GENERAL CONSIDERATIONS

Despite its potential advantages as anode material, aluminum has not yet been utilized in a commercially important battery. The protective oxide film to which aluminum owes its stability is, in general, detrimental to battery performance, contributing to failure of the anode to achieve its reversible electrode potential and resulting in a cell voltage considerably lower than the theoretical. In fact, aluminum is actually cathodic to zinc in many neutral or acid electrolytes(24) so that a zinc cell might show a higher voltage than the analogous aluminum unit in such electrolytes. The oxide film also causes the phenomenon of "delayed action"; that is, the time lag before the cell reaches its maximum operating voltage when the circuit is closed. Although this can be minimal for freshly prepared cells, it can be substantial for partially discharged cells or those which have been stored for some time.

The oxide layer can, of course, be removed by dissolution in concentrated alkali solutions or by amalgamation; but, as already noted, any gain in anode potential through such means is accompanied by accelerated wasteful corrosion and poor shelf life.

These and other difficulties have long delayed the development of a satisfactory aluminum dry cell. As might be expected, simple substitution of aluminum for zinc in the conventional Lechanché system results in a cell of virtually no shelf life(25). The Sargent alkaline $Al:MnO_2(C)$ unit(19) had poor shelf life and low capacity, and the Ruben $Al:MnO_2(C)$ system(20) with manganese chloride electrolyte failed to match the conventional dry cell in these respects. Nevertheless, as the discussion to follow will show, an $Al:MnO_2$ dry cell(21)(22)(23) competitive with its zinc and magnesium analogues, is still a reasonable objective(26).

THE ALUMINUM:MANGANESE DIOXIDE DRY CELL

The current discussion is limited to cylindrical "D" cells, though other sizes have successfully been made and operated. The overall construction—anode (can): paper or equivalent separator: electrolyte-wet MnO_2, carbon cathode mix: central carbon electrode—is analogous to that of the paper-lined $Zn:MnO_2$ (Chap. 1) or $Mg:MnO_2$ (Chap. 2) cells (Fig. 3.1). It should be recognized at the outset that the aluminum dry cell

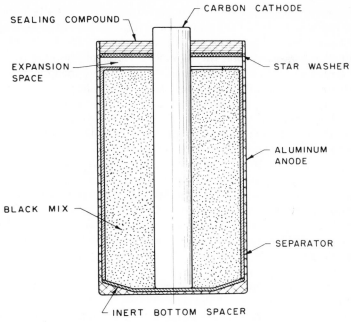

Figure 3.1 Cross section of an aluminum dry cell.

is still in the experimental stage and that much additional refinement in cell formulation and further development and testing may well be required before entry into the highly competitive dry battery market can be established.

ANODE

Aluminum as anode metal suffers several handicaps: (a) due to its oxidic surface film, its electrode potential (about -0.4 V on the hydrogen scale) may actually be less anodic (by about 0.3 V) than that of zinc in chloride electrolyte; (b) the disruption of that surface during service leads to excessive corrosion, with liberation of hydrogen and poor anode efficiency; (c) as with magnesium (Chap. 2), there may be a time lag of several seconds after the circuit is closed before an aluminum cell reaches operating voltage.

A further complication is the tendency of ordinary aluminum sheet to corrode unevenly, leading to premature failure of conventional cylindrical cells due to can perforation. This difficulty can be minimized by using a two-layer anode composed of aluminum alloys of different electrode potential, as described by Stokes(21). Thus, cell action can be confined to the inner, more anodic layer which, until it is substantially consumed, protects the outer shell from corrosive attack. The outer layer should constitute as little of the total thickness as practical, about 20% being sufficient. It is important that the two alloys be firmly bonded together to prevent any separation or scaling during operation of the cell. This is best accomplished by hot rolling together slabs of the alloys, after which the composite material is reduced to the desired thickness.

Many alloy combinations are acceptable, but the one used in much of the work described consisted of an inner layer of Al-5 percent Zn and an outer layer of commercial 3003 alloy; that is, Al-1.2 percent Mn, which differ in potential by about 0.1 V in the electrolytes used. A "D" size Al:MnO₂ cell with an anode can of this composition and a wall thickness equal to that in a commercial Zn:MnO₂ cell (0.016–0.018 in.) had a perforation resistance comparable to that of the zinc cell. Figures 3.2 and 3.3, cross sections of typical cell walls after discharge, show that the outside layer of Al 3003 alloy was not attacked while the inside layer of Al-5 percent Zn alloy had served as the anode metal and was more or less consumed.

Considerable work has been done in recent years on the development of aluminum alloys more anodic and corrosion resistant than the base metal. In a survey of the subject(27)(28), Reding and Newport noted that the electrode potential in chloride electrolyte was made 0.1 to 0.3 V more anodic by small additions of zinc, cadmium, magnesium, or barium; 0.3 to

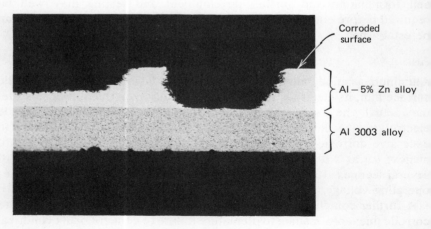

Figure 3.2 Photomicrograph (70×) of aluminum anode from dry cell after discharge. Section taken at air paste interface. Upper layer: Al-5% Zn active material. Lower layer: 3003 Al alloy.

Figure 3.3 Photomicrograph (70×) of aluminum anode from dry cell after discharge. Section taken near bottom of anode can. Upper layer: Al-5% Zn active material. Lower layer: 3003 Al alloy.

0.9 V by gallium, mercury, tin, or indium. However, anode efficiencies for single element additions were usually undesirably low. Success was greater with two element additions, notably zinc-mercury [cf. also (29)(30)], zinc-tin [cf. also (31)(32)], and zinc-indium [cf. also (33)(34)]. By way of examples, an alloy containing 0.45 percent Zn, 0.45 percent Hg, the balance aluminum of 99.9 percent purity, had a potential of −0.81 V and a current efficiency, under load, of 95 percent[1] in synthetic seawater.

In a similar electrolyte, an Al-7 percent Zn-0.12 percent Sn alloy had a potential of about −0.90 V and a current efficiency of about 90 percent(35). An even more anodic alloy (−1.25 V) containing 6 percent Mg and 0.066 percent Hg has been patented(36). A number of other alloys and optimum heat treatments for these alloys have been described(37)(38)(39). The future of the aluminum cell may well depend on the extent to which these anodic, corrosion resistant alloys are adaptable to battery technology.

ELECTROLYTE

The currently preferred electrolyte salts are $AlCl_3 \cdot 6H_2O$ or $CrCl_3 \cdot 6H_2O$ (21), these in themselves being less corrosive than the NH_4Cl-$ZnCl_2$ used in conventional $Zn:MnO_2$ dry cells. However, as in the case of its magnesium counterpart (Chap. 2), a chromate inhibitor (Na_2CrO_4, K_2CrO_4, or preferably, $(NH_4)_2CrO_4$) should be added to reduce

[1]Magnesium, under similar conditions shows about 60 percent current efficiency.

wasteful corrosion(22). Cells made with such electrolytes have oper circuit potentials of 1.6–1.7 V and a shelf life of at least a year. Again, as with magnesium, the protective film formed by addition of inhibitor is broken down during service and hence may lose its effectiveness in partly discharged cells. The preference for $(NH_4)_2CrO_4$ is due to the fact that i does not increase "delayed action" (the time required for a cell to become operable after the circuit is closed) (cf. Chap. 2).

SEPARATORS

The separator, in addition to forming a barrier preventing short circuit of anode and cathode, must absorb electrolyte and provide uniform we contact over the active anode surface. To improve that contact and ensure uniform anode corrosion, a gel or paste, either separately or as a coating on paper, is employed in conventional dry cells. This expedient is not immediately applicable to the aluminum system because the colloids (e.g., cereal pastes) generally used are unstable in the relatively acid $AlCl_3$-$CrCl_3$ electrolyte. Most of the work has, therefore, been done with uncoated (e.g., filter) paper, an expedient adequate for the evaluation of experimental cells, but likely to cause difficulty in commercial assembly where poor wet strength, excessive porosity, or nonuniformity migh cause service irregularities or even destructive short circuits.

A new method of producing an effective coated-paper separator has been described by Cahoon and Korver(40). The preferred colloids, locus bean or karaya gum, too insoluble in water or the ordinary organic solvents to be used directly, are first suspended in an acetone solution of polyvinyl acetate, which is then applied to one side of a special high wet-strength paper and dried. During cell assembly, electrolyte contacts the coating, the polyvinyl acetate is hydrolyzed, and the colloid is released. The latter swells in the electrolyte to furnish the adhesive layer that fills the voids between the wet paper and the aluminum surface. This procedure improved shelf life and delayed service.

CATHODES

Cathode manufacture is much the same as that for the conventiona Leclanché cell. Powdered manganese dioxide ore and acetylene black are mixed thoroughly with the pulverized electrolyte salts and the appropriate amount of water is added to provide the desired consistency. Examples of successful cathode formulations are shown in Table 3.2. A weighed amount of the cathode mix is tamped into the aluminum can after it has been lined with a plain or treated paper or other suitable liner in the conventional manner. The central carbon electrode is then driven into the cathode mix to serve as positive terminal.

TABLE 3.2. CATHODE COMPOSITIONS SUITABLE FOR ALUMINUM DRY CELLS

(1)	13.6 g Aluminum chloride	$AlCl_3 \cdot 6H_2O$
	5.7 g Chromic chloride	$CrCl_3 \cdot 6H_2O$
	9.1 g Ammonium chromate	$(NH_4)_2CrO_4$
	71.6 g Cathode mix [7 parts natural (African) MnO_2—1 part Shawinigan acetylene black]	
	30 cm³ water per 100 g of dry constituents	
(2)	6.1 g Chromic chloride	$CrCl_3 \cdot 6H_2O$
	6.1 g Ammonium chloride	NH_4Cl
	4.5 g Ammonium chromate	$(NH_4)_2CrO_4$
	83.3 g Cathode mix [5 parts natural (African) MnO_2—1 part Shawinigan acetylene black]	
	42 cm³ water per 100 g of dry constituents	

CELL REACTION

The end products have not been definitely identified and may vary with different electrolytes and operating conditions. The following overall reaction might be expected during cell discharge.

$$2Al + 6MnO_2 + 6H_2O \rightarrow 6MnO \cdot OH + 2Al(OH)_3$$

SERVICE CHARACTERISTICS

Because the electrolyte is acidic (pH ~ 3.5), it is not surprising that, as indicated in Fig. 3.4, the aluminum cell shows a higher operating voltage than its zinc counterpart and, therefore, a greater ampere-hour output on constant-resistance, continuous-drain test. Figure 3.5 shows the condition of the cell anodes after the 52 hours of discharge.

The performance of aluminum cells on standard tests(41) is shown in Table 3.3. Except for the long-term tests, the aluminum cells gave more service than the zinc units. Even on 4-ohm continuous test, anode efficiency, as determined from ampere-minutes of service and hydrogen evolution, was initially only of the order of 85 percent, dropping to an overall figure of 70 percent at a 0.9 V cutoff. On intermittent or long-extended service, the losses were even greater. The results shown are with natural MnO_2, but it should be noted that, as with zinc cells, heavy-duty service is improved with electrolytic MnO_2. For example, on the heavy industrial flashlight test (4 ohm), an aluminum cell with electrolytic MnO_2 had an initial closed circuit voltage 0.03 V higher and a capacity 35 percent greater than a cell with natural MnO_2.

Belitskus has found(42) that at least two factors limit capacity of these aluminum cells. The anode corrosion reaction consumes hydrogen ions to such an extent that the buffer capacity of the electrolyte (hydrolysis of

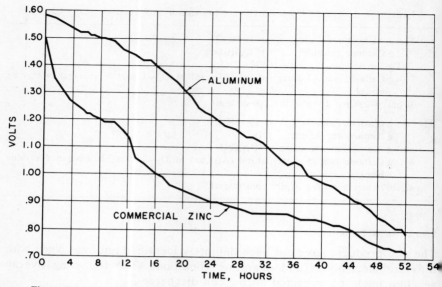

Figure 3.4 Discharge curves for "D" size dry cells (15 ohm continuous drain).

aluminum and chromic ions results in a pH of about 3.5) is exceeded and the pH of the cathode mix rises to a much higher level than that of a zinc cell during discharge. This increase in pH reduces the voltage of the MnO_2 half-cell reaction (cf. Chap. 1), so decreases capacity to a given cutoff voltage. Since aluminum corrosion continues during periods on open

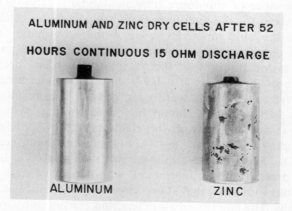

Figure 3.5 Aluminum and zinc dry cells after 52 hours continuous 15 ohm drain.

TABLE 3.3. COMPARATIVE PERFORMANCE OF ALUMINUM AND COM-
MERCIAL GENERAL PURPOSE ZINC CELLS[a]

Type of Test[b]	Zinc Cell Performance (minutes)	Aluminum Cell Performance (minutes)
Continuous drain		
(4-ohm to 0.90 V)	200	375
(2.25-ohm to 0.65 V)	200	275
Heavy industrial flashlight		
(4-ohm to 0.90 V)[b]	425	450
Light industrial flashlight		
(4-ohm to 0.90 V)[b]	700	500
General purpose		
(2.25-ohm to 0.65 V)[b]	475	225

[a]Cells made with two-layer anode, plain filter paper liner, $AlCl_3$-$CrCl_3$-$(NH_4)_2CrO_4$ electrolyte, and MnO_2 (natural African ore)—acetylene black cathode mix.
[b]Standard tests for "D" size Leclanché cells(41).

circuit, it is particularly detrimental during very intermittent service. The pH of the cathode mix of a zinc cell levels off (at ~6) during this type of service, and capacity is much greater than during heavy-duty service.

A second factor limiting capacity of these aluminum cells is internal resistance. Whereas the internal resistance of a "D" size zinc cell remained fairly constant (at about 0.4 ohm) during discharge (3-ohm continuous drain), the resistances of these aluminum cells were about 1.0 ohm initially, decreased to 0.5–0.6 ohm in the first one to two hours of drain (probably because of dissolution or modification of the original film), then increased to as high as 1.2 ohms after six hours (probably because of a buildup of $Al(OH)_3$, from both the cell reaction and corrosion reaction).

The aluminum alloys for which more anodic operating potential and higher current efficiency have been claimed offer hope for significant improvement but seem not to have been adequately evaluated in operating cells. The Raclot patents(36), for example, covering Al-Mg-Hg anodes, describe a dry cell which, with MnO_2:(C) cathode, showed an open circuit potential of 2.1 V. This unit was said to have delivered more than twice the ampere-hour output of its zinc counterpart (to 0.75 V cutoff), but its behavior on other than continuous, constant current service, or its shelf characteristics, was not indicated.

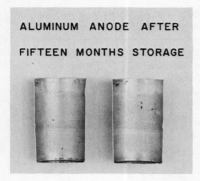

Figure 3.6 Interior of sectioned aluminum dry cell anode after 15 months storage.

Figure 3.7 Interior of sectioned aluminum dry cell anode after 7 years storage.

SHELF LIFE

This is an area that requires further study, since delay tests, as might be expected with experimental cells, have given erratic results. Enough has been done, however, to indicate that a properly made and adequately sealed aluminum cell should have an acceptable life expectancy. Thus, some of the better lots after three years of storage had capacities of 95 to 100 percent of freshly prepared cells and after seven years retained 60 to 75 percent of their initial capacity.

Figures 3.6 and 3.7 show the interior of composite anodes taken from cells stored for 15 months and seven years, respectively. After 15 months there was virtually no attack on the container. A small amount of attack appeared after seven years; at its worst, this amounted to only 25 percent of the total wall thickness of the anode. There was no perforation of the cans during this period.

DELAY

As with the magnesium cell (Chap. 2), "delayed action" experienced with the $Al:MnO_2$ system is still a problem, its magnitude a function of anode alloy, electrolyte composition, storage time, and degree of discharge. With the anode alloys and electrolyte compositions mentioned, freshly made cells had delay times of less than a second. With older or partially discharged cells, delay time increases and may reach 15 seconds or more.

Figure 3.8 is an oscillograph showing the voltage changes produced by alternately opening and reclosing the circuit of an $AlCl_3$-$CrCl_3$-$(NH_4)_2CrO_4$ type aluminum cell under 3-ohm load. On opening the circuit, a film apparently formed in about 0.1 second, as indicated by a slight drop in the

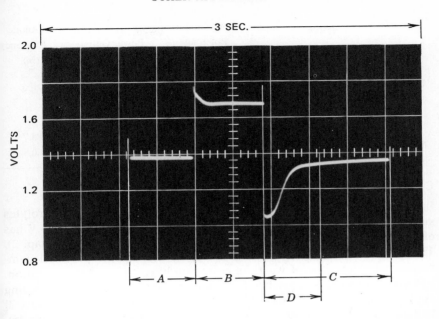

Figure 3.8 Al-MnO$_2$ cell voltage changes produced by interrupting a 3 ohm discharge for about $\frac{1}{2}$ second. First heavy line A shows working voltage under load; second line B shows open circuit voltage; third line C shows voltage after reapplying the load; major part of delay shown by D.

open circuit voltage; on reclosing the circuit, about a one-second delay occurred before the working voltage was reestablished. The major part of the delay disappeared in about 0.4 second.

It has been found(43) that some tin-containing aluminum alloys do not show this delay characteristic.

OTHER APPLICATIONS

The development of the more anodic aluminum alloys has suggested their application to batteries other than the Al:MnO$_2$ system.

ALUMINUM: SILVER CHLORIDE SYSTEM

Outstanding results have been claimed(44) for the substitution of an aluminum-tin alloy for magnesium in the Mg:AgCl system, as shown in the following table.

Anode	Open Circuit Potential (volts)	Power Output (kw-min/lb)	Anode Efficiency (%)
Base aluminum (99.5% purity)	0.78	28.4	69.8
Alloy			
(0.30% Sn, remainder Al of			
99.5% purity)	1.49	64.2	85.7
Magnesium	1.59	38.7	60.8

CATHODIC PROTECTION

In the past, the fact that commercial grades of aluminum in electrolytes such as seawater are actually more cathodic than zinc by about 0.3 V has effectively barred their application to cathodic protection (cf. Chap. 2). The aluminum alloys (described earlier) of high anodic potential combined with a low rate of wasteful corrosion have changed this situation. After a cost analysis, it has been concluded that aluminum anodes having current capacities of at least 1100 amp-hr/lb (80 percent efficiency) will largely replace magnesium and zinc anodes for seawater applications (45). An Al-7 percent Zn-0.12 percent Sn alloy has been used since 1962 for cathodic protection of offshore oil field structures (35). The ballast compartments of a 71,000 ton deadweight bulk carrier ship have also been protected with anodes of this alloy. Total anode weight was 26 tons, half of that required with magnesium anodes. Other successful field testing of aluminum alloy anodes has been carried out (46, 47, 48).

CURRENT STATUS AND PROSPECTUS

The aluminum system is still in the experimental stage and until a reproducible cell combining optimum components and refinement of construction has been developed and evaluated its future prospects, in competition with conventional systems, will remain in doubt. It is definitely indicated that an aluminum cell can have a higher operating voltage and greater watt-hour capacity on short-term, heavy-drain test than commercial dry cells. Wasteful anode corrosion is still a problem, lowering service capacity on intermittent and long-term tests and causing irregularities in shelf quality. The time lag or delay when circuit is closed must also be corrected to ensure commercial acceptability. Nevertheless, the fact that similar problems have been brought under control for the magnesium dry cell (Chap. 2) augurs well for the future of the aluminum system.

Valid cost comparisons are not yet available but should not prove unfavorable, provided that wasteful corrosion can be controlled and that use of the duplex anode can be avoided. Aluminum, to be sure, costs about twice as much per pound as zinc, but its higher electrochemical equivalent could bring its unit cost down to only slightly more than half that of the zinc anode of equal ampere-hour capacity.

REFERENCES

1. Latimer, W. M., *Oxidation Potentials*, Prentice-Hall, Englewood Cliffs, N.J. (1952).
2. Hulot, M., *Compt. rend.*, **40**, 148 (1855); cf. (4), p. 93.
3. Cazin, A., *Piles Électriques*, Gauthier-Villars, Paris (1881), p. 131.
4. Tommasi, D., *Traité des Piles Électriques*, Georges Carré, Paris (1889), p. 131.
5. Mennons, M. A. F., Brit. Pat. 296 (1858).
6. Eager, H. T. and Milburn, R. P., Brit. Pat. 6924 (1891).
7. Id., Brit. Pat. 899 (1891).
8. Brown, C. H., U.S. Pat. 503,567 (1893).
9. Anderson, E. L., U.S. Pat. 706,631 (1902).
10. Noble, J. and Anderson, E. L., U.S. Pat. 759,740 (1904).
11. Peek, F. W., U.S. Pat. 1,008,860 (1911).
12. Paine, A. J., *The Electrician*, **72**, 238 (1913).
13. Vince, C. H., Brit. Pat. 399,561 (1933).
14. Pennock, A. G. L. and Lee, S., Brit. Pat. 437,536 (1935).
15. Heise, G. W., Schumacher, E. A., and Cahoon, N. C., *J. Electrochem. Soc.*, **94**, 99 (1948).
16. Lozier, G. S., Glicksman, R., and Morehouse, C. K., U.S. Pat. 2,874,079 (1959).
17. Zaromb, S., *J. Electrochem. Soc.*, **109**, 1125 (1962).
18. Bockstie, L., Trevethan, D., and Zaromb, S., *J. Electrochem. Soc.*, **110**, 267 (1963).
19. Sargent, D. E., U.S. Pat. 2,554,447 (1951).
20. Ruben, S., U.S. Pat. 2,638,489 (1953).
21. Stokes, J. J., Jr., U.S. Pat. 2,796,456 (1957).
22. Id., U.S. Pat. 2,838,591 (1958).
23. Id., U.S. Pat. 3,307,976 (1967).
24. Mears, R. B. and Brown, C. D., *Corrosion*, **1**, 113 (1945).
25. Tosterud, M. and Taylor, C. S., Aluminum Company of America (in house) Report R-98, "Aluminum Dry Cells" (1927).
26. Stokes, J. J., Jr., *Electrochem. Tech.*, **6**, 36 (1968).
27. Reding, J. T. and Newport, J. J., *Materials Protection*, **5**, 15 (Dec. 1966).
28. Reding, J. T. and Newport, J. J., U.S. Pats. 3,281,239 (1966); 3,321,306 (1967); 3,337,332 (1967); 3,337,333 (1967).
29. Rohrman, F. A., U.S. Pat. 2,758,082 (1956).
30. Schreiber, C. F., U.S. Pat. 3,537,963 (1970).
31. Rutemiller, H. C., U.S. Pat. 3,227,644 (1966).
32. Rutemiller, H. C. and Montgomery, A. M., U.S. Pat. 3,274,085 (1966).
33. Sakano, T. and Toda, K., U.S. Pat. 3,172,760 (1965).
34. Rutemiller, H. C., U.S. Pat. 3,418,230 (1968).
35. Ponchel, B. M. and Horst, R. L., *Materials Protection*, **7**, 38 (March 1968).
36. Raclot, B., U.S. Pats. 3,257,201 (1966); 3,318,692 (1967).
37. Pryor, M. J., Keir, D. S., and Sperry, P. R., U.S. Pats. 3,180,728 (1965); 3,186,836 (1965); 3,189,486 (1965); 3,241,953 (1966); 3,368,958 (1968); cf. also (43, 44).

38. Keir, D. S., Pryor, M. J., and Sperry, P. R., *J. Electrochem. Soc.*, **114**, 777 (1967).
39. Id., *J. Electrochem. Soc.*, **116**, 319 (1969).
40. Cahoon, N. C. and Korver, M. P., *J. Electrochem. Soc.*, **106**, 469 (1959); cf. U.S. Pat. 3,048,647 (1962).
41. ANSI, Specifications for Dry Cells and Batteries, C18-1969, American National Standards Institute, New York (1969).
42. Belitskus, D., *J. Electrochem. Soc.*, **119**, 295 (1972).
43. Pryor, M. J., Keir, D. S., and Sperry, P. R., U.S. Pat. 3,250,649 (1966).
44. Id., U.S. Pat. 3,368,952 (1968).
45. Ponchel, B. M. and Horst, R. L., *Materials Protection*, **6**, 39 (June 1967).
46. Sakano, T., Toda, K., and Hanada, M., *Materials Protection*, **5**, 45 (Dec. 1966).
47. Grandstaff, C. M. and Craig, H. L., Jr., *Materials Protection and Performance*, **10**, 29 (May 1971).
48. Reding, J. T., *Materials Protection and Performance*, **10**, 17 (Oct. 1971).

4.

Organic Cathodes and Anodes for Batteries

J. S. Dereska

This chapter is primarily directed to the classical inorganic-oriented battery technologist seeking a simplified introductory perspective on the role of electrode-active organic materials (electrophores) as electrode depolarizers; and secondarily, to the organic chemist interested in characterizing galvanic cell organic processes as extensions of electrolytic cell syntheses. Hopefully, the chapter can also serve as an introductory guideline and digested review integrating some modern elementary theoretical organic chemistry with some electrochemical technology as they relate to the use and limitations of organic materials as electrodes in practical galvanic primary cell systems.

Arbitrarily, predominant attention is given to the cathodic process electrodes and to cell systems using aqueous electrolytes. Organic anodes have been discussed to a limited extent under "Fuel and Continuous-Feed Cells," (Chap. 9, Vol. I), and some organic electrolytes have been described under "Low Temperature Nonaqueous Batteries," (Chap. 10, Vol. I).

Academic and industrial electroorganic chemistry publications in the past century have emphasized the syntheses of organic electrode products from electrolytic cells more than the conversion of organic chemical reaction energy to useful electrical energy in galvanic cells. The past generation, however, has witnessed expansion in both areas and a closer integration of organic electrochemistry with physics and several other disciplines.

Energy economics, organic fuel conservation, and ecology at the present time are stimulating even greater interest in new galvanic power sources including exploitation of all types of organic materials in smaller galvanic cell systems, and in larger systems combining electrolytic and galvanic cell functions.

While, mechanistically and otherwise, the electrolytic and galvanic cell considerations are closely related, it must be unequivocally understood they are by no means synonymous. Hence, much of the vast information available on organic syntheses requires reappraisal and modification before applying it to galvanic cell systems delivering electrical power rather than using it.

For practical galvanic cell system applications, it is widely recognized that a number of cathodic organic depolarizer types intrinsically possess such electrochemically and commercially attractive features as (1) relatively high coulometric capacity, (2) reasonably useful range of operating potentials, (3) acceptably low polarization losses and flat discharge voltage levels, (4) competitive costs, and (5) availability within the continental United States. Consequently, it is not surprising that organic compounds should have been considered for use as single or both electrodes in primary, secondary, and fuel cells ever since the advent of organic electrochemistry itself. A casual review of the pertinent literature reveals that essentially every combination of elements and compounds, both organic as well as inorganic in aqueous, nonaqueous, and many molten and solid electrolytes has been considered or empirically evaluated.

Unfortunately however, unlike the precedent established by many of the inorganic cell systems previously described in this book, no widescale, commercially established primary battery with at least one organic electrode in aqueous electrolyte has yet been attained despite a great deal of research and development effort. Although this represents the present status, it is not unlikely that present and future work can provide a completely organic battery as well as various successful hybrids.

Some of the questions of general interest usually raised include: Why organic depolarizers and batteries? What is their history? How do they function fundamentally? What do they comprise and how are they fabricated? What are some of the commercial limitations?

An attempt to at least partially answer some of these questions will be made in more-or-less sequential order in the six sections comprising this chapter.

BRIEF HISTORY

Historically, the introduction of organic materials as "electrophores" in both electrolytic and galvanic electrochemical cells, seems to have coincided with the development of the broad subject of electroorganic chemistry itself, and in the early part of the 19th century perhaps highlighted by the synthesis of ethane by Kolbe(1) in 1849. Earlier, Rheinhold and Erman(2) had established the oxidizing and reducing properties of organic compounds. Haber(3) was the first to note the very important, presently accepted relationship between cathode potentials and organic electrode products. He also first recorded the study techniques and the mechanisms for the stepwise reduction of nitro compounds, a subject of considerable importance to this chapter because of the illustrative consideration given later to some of them as organic cathodes. The extensive literature of the electrolytic field has been reviewed notably by Swann(4), Brockman(6), and others(5)(49) while Allen(52) and, recently Baizer(38) have presented excellent texts concerning electrolytic cell processes(40–54).

As early as 1802, Sir Humphry Davy attempted (unsuccessfully, to be sure) to oxidize carbon anodically in a galvanic cell containing nitric acid as both the electrolyte and the cathodic reactant(7). In 1843, Grove, who made the first hydrogen-oxygen cell, showed that ethylene or carbon monoxide could be substituted for hydrogen(8); Gaugain is reported(9) to have used alcohol or ether vapors as anodic reactants in cells with molten electrolytes. Maiche and others(10) worked with illuminating, producer, or water gas.

Numerous electrochemical cell examples illustrating the use of organic electrolytes, mixtures, and aqueous solutions and suspensions can also be found, but generally the all-organic electrolytes have been used with inorganic electrodes. Propylene carbonate electrolyte(11) in combination with lithium-copper fluoride represents one such example which features a high energy-to-weight ratio and high voltage, and is described in more detail in the nonaqueous chapter in Volume I.

Aqueous primary galvanic cell systems incorporating nitro compounds as depolarizers were patented by Bauer(12) in 1915. Arsem(13) in 1942, widely expanded the patented organic depolarizers to include neutral, acid, and alkaline electrolytes as well as the nitro compounds and their reduction products, positive halogen types and the quinone compounds. Bloch(14), in 1951, described the use of quaternary alkylhalide mixtures while Sargent(15), in the same year, patented the use of aluminum anodes with organic cathodes.

Morehouse and Glicksman(16), in 1958–59, and later Lozier of RCA patented the use of magnesium anodes in a series of galvanic cell systems using many of the organic cathodic depolarizer classes introduced earlier by Arsem who had used zinc anodes. The cathodic (for example, oxyiodine types) of organic depolarizers were also patented(17) by these RCA investigators who also described screening tests for a number of anodic organic depolarizer types including hydrazine, hydroxylamine, hydroxy and aminophenol anode compounds.

Attention was directed to polyhalides and interhalogens as a general class of organic and inorganic cathodic depolarizers by Dereska et al.(18) in 1962, while Cohn(19) patented cells using hydroxylamine as the cathodic depolarizer. Klopp(20) cathodically reduced nitro compounds in cells using nonaqueous liquid ammonia electrolyte and magnesium anodes.

Bioelectrochemical cells involving microbiol and catalytic enzyme anodic processes such as the indirect oxidation of organic carbohydrates, alcohols, fatty acids, and urea are well known and predictions for hydrocarbon bioelectrodes have been described(21–28).

Pinkerton(29) in organic electrochemical processes reminiscent of organic chemistry's Friedel-Crafts and Grignard synthesis(5)(46)(47)(61) patented high energy cells using organic electrolytes which participated in both the charge carrier and electrode processes. High potential inorganic anodes such as Na, Ca, Mg, and Al were generally used along with composite organic intermediate cathodic "systems" comprising an oxidizer and an acceptor. Oxidizers included the organic electrolyte intermediates formed with $AlCl_3$, BF_3, LiF, esters, halogen ethers, and metalloids while the acceptors were largely drawn from the organometallic compounds.

Among the numerous more recent patents, some of which illustrate cell constructions and secondary cell applications, are those relating to azodicarbonamides(30)(70), biurea(31)(70), azobis formamides(32), nitrothiazol derivatives(33), benzofurazan oxide(34), polynuclear aromatic charge transfer complexes(35), benzopheroximine thiazole and redox-polyester polymers(36).

In his substantial studies of electrochemical energy conversion systems Foley(37) reports on various electrolytes and includes a number of organic depolarizers. Perhaps one of the most complete and up-to-date texts on organic electrochemistry which is primarily directed to the organic chemist, but one which appears most instructive for the inorganic battery technologist interested in organic electrodes is one edited by M. M. Baizer(38). Some additional earlier publications on organic electrodes can be found in Volume I of this book and in references (39–66).

FACTORS LIMITING PERFORMANCE OF ORGANIC DEPOLARIZERS

The electrochemically and commercially attractive features of organic depolarizers and batteries such as high theoretical coulometric capacities, service times per unit weight, competitive costs, competitive unit cell voltages when used with high potential anodes, and availability within the continental United States have been pointed out earlier.

One or more of the following factors appear to have restricted the widespread galvanic cell use of organic electrophores and depolarizers in particular.

1. Present status relative to the advantageously competitive inorganic depolarizers.
2. Environmental instability in relation to the restricted choice of high potential cathodes and anodes.
3. Solubility and colloidal dispersion problems.
4. The generally low electrical conductivity of organic depolarizer mixes.
5. Secondary mass transfer limitations.
6. The low specific gravity of organic compounds.
7. Relatively greater electrode process complexities.
8. Incomplete fundamental understanding of electrodes processes.

PRESENT STATUS RELATIVE TO THE ADVANTAGEOUSLY COMPETITIVE INORGANIC DEPOLARIZERS

Perhaps one of the greatest restrictions to exploiting organic electrophores commercially is attributable to the present economic uncertainty and the understandable reluctance to replace stable, well established, well developed, and generally acceptably priced inorganic depolarizers such as the oxides of Mn, Hg, Ag, Ni, Pb, Cu, Fe and such metallic anodes as Zn, Mg, Cd, Pb with the less developed, problematical organic electrophore substitutes—at least at the present time.

ENVIRONMENTAL INSTABILITY IN RELATION TO THE RESTRICTED CHOICE OF HIGH POTENTIAL ANODES AND CATHODES

Many electrode-active organic materials are unstable in the protic electrolyte cell environments, reacting for example with water, aqueous salt solutions, or with aqueous-solvent suspensions or mixtures via hydrolysis, decomposing at elevated temperatures, etc.

Some separator materials and gelling agents oxidize or hydrolyze in the presence of oxidants including air, and many electrolytes including the aqueous-organic compositions are subject to similar shelf deterioration. Thus instability of organic electrode, separator, and electrolyte, and materials of construction can be responsible for a significant loss in cell discharge efficiency.

Electrode instability limitations for primary cell systems using organic depolarizers is reflected as an inverse relationship existing between the oxidation potential and the stability in aqueous electrolytes. Many organic materials hydrolize or otherwise decompose in water, hence many of the more stable organic cathodes are characterized by relatively low cathodic operating potentials at practical current drains. To meet the 1.4–1.6 V criteria commercially established in past applications of the Leclanché (MnO_2/NH_4Cl, $ZnCl_2/Zn$) primary dry cell, an anode (the cell's counter electrode) with a relatively high anodic potential must be used. High potential anodes stable in aqueous environment are quite limited. Anodes of magnesium, aluminum, titanium, and alkaline earth, alkali metals Zn, Cd, In, Pb, Sn, Cu, and H_2 organic compounds have all been suggested for use with organic cathodic depolarizers. The introduction of magnesium anodes has been essentially responsible for the relatively recent attention given to organic depolarizers for primary cells (16). Seriesed cell multiplicity is the other alternative to high potential anodes but processing is generally more involved and costly, hence is a less desirable choice.

Attainment of practically useful cell voltages with organic cathode depolarizers exhibiting very low cathodic potentials in aqueous electrolytes is often limited by the instability of the complementary high anodic potential anodes required.

In an analogous way, the relatively low anodic potentials of organic anodes emphasize the complementary use of unstable high cathodic potential oxidants whether they be organic or inorganic. Thus, a compromise between the stability of one or both electrodes consistent with presently specified cell voltage requirements has to be constantly faced in galvanic cell designs. Perhaps environmental instability represents one of the major restrictions to the use of organic electrophores in galvanic cells.

SOLUBILITY AND COLLOIDAL DISPERSION PROBLEMS

The third factor deserving additional comment here concerns the solubility and colloidal dispersability of the depolarizer and the consequent depolarizer-anode chemical reactions causing shelf problems via coulometric capacity loss and fouling of electrodes.

While homogeneous processes and soluble organic depolarizer species pose no special problem when suitable diaphragms are used in industrial electrolytic cell syntheses and perhaps in some reserve-action, fast-discharge galvanic cells, the relatively insoluble types are preferred for galvanic cell with some reasonable shelf life. Thus most of the soluble organic cathode depolarizers like their inorganic counterparts can diffuse or be electrically transported through the diaphragm separator to the anode and be locally reduced during shelf storage. Colloidally dispersed organic solids and liquids act in a similar way.

Examples of organic substances dissolved or colloidally dispersed in the electrolyte which are reduced by contact with metals in acids, include the Clemmensen reductions of the carbonyl groups. Nitro compound reductions by contact with electrolyte and with various metals and their amalgams, such as zinc, tin, iron, magnesium and aluminum, also have long been known. Reductions by metals in acidic, neutral, and alkaline electrolytes in which amalgams and alcoholates of even the alkali metals serve anodic functions are also well described in classical organic chemistry texts, (5)(45)(46)(61–63).

Thus, it should be anticipated that even slightly soluble organic cathodic depolarizers such as m-DNB dissolving from the cathodic mix in galvanic primary cell systems can diffuse through separators to the anode metal surface, and then be reduced partially to intermediates or even to final products. This environmental shelf process represents wasteful consumption of both anode and cathode materials and often electrolyte as well; and to the formation of resistive reduction product films adhering to the anode surface and reducing its effective area. Both of these effects are then reflected in lowered coulometric capacity and of lower operating voltages during discharge.

The solubility factor has been responsible for the failure of an appreciable number of organic depolarizers to be more rapidly exploited.

The Generally Low Electrical Conductivity of Organic Depolarizer Mixes

The characteristic electrical conductivities of many inorganic types of solid oxide depolarizers in polycrystalline or particulate form are generally too low for acceptable internal cell charge transport rates. Consequently, milled mixtures of these oxide depolarizers with some type of supplemental electronic conductor such as graphite, acetylene black, or inert metal must be used to minimize the iz voltage losses.

Because of the characteristically lower density of organic depolarizers, because of their relatively low conductivity(55), and because typical heterogeneous electrode processes function primarily at surfaces, a greater amount of the supplemental conductor which occupies critically needed space and contributes nothing coulometrically, is typically required for organic depolarizer mixes. Similar reasoning applies to the anodic electrophores.

Inorganic oxides such as those of manganese and nickel in the solid state are semiconductors and it is believed that this property materially participates in the electrode mechanism, particularly when the oxide is being completely reduced electrochemically. The participation of organic semiconductor charge transport processes in electrode reactions has not yet been widely studied.

SECONDARY MASS TRANSFER LIMITATIONS

Practical galvanic cells, particularly the dry and miniature cell designs, are volume-limited systems. To maximize discharge service time, solid electrophores are used and the amount of electrophores are usually maximized, often at the expense of the less dense aqueous electrolyte. Thus, in the spatially limited electrolytes, higher concentrations of electrode products, and greater pH changes occur during discharge. This not only restricts the rate of removal of products and the rate initial reactant electrophores can be mass transferred to the electrode surface, but also introduces a secondary complication with organic electrophores. This is manifest as side-reactions in the form of rearrangements, couplings, condensations, and hydrolyses in the anode and cathode compartments. The resulting iz voltage losses decrease the service time and discharge efficiency to a given cutoff.

As an example, the hydrolysis of anodically produced Zn^{++} ions at the higher localized concentrations prevailing at the spatially limited anode surface, can lower the pH to values which can cause rearrangements of nitro compound reduction intermediates such as phenylhydroxylamine to p-amino-phenol. An increase in pH is typical for the cathode compartment or mix which can lead to couplings and condensations to form azoxybenzene, and so forth.

THE LOW SPECIFIC GRAVITY OF ORGANIC COMPOUNDS

Practical galvanic cell size specifications impose volume limits on the cell contents making the denser electrode materials preferred.

Most organic compounds have a specific gravity much lower than the corresponding inorganic compounds that the experimenter may be trying to replace. A conventional inorganic cathode material, that is, MnO_2, has a specific gravity of from 4.5 to 4.8. Thus it serves as a very compact source of the available oxygen in the compound. In the design of batteries, one objective is to build into the package volume the greatest amount of capacity possible. With materials of low specific gravity this becomes more difficult with organic materials than with the conventional oxides.

RELATIVELY GREATER ELECTRODE PROCESS COMPLEXITIES

With their characteristic electrical-chemical step sequences, the organic electrode processes in general appear to be more complex than their inorganic counterparts. Thus the greater sensitivity of electrode process and product to electrode potential changes becomes apparent. And as the organic electrode potential progressively changes during discharges into constant load resistors, a variety of products often results rather than the usually single or relatively few products found for the inorganic electrophores. And, of course, the service and coulometric capacity reflect the products formed.

The catalytic nature of the current collector surface whether metal or carbon also influences the discharge processes and products formed.

The presence of oxygen or hydrogen "carriers" acting as intermediate agents in the overall electrode process also can alter galvanic cell discharges, and the secondary pH effects on rearrangements and condensations has already been mentioned.

INCOMPLETE FUNDAMENTAL UNDERSTANDING OF ELECTRODE PROCESSES

While the past three decades have witnessed somewhat profound advances in electroorganic syntheses, instrumentation, equipment, materials, techniques, and some new fundamental information on organic electrode processes in electrolytic cells, more such basic knowledge is still needed for the quantitative interpretation not only of electrolytic but practical galvanic cells (55–57). Although the more quantitative but complex relationships between structure and oxidation potential, relating potentials to molecular orbital energies and to the kinetic or non-thermodynamic changes, are currently under study (38), the broader generalizations appear somewhat obscure, making the earlier qualitative and quasi-quantitative information still of practical value to the battery technologist.

At present, simplified concepts integrating the organic and electrochemical disciplines are still widely used.

Studies of the electrical properties of organic solids, (55) liquids, and gases useful as electrophores are complex and are still in somewhat of a state of transition or flux despite the accelerated efforts of the past several decades. Further detailed knowledge and discipline-integrated theories are needed in areas relating states of matter, electrical charge generation, transfer and transport within and between phases and molecular structures.

Quantitatively defined theories explaining the concurrent as well as the sequential electron charge distributions between the covalently bonded organic species and the current collector electrode, as well as defining the intermediate organic-collector species may serve to simplify and more clearly relate the oxidation potentials of organic and inorganic electrophores.

Despite the enormous attendant complexity, and in view of the recent accelerated sophistication and speeds in computers, further rigid solutions might again be attempted of the Schrodinger relationship between total system energy, potential energy in a given force field, and the orbital positions of the electrons which are applicable to polyelectronic and molecular species.

COULOMETRIC EVALUATION OF ORGANIC ELECTRODE PROCESSES

Despite the previously reviewed complexities and the still remaining inadequacies in the rigid, fundamental understanding of organic electrode processes, some highly oversimplified but useful guidelines can be conceived for battery technologists evaluating theoretical coulometric capacities.

The simplification for aqueous electrolyte cells stresses the role of organic chemistry's "functional groups" in electrode processes, and the direct or indirect electrical transfer between the organic electrophore and either the *electron source* current collector surface (cathode) or the *electron sink* current collector surface (anode).

Thus, as an example of the "indirect" reduction of an organic depolarizer as a sequential electrical charge transfer—chemical reaction (an E-C sequence), the intermediate atomic, radical, or surface species resulting from the primary electrochemical charge transfer step such as $H^+ + e \rightarrow H°$ can subsequently react chemically with the more reactive $H°$ species. This makes possible predictions for the net coulometric capacities as limited by the "saturation" levels of electrons and atoms corresponding to the end products formed. However, it must be clarified here that this coulometrically related charge transfer does not include the additional inter- and intramolecular electron and proton charge transfers accompanying such an over all "electrode mechanism", usually included in various electrical-chemical sequences.

As a hypothetical generalization of organic electrode cathodic processes in aqueous electrolytes, the "saturating" steps have been occasionally pictured as in Fig. 4.1 as a sequential series of additions of activated surface state hydrogen atoms (hydrogenations) and losses of H_2O (dehydrations) to form various intermediate products up to complete "saturation" levels such as exemplified by aniline and methane starting with propane and carbon dioxide respectively. It should be recognized that not all of the specific examples shown have yet been demonstrated.

Analogously, the anode processes are considered to involve sequential dehydrogenation and hydration. The net processes can involve sequential charge and mass transfer steps.

Many organic anodic processes are simplified in illustrations as the reverse of the reductions above, that is, dehydrogenation processes are characteristic.

Exceptions to the preceding of course abound, and as an example of the Kolbe reaction at an anode shows a net loss of electrons resulting in a free radical mechanism. Here carboxylic anion *sources* transfer electrical charge to the electron *sink* anode forming reactive carboxylic radicals which subsequently lose CO_2 and couple to form higher molecular weight products.

$$2\,RCO_2^- \xrightarrow{\text{charge transfer}} 2\,RCO_2^{\cdot} \xrightarrow{\text{chemical}} 2\,R^{\cdot} + CO_2$$
$$\searrow$$
$$R\text{-}R$$

Consider now, a specific example of this "overall" or indirect charge transfer which involves a sequence of a charge transfer step, followed by a chemical step, for a nitro compound reduction in aqueous electrolyte. Aqueous electrolyte charge carriers such as hydrated H^+ and OH^- are usually available from the dissociation of H_2O. Approximately 1.6×10^{-19} coulombs of charge can be transferred as an electron passing from the cathodic current collector to one H^+ ion sink to form, for example, a neutral intermediate (atomic or activated surface hydrogen specie) at the electrode surface. This is equivalent to 96,487 coulombs or 1608 amp-min/g atom of hydrogen. This unstable atom subsequently can undergo chemical reaction with a covalently bonded organic species increasing its degree of "saturation."

Thus this overall cathodic mechanism illustration suggests some resemblance to the more conventional catalytic chemical hydrogenation processes described in earlier organic chemistry literature (5)(48)(61)(63), and can proceed coulometrically at one unit of charge per hydrogen ion consuming charge to various levels of saturation up to a maximum saturation point. Figure 4.1 can be further detailed as follows.

CHARGE TRANSFER STEP AT THE CATHODE

As shown, a single hydrogen ion in the electrolyte near the electron source accepts a single electron thus transferring a single charge across the interface and forming an excited state neutral atom or atomic surface specie.

DEPOLARIZER REDUCTION STEP

When favorably oriented close to the electron source cathode surface, the unsaturated, m-directing, electron attracting nitro functional group in one molecule of nitrobenzene (NB) either directly transfers charge from the

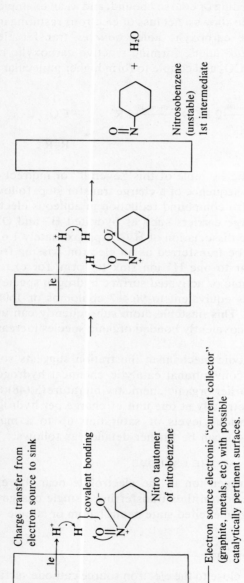

(a) First two electron cathodic reduction step of nitrobenzene.

Charge transfer from electron source to sink

Nitro tautomer of nitrobenzene

Nitrosobenzene (unstable) 1st intermediate

covalent bonding

Electron source electronic "current collector" (graphite, metals, etc) with possible catalytically pertinent surfaces.

(b) Overall cathodic reduction of nitrobenzene to phenylamine.

Nitrobenzene $\xrightarrow{2\begin{Bmatrix}H^+\\e\end{Bmatrix}}$ Nitrosobenzene $+ H_2O$

Nitrosobenzene $\xrightarrow{2\begin{Bmatrix}H^+\\e\end{Bmatrix}}$ Phenylhydroxylamine

Phenylhydroxylamine $\xrightarrow{2\begin{Bmatrix}H^+\\e\end{Bmatrix}}$ Phenylamine $+ H_2O$

198

(c) Overall cathodic reduction of carbon dioxide and anodic oxidation of methane.

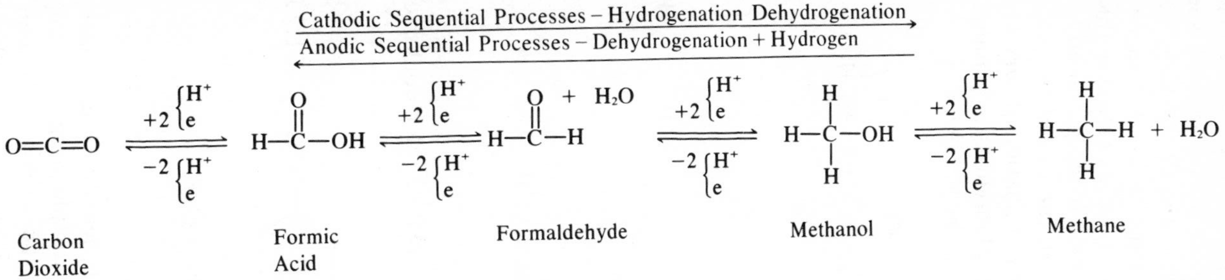

Cathodic Sequential Processes – Hydrogenation Dehydrogenation →

Anodic Sequential Processes – Dehydrogenation + Hydrogen ←

$$O=C=O \underset{-2\begin{Bmatrix}H^+\\e\end{Bmatrix}}{\overset{+2\begin{Bmatrix}H^+\\e\end{Bmatrix}}{\rightleftarrows}} \underset{\substack{H-C-OH\\\|\\O}}{} \underset{-2\begin{Bmatrix}H^+\\e\end{Bmatrix}}{\overset{+2\begin{Bmatrix}H^+\\e\end{Bmatrix}}{\rightleftarrows}} \underset{\substack{H-C-H\\\|\\O}}{} + H_2O \underset{-2\begin{Bmatrix}H^+\\e\end{Bmatrix}}{\overset{+2\begin{Bmatrix}H^+\\e\end{Bmatrix}}{\rightleftarrows}} \underset{\substack{H\\\|\\H-C-OH\\\|\\H}}{} \underset{-2\begin{Bmatrix}H^+\\e\end{Bmatrix}}{\overset{+2\begin{Bmatrix}H^+\\e\end{Bmatrix}}{\rightleftarrows}} \underset{\substack{H\\\|\\H-C-H\\\|\\H}}{} + H_2O$$

Carbon   Formic   Formaldehyde   Methanol   Methane
Dioxide   Acid

Figure 4.1 Schematically illustrative covalent surface state–charge transfer processes.

199

electrode current collector to the charged specie or an ion radical, or chemically reacts with the activated atomic atom or surface species initially formed via charge transfer at the catalytic surface. Direct chemical reduction to the neutral molecule by the atomic hydrogen may also be a possibility. Chemical reduction then proceeds stepwise through various intermediates, some concurrently formed if the potential is high enough, and the coulombs delivered depend on the extent of reduction products formed for the time permitted by the termination of service. The endpoint indicated in the following reaction is phenylhydroxylamine, but further reduction and coulometric capacity can be achieved.

$$\langle\bigcirc\rangle\!-\!NO_2 \; + \; 4H^{\cdot} \; \longrightarrow \; \langle\bigcirc\rangle\!-\!\overset{\overset{\displaystyle H}{|}}{N}\!-\!OH \; + \; H_2O$$

OVERALL CATHODIC ELECTRODE PROCESS

$$\langle\bigcirc\rangle\!-\!NO_2 \; + \; 4H^+ \; + \; 4e \; \longrightarrow \; \langle\bigcirc\rangle\!-\!\overset{\overset{\displaystyle H}{|}}{N}\!-\!OH \; + \; H_2O$$

Specific coulometric capacity

$$q_s = \frac{4 \times 96,480}{123.11} = 3135 \text{ coulombs/gNB}$$

$$q_s = 52.3 \text{ amp-min/gNB}$$

Cathodic reduction to aniline (phenylamine) yields greater coulometric capacity and service.

$$\langle\bigcirc\rangle\!-\!NO_2 \; + \; 6H^+ \; + \; 6e \; \longrightarrow \; \langle\bigcirc\rangle\!-\!NH_2 \; + \; 2H_2O$$

$$q_s = \frac{6 \times 96,480}{123.11} = 4702 \text{ coulombs/gNB}$$

$$q_s = 78.4 \text{ amp-min/gNB}$$

Subsequent and concurrent side reactions, pH sensitive condensations, rearrangements, and so forth, often occur complicating the simpler interpretations. The importance of hydrogen ion activity, pH, can be inferred from the above equation and previous comments as well as the catalytic influence of the current collector metal or graphite.

Specific mention must be made of the enormous differences made by the type of electrolyte or media on the behavior of depolarizers(4)(38).

CLASSIFICATION OF ORGANIC ELECTROPHORES

A single rigid, comprehensive, and unique classification of organic anodic and cathodic compounds is not a simple thing to achieve with respect to primary cell objectives.

Classification of the organic compounds as anodes or cathodes is complicated because many contain several functional groups in one molecule, many are intermediates and can be utilized in either electrode function, and a number of them can be used in secondary or rechargeable cell applications in which case they can alternately serve as anodes and cathodes.

Choices of organic electrophore classes in galvanic cell functions can assume various arbitrary forms. These may be made according to (a) generally known organic chemical reactions, (b) functional groups in the organic molecule, (c) relative anodic and cathodic activity, (d) electrochemical mechanism, and (e) electrophore specie whether ionic, radical, radical ion, atomic, molecular, intermediate surface specie, etc.

The parallelism between an electrochemical process and the three fundamental aspects of organic chemistry: structure, reactions, and mechanisms suggest several possible classification schemes. Following the organic chemistry precepts, classification can be made to cover the major types of well-known chemical reactions and processes.

Organic Reaction Types

1. Oxidations and reductions, including irreversible and redox processes.
2. Substitutions, aliphatic and aromatic, nucleophilic, electrophilic, free radical, etc. in which an atom or group attached to a carbon atom is removed and another enters its place. No change in degree of unsaturation of reactive carbon occurs.
3. Additions of molecules, atoms, or ions to carbon-carbon, and to carbon-hetero multiple bonds to other element unsaturated groups, and so on. These involve an increase in number of groups attached to carbon thus increasing the degree of saturation.
4. Eliminations which involve a decrease in the number of groups bound to carbon. The degree of unsaturation increases.
5. Rearrangements, condensations, inversions in which the carbon skeleton of the molecule is internally rearranged.
6. Proton-transfer.

7. Reordering in which one reactant contributes an unshared pair electron to another leaving an empty orbital available. This is simply a Lewis acid-base reaction.

8. Single cleavage.

LEAVING GROUPS

Complementing the above, another useful grouping drawn from reported organic chemistry reactions(5)(46)(47), which is useful in qualitatively predicting primary electrode products focuses attention on the somewhat limited number of characteristic, displaced substituents called "leaving groups" or "residual" secondary products of the reaction rather than on the substrate itself.

Thus in principle (and analogous to the Raney nickel catalytic desulfurization syntheses), alkanethiols may conceivably be reduced in suitable media at nickel cathode current collectors and hence hydrogenated to their respective alkane products (substitute group) and a characteristic leaving group such as H_2S.

FUNCTIONAL GROUPS

Organic electrophore classification according to the relative cathodic or anodic activity can be conceived which is based on the characteristic relationship between the observed electrode potentials and the functional group and structure as pioneered by Haber(3) and Heyrovsky(65).

Shikati and Taschi(66), working with nitro compounds, introduced the electronegativity rule observing that reduction potentials became more positive the greater the degree of electronegativity of the substituents in the molecule. Implementing this concept with quinones Fieser(5) found experimental support that the cathodic reduction potential depends on the class of compounds, and the type of position of substituted groups in any class; thus Class II organic compounds comprising such unsaturated or electron attracting groups as $-NO_2$, $-COOH$, $-CHO$, $-SO_3H$, $-CN$, $-CONH_2$, etc. decrease the electron density, hence energy levels in the aromatic ring for all positions but in different degrees depending on the position. Therefore, the introduction of one of these functional groups raises the potential of the parent quinone up to about 200 mV, whereas about a 0 to 100 mV potential lowering effect is exerted by the electron repelling Class I groups such as $-NHR$, $-NH_2$, $-N(CH_3)_2$, $-OH$, $-OR$, $-CH_3$, etc.

Publications, notably by Allen(52), S. Swann, Jr.(4), and others(38)(40)(49)(57) provide much more specific and inclusive electrolyte cell classifications and listing of organic compounds useful as cathodic and anodic electrophores in galvanic cells than have been attempted here.

However, based on arbitrary classification choices (b) and (c) above which reflect the demonstrated influence of organic compound structure, type and position of substituent group, nature of electrolyte, type of catalyst or carrier agent present, pH, and the like on such electrochemical parameters as operating electrode potentials, coulometric capacities, a convenient if arbitrary classification is presented here which may be of some use to the battery technologist.

Numerous specific compounds, derivatives, analogues and combinations of groups have been omitted for the sake of brevity and only rough guidelines are intended. It must be emphasized that all the depolarizers indicated are not presently used in commercial primary cell systems and that the list is neither comprehensive nor inclusive. Many of the specific compounds listed are unstable and several are explosives. However, a considerable number of examples of each category have been actually demonstrated to be electroactive as anodes and cathodes in either an electrolytic or galvanic cell or both. Seven potential or demonstrated categories of cathodes and seven categories of anodes have been listed in Tables 4.1 to 4.3 with the first showing general classes and Tables 4.2 to 4.3 listing several illustrative examples of each of the general classes.

While not exhaustive, the limited number of groupings or classes of representative compounds containing such elements as carbon, hydrogen, oxygen, nitrogen, sulfur, the halogens and various metal combinations are included. Some of the common structural characteristics such as unsaturation, conjugation, and resonance are evident from the illustrative class examples shown in these tables.

TABLE 4.1. AN ARBITRARY GALVANIC CELL CLASSIFICATION FOR SOME DEMONSTRATED AND POTENTIAL ORGANIC DEPOLARIZERS ACCORDING TO FUNCTIONAL GROUPS

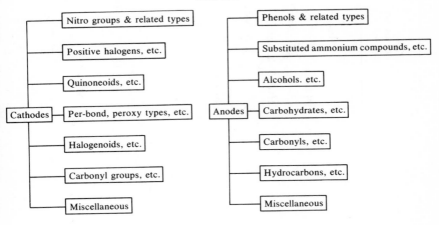

TABLE 4.2. CATHODE ASSOCIATED DEPOLARIZER TYPES

I Nitro Group and Related Types	II Positive Halogen and Related Types	III Quinonoid and Related Types
Nitro	Amines	Quinones
Nitroso	Amides	Benzo, Naptho Anthro, Dipheno, etc.
Hydroxylamines	Imides	Imines
Azoxy	Cyclic Ureides	Quinone Imines
Azo	Amidines	Quinone Mono and Dioxines
Polyazo Hydrazo		Redox Resins and Macromolecules Ion Exchange Polymers
Nitramine		

204

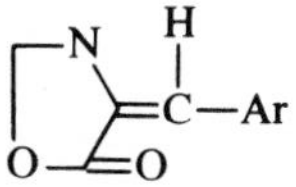

Oxyiodine Compounds

Iodoso—IO
Iodoxy—IO₂
Iodyl—salts

Polyhalogen Types

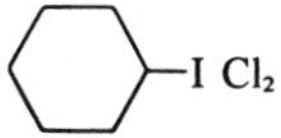

Special Halogen

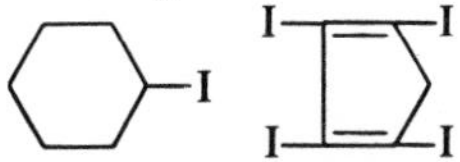

Halogen atoms on C next to carbonyl or phenyl gp. or those attached to metals in metallo-organics

Pyrollidone mixture, polyvinyl with 10% Iodine.

Nitrosoamines

$$H-N-N=O$$

Alkyamine Oxides

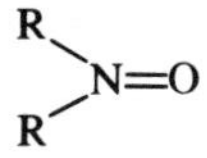

Oximes

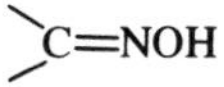

Nitrile Oxides

$$R-C\equiv N=O$$

Diazonium

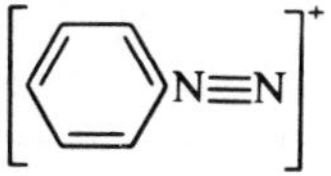

Quinazolones

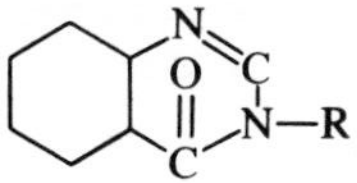

Imidazoles

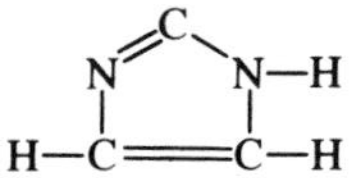

Azlactones

$$\text{TABLE 4.2. } (\textit{Contd.})$$

I Nitro Group and Related Types	II Positive Halogen and Related Types	III Quinonoid and Related Types
Phenylhydrazones **Azodicarbonamides** **Azobisformamides** 		

Cathode Associated Depolarizer Types

IV Per-Bond, Peroxy, and Related Types	V Halogenoid and Related Types	VI Carbonyl Group and Related Types
Peroxides, peroxyethers	Nitriles, cyanides, cyanogen —C≡N (CN)$_2$	Aldehydes —C=O with H
Hydroperoxides Peroxyhydrates Peroxyacids	Carbylamines, isocyanides —N≡C	Ketones —C=O
Ozonides	Cyanimides, cyanalkines, aminopyrimidines HN H$_2$ —C—C—C≡N	Carboxylic acids and anhydrides —C=O with OH, O=C=O, CO
Disulfides N—S—S—N	Cyanates R—O—C≡N	Esters —C=O with OR
Persulfides, polysulfides S=C—S—S—C=S	Isocyanates, isothiocyanates —N=C=S —N=C=O	—O—N=O O —O—N=O
Perhalo compounds	Azides	Amides —C=O with NH$_2$
Percarbonates Peroxyphosphates	Thiocyanates (SCN)$_2$ (SCN)— (NCS)	Imides —C—C=NH with —C=O
Peracetates Persulfates		

207

TABLE 4.2. (*Contd.*)

IV Per-Bond, Peroxy, and Related Types	V Halogenoid and Related Types	VI Carbonyl Group and Related Types
Perborates	Azido carbon disulfide $(S$—C—$SN)_2$ $(SCSN_3)$	Imido esters
Transannular types and photooxides	Fulminates ONC	
	Cyanogen, and thiocyanogen halides $SCNCl_3$ $I(OCN)_3$	Amidines
	Cyanines, pthalocyanines, chlorophyl, hemin	
	Polyhalide halogenoids Schiff's bases and hydrazones	Aldimides
	Selenium, tellurium analogues $(Se\ CN)$, $(TeCN)^{-2}$	Carbon suboxides
	Nitrothiazole	

TABLE 4.2. (*Contd.*)

VII

Miscellaneous Types and Processes

Alkaloids

Sulfur, sulfonylhalides, nitrothiazol derivatives sulfinic acid derivatives, dehalogenations, desulfonations, deoxygenations

Additions of H to unsaturated C-C bonds, acetylenes, acetylides, ketenes, carotene etc., diphenylpolyenes, conjugated, cumulative bond systems

Organic dyes (redox)

Organo-metallics As, Hg, diphenyl beryllium + $TiCl_4$, intermetallics

Oxygen cycling complexes, salcomine, oxyhemoglobin, chlorophyll

Sugars, hexahydric alcohols

Carbonyl adducts of persulfides, polysulfides

Bacteria, enzymes + nutrients, nitrates, sulfates (desulfovitrio desulfuricans) carbonates + catalysts

Charge-transfer complexes (as for organo metallics, intermetallics, etc.)

Silicone analogs and hybrids of reducible and unsaturated organic compounds

Oxygenated, chlorinated, nitrated graphite, rubbers, polymers, carbon suboxides

Polymeric oxidants, polymolecular compounds, perylene, etc.

DESCRIPTION OF SEVERAL CELL SYSTEMS USING ORGANIC ELECTROPHORES

Engineering and technological nomenclature generally refers to a functionally integrated, constructionally completed galvanic cell as a cell "system." Constructional designs and cell performance represent two of the more critical concerns of the battery technologist.

CELL CONSTRUCTIONS

No single cell construction design appears uniquely best for all types of depolarizers and various designs have been described and used in various patents. Two basic types, distinguished as the round and flat cell constructions, are schematically illustrated in Fig. 4.2.

The round cell type often used with both stable and reserve action depolarizers is shown in Fig. 4.2a. A graphite, titanium or other compatible metal current collector rod (1) with cap, (2) is inserted into the dry or electrolyte-wetted depolarizer-carbon mix, (3) forming a particulate cathode called a "bobbin." The bobbin is prevented from making electronic contact (shorting) with the anodic zinc or magnesium can (4) by the intervening paper or polymeric gel separator (5). Asphalt or polymer sealant (6) is poured on top of the paper or polymer washer (7) above the mix bobbin. A number of the solid and liquid nitro compound depolarizers can be utilized in such or similar constructions and designs.

TABLE 4.3. ANODE ASSOCIATED ELECTROPHORE TYPES AND PROCESSES

I Phenols and Related Types	II Substituted Ammonia and Related types	III Alcohols and Related Types	IV Carbohydrates and Related Types
Mono and Polyhydroxy Compds.	Hydrazines	Alcohols	Simple Sugars
Aminophenols	Hydrazine derivatives	Glycols	Glycosides Sucrose Maltose, etc.
Quinones	Hydroxylamine and derivatives	Alkanethiols R—S—H	Starches
Thiophenols	Amines		
	Nitroso and other Nitro compd.		

210

V	VI	VII
Carbonyl and Related Types	Hydrocarbons	Miscellaneous Types and Processes

V

Carbonyl and Related Types

Aldehydes

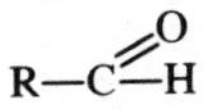

$$R-\overset{\overset{\displaystyle O}{\|}}{C}-H$$

Ketones

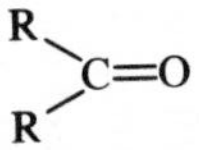

$$\overset{\displaystyle R}{\underset{\displaystyle R}{\diagdown}}C=O$$

Carboxylic Acids

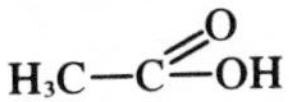

$$H_3C-\overset{\overset{\displaystyle O}{\|}}{C}-OH$$

Esters

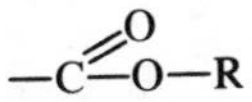

$$-\overset{\overset{\displaystyle O}{\|}}{C}-O-R$$

Ureas, Thioureas

$$H_2N-\overset{\overset{\displaystyle O}{\|}}{C}-NH_2$$

$$H_2N-\overset{\overset{\displaystyle \|}{\underset{\displaystyle S}{}}}{C}-NH_2$$

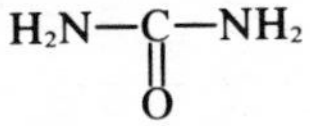

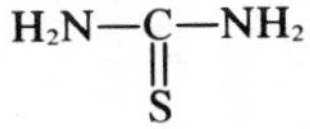

VI

Hydrocarbons

Aliphatics

CH_4, C_3H_8, etc.

Alicyclics

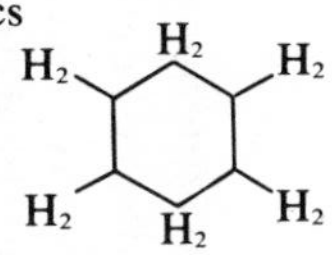

Aromatics

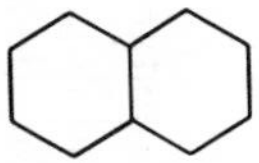

Unsaturated Drying Oils

VII

Miscellaneous Types and Processes

Substitution Processes
Oxygenations
Halogenations
Nitrations
Thiocyanations
Alkylations
Alkoxylations
Acetoxylations

Bacteria and Enzymes
Fuels
Carbohydrates
Fatty or amino acids
Alcohols
Ureas
NH_3, S, Fe^{++}, H_2, CO_2

Microprobes and Biocatalysts
Nitrosomonas, nitrobacter,
Thiobacillus thiooxidans
Ferroxidans
Glucose oxidase
Escherida Coli
Clostridium butyricum

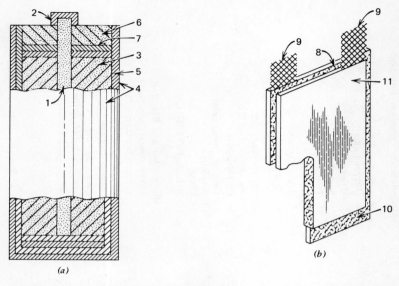

Figure 4.2 Two basic types of cell construction. (*a*) Round cell. (*b*) Flat cell.

For positive halogen types which are unstable to aqueous environmen and cause excessive metals corrosion and degradation of cathode materi als, reserve activation or deferred action designs have been frequently employed. The carbon-depolarizer mix can be assembled dry or in sealed water soluble film containers. Electrolyte or water can be added at the moment of activation. There are numerous other variations on this design theme.

The basic flat cell construction is schematically illustrated in Fig. 4.2*b* The cathodic depolarizer mix or paste (8) can be applied to one or both sides of a porous carbon or titanium grid or plate which serves as cathodic current collector (9). Paper, starch, or polymeric gel separator (10 isolates depolarizer paste from the sheet or powdered zinc anode (11).

In a "cathode envelope" version (73) of the flat cell shown as a cutaway illustration in Fig. 4.3, a paper covered sheet of zinc is coated on both sides with wetted depolarizer carbon mix and a sheet of carbon or metal coated polymeric film is folded around one end of the combination to serve as cathode current collector.

Various other combinations of porous metal or carbon rods, plates plaques, sinters, expanded grids, and so on, have been described for use with gaseous, liquid, and solid organic depolarizers in developmental cel constructions.

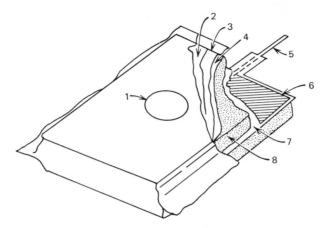

Figure 4.3   Cathode envelope flat cell. Cutaway of typical cathodic envelope cell(73). (1) positive contact with metal foil; (2) plastic film wrapper; (3) metal foil; (4) conductive film; (5) negative collector lead; (6) anode; (7) separator; (8) depolarizer mix cakes.

CELL PERFORMANCE CRITERIA

Cell performance usually refers to discharge service time, but in a more general sense, the cell can be viewed as a rate sensitive, imperfect electrochemical energy converter or transducer changing potentially available chemical free energy to joules of electrical energy ($E_{joules/coulomb} \times i_{coulombs/sec} \times t_{sec}$) in varying degrees of efficiency depending on the discharge rate and other conditions.

In this view involving voltage, current, and time as variables, characteristics of interest include overall cell and individual electrode limitations in discharge time to cutoff; coulometric and energy efficiencies; energy and current per unit weight, volume, and area; discharge rates or drains, voltage and electrode "losses" in the form of resistive and reactive impedances, (electrochemically designated as representing ohmic, activation, and concentration polarization); the environmental stability and a few others. Several of these above characteristics will be covered when describing specifically selected organic cell examples.

EXAMPLES OF CELL SYSTEMS USING ORGANIC CATHODES

As to the number and types of organic depolarizers that have received some degree of developmental or contract effort, no quantitatively reliable answer is known. However, a large number of depolarizer types listed in the previous organic literature classifications have been electrolytically prepared in the laboratory, fewer have achieved importance in commercial electrolytic synthesis, and perhaps still fewer yet hav qualified for use in galvanic cell development as recognized powe sources.

Metadinitrobenzene (m-DNB) in actual experimental cells has bee investigated from many different viewpoints by Morehouse an others(16). Figure 4.4 shows discharge curves of "A" size cells on a 16-ohm drain for cells using magnesium anodes and m-DNB cathod depolarizers compared with their Leclanché counterparts(72). It is clea from this data that the Mg/Mg(ClO₄)₂/m-DNB system provides a rela tively flat discharge curve that provides more hours of service than eithe of the other two systems. It should be pointed out that the workin voltage of the Mg/m-DNB cell is substantially lower than that of the othe units shown. In spite of this handicap, the Mg/m-DNB cell has provide up to 55 Whr/lb at the 25 hour rate, as compared with 45 for the equivalen Mg/MnO₂ cell. The higher output of the Mg/m-DNB system has promp ted a considerable amount of technical effort to explore the advantages o this system.

A number of the organic gases, liquids and solids in molten, nonaque ous, and aqueous electrolytes have had extensive study as anodes an cathodes in fuel cells. An impressive number of experimental primary an secondary cells using organic cathodes have been reported, and th numbers are increasing.

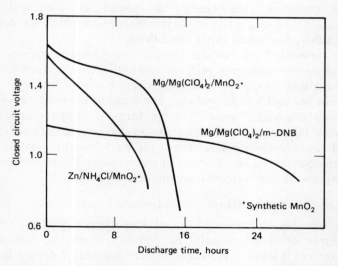

Figure 4.4 Comparison of MnO₂ versus m-DNB depolarizer in "A" cells using magnesium anodes and discharging continuously into 16-⅔ ohm loads.

However, while an appreciable number of anodic and cathodic developmental samples of primary, secondary, biochemical, and electrolytic cells have been described, space here permits no more than a brief consideration of only a few selected cell systems emphasizing organic cathodes. These cover the nitro, N-halogen, quinone and other cathodic types used in primary and secondary cells. The nominal reactions for some of these are shown in Table 4.4.

SPECIFIC ORGANIC CATHODE CONSIDERATIONS

Electrically, organic electrophores serve the function of electron sinks for the cathode current collector sources. Thus, electric charge transfer occurs and electrode products are formed.

Among the cathode studied, the nitro compounds in various construction designs have received the greatest amount of attention and will be considered in some detail, including some previously reported (18)(68) and some unreported experimental confirmations related to mechanisms and technology.

In this section, the various experimentally examined aspects of the behavior of m-DNB as an electrophore are discussed under the headings: theoretical coulometric capacities; illustrative capacity-cost index for nitro compounds; electrolyte pH influence in polarographic and developmental cells; typical aqueous primary cells; an experimental comparison of MnO_2 and m-DNB polarization characteristics; stepwise reduction intermediates isolated and identified after galvanic cell discharge; coulometrically indicative iz voltage drops associated with specific reduction intermediates; and hypothetical cell design considerations.

Theoretical Coulometric Capacities

Figure 4.5 illustrates the dependence of the theoretical (maximum saturation) capacity on the number of nitro groups per organic molecule. From a weight standpoint the advantages of the lower molecular weight alkyl types over the aryl nucleus types will be noted. Several indications of volumetric comparisons have also been included. The coulometric advantages of some of the organic compounds per unit weight over MnO_2 in acidic and alkaline electrolytes are also apparent. The catalytic nature of the cathode surface has long been known to influence reduction products, hence capacity.

From Fig. 4.5 it will be noted that the stability of the individual compounds varies with the degree of substitution in the molecule. Since stability, and safety are important considerations in battery application, it becomes apparent that the cathodic depolarizer choices might often be

TABLE 4.4. A COMPARISON OF CATHODE PROCESSES, USEFUL OPERATING POTENTIALS, AND THEORETICAL COULOMETRIC CAPACITIES

Cathode Process	Electrolyte pH and Current Drain	Operating Potential[a] volts vs. NHE	Theoretical Capacity Amp-Min/g
Manganese dioxide			
$MnO_2 + 4H^+ + 2e \rightarrow Mn^{++} + 2H_2O$	(low pH)	+1.15	37.0
$2MnO_2 + 2H^+ + 2e \rightarrow 2MnOOH$	(intermediate pH)	+0.6	18.5
$3MnO_2 + 2H_2O + 4e \rightarrow Mn_3O_4 + H_2O$	(high pH)	+0.15	24.7
Nitro compounds			
$Ar\text{-}NO_2 + 4H_2O + 6e \rightarrow ArNH_2 + 6OH^-$		−0.3	78.4
Nitrobenzene	(max. light drain, low pH)	−0.15	115
m-dinitrobenzene	(heavy drain, neutral pH)	−0.15	76
Tetranitropropane		+0.85	196
C-Nitroso compounds			
$Ar\text{-}NO + 3H_2O + 4e \rightarrow RNH_2 + 4OH^-$		0.0	52.2
p-nitrosophenol			
N-Halogen compounds			
$RNHX + H_2O + 2e \rightarrow RNH_2 + OH^- + X^-$		+0.93	32.6
N,N'-Dichlorodimethylhydantoin			
Pyridinium tetrachloroiodate		+1.0	27.7
Quinone compounds			

$$O=\!\!\bigcirc\!\!=O \;+\; 2H_2O \;+\; 2e \;\rightleftharpoons\; HO-\!\!\bigcirc\!\!-OH \;+\; 2OH^-$$

p-quinone		+0.10	29.8
1,4-Naphthoquinone		−0.15	20.4

[a]These refer to the closed circuit conditions and are not the open circuit nor thermodynamic potentials.

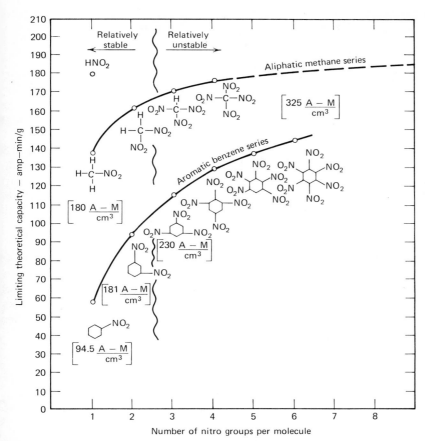

Figure 4.5 Theoretical coulometric capacity ranges for some organic nitro compounds. The nominal theoretical coulometric capacity of some inorganic oxide cathode depolarizers used in commercial galvanic cells and batteries which can be compared to the above include:

	Depolarizer		Electrons/mole	Amp min/g
1.	MnO_2	(a) In strongly acidic electrolytes	2	37.50
		(b) In strongly alkaline electrolytes	$\frac{4}{3}$	24.67
		(c) In MnO_2/Zn batteries with Leclanché or $ZnCl_2$ electrolyte (pH 3-7)	1	18.50
2.	Ag_2O		1	13.91
3.	HgO		2	14.88
4.	$NiOOH$		1	17.52

compromised between stability and highest ratio of functional nitro
groups per molecules.

Illustrative Capacity-Cost Indices for Nitro Compounds

Among other practical considerations, the relative coulometric capacity
per unit cost of nitro compounds is always important and can vary
profoundly with economic conditions. Table 4.5 shows a listing of
theoretical capacities according to the July 1971 costs along with a rough
index based on MnO_2 unity.

It is clear that many of the nitro compounds listed in Table 4.5
reflecting a range of stability and water solubility, are more costly than the
natural MnO_2 ore that is included for comparison. Even if compared with
the more expensive synthetic MnO_2 used widely today, several of the
nitro compounds are still more expensive.

Electrolyte pH Influence in Polarographic and Developmental Cells

Figure 4.6 illustrates the polarographic reduction of aqueous solutions of
m-DNB in essentially two 4-electron stages for acidic to neutral electrolytes as might be representative for subsequently intended primary cell
operation(68). Supplemental electron spin resonance data(71) suggest
that each polarographic 4-electron stage comprises three or four intermediate individual steps such as:

Potentiostatic reductions of the same aqueous m-DNB solutions at
stationary cathode graphite (cathode current collectors) confirmed the
coulometrically easily achievable two 4-electron stages and permitted a
more convenient identification of corresponding products. In general, the
acidic supporting electrolytes favored polarographic and potentiostatic
reductions to reasonable well defined products such as nitrophenyl
hydroxylamine.

TABLE 4.5. ILLUSTRATIVE CAPACITY COST INDEX OF SOME COMMON, COMMERCIALLY AVAILABLE POTENTIAL NITRO COMPOUND CATHODE DEPOLARIZERS

Nitro Compound	Mol. Wt.	Capacity[a] Amp-Min/g	O.P.D.[b] Price Dollars/lb	Amp-Hr/ Dollar	Relative Index
Nitrobenzene	123.11	78.37	0.085	6978	3.59
1-Nitropropane	89.09	108.28	0.16	5123	2.63
2-Nitropropane	89.09	108.29	0.35	5103	2.63
2,2-Dinitrochlorobenzene	202.56	95.26	0.1625	4436	2.28
o-Nitrotoluene	137.13	70.35	0.14	3903	1.96
2,4-Dinitrotoluene	182.13	105.94	0.220	3564	1.83
Nitromethane	61.04	158.06	0.81	3559	1.83
o-Nitrochlorobenzene	157.56	61.23	0.15	3090	1.59
Nitroethane	75.07	128.52	0.35	2779	1.43
p-Nitrotoluene	137.13	70.35	0.20	2662	1.37
m-Dinitrobenzene	168.11	114.78	0.86	2413	1.24
p-Nitrochlorobenzene	157.56	61.23	0.20	2317	1.19
Manganese dioxide (African Ore)	86.93	18.50	0.072	1944	1.00
2,4-Dinitrophenol	184.11	104.80	0.45	1763	0.91
o-Nitrophenol	139.11	69.35	0.43	1221	0.63
Picric acid	229.11	126.23	0.85	1125	0.58
p-Nitroaniline	138.11	69.35	0.48	1101	0.57
2,4-Dinitroaniline	183.12	105.37	0.90	887	0.46
p-Nitrobenzoic acid	167.12	57.73	0.50	874	0.45
m-Nitroaniline	138.11	69.85	1.33	398	0.20

[a]This depends on conditions and is not necessarily fixed.
[b]Data from Oil, Paint and Drug Report.

219

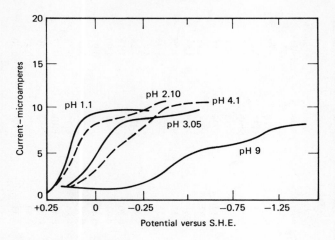

Figure 4.6   Polarograms illustrating pH-related differences in cathodic reduction processes of aqueous *m*-DNB solutions (68).

Typical aqueous electrolyte primary cells

Usually typical galvanic primary cells using aqueous electrolyte are discharged at constant resistance which precludes the simpler results and interpretations applicable to cathodic reduction at either constant potential or pH. Generally, the depolarizer potential shifts to more anodic values and the catholyte pH increases causing a lowering of potential values and resulting in mixtures of products rather than single selective species.

This "average" value of coulometric output to a given cell cutoff voltage assumes more practical significance than the ultimate reduction products and capacities projected from electro-organic syntheses. As a preliminary indication preceding actual prototype developmental cell tests, Fig. 4.7 shows cathodic potentials of discharging experimental $Mg/MgBr_2/m$-DNB "beaker" cells under "flooded" electrolyte conditions, as well as a reasonably close comparison of the cell's cathode potential with concurrently completed polarograms illustrating effects of electrolyte on the cathodic reduction potentials shown in Fig. 4.6.

As noted for the 1 M $MgBr_2$ electrolyte example, perhaps more closely approximating the developmental cell environment, the readily attained 8 electron capacity coincides with the corresponding polarographic 8 electron slope capacities experimentally observed. However an overall

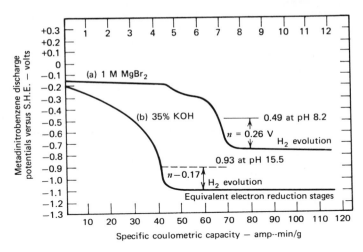

Figure 4.7 The influence of electrolyte pH on cathode discharge of m-DNB in Mg/MgBr$_2$/m-DNB cells. Cell and test features:
Cathode mix: 0.250g m-DNB
               0.250g acet. black
Electrolytes: 300 ml used
        (a) 1M MgBr$_2$ (pH 7.2–8.0)
        (b) 35% KOH
Discharge current: 100 ma cont.
Reference electrode via KCl bridges versus SCE at 25°C.

cell value of about 9 electrons capacity to H$_2$ evolution can be approached even with the relatively high constant current drain used.

For the more alkaline 35% KOH electrolyte example, less definition and greater obscuration of cathode potentials and less capacity was experienced which was attributed to the alkaline condensation step following the 4 electron reduction to the pH sensitive phenylhydroxylamine intermediates.

The complications associated with the isolation and analyses of practical cell intermediates here prevented the extraction of significant amounts of the diphenylhydroxylamine compounds which were more easily recovered in the corresponding polarographic cell test under nitrogen atmosphere. Concurrent oxidation of the more reduced intermediates by O$_2$ was noted where the practical cells were not hermetically sealed or discharged under inert atmospheres. This, of course, not only complicates the cell product and process analyses and interpretations, but also can lead to subsequently increased discharge service from the air-oxidized species.

An Experimental Comparison of MnO₂ and m-DNB Polarization Characteristics

The characteristic capability of the m-DNB to retain higher discharge potentials at higher current densities is experimentally illustrated in Fig 4.8. At the lighter current drains the MnO₂ discharge potential was more cathodic than m-DNB. The reverse was true for the higher drains.

Perhaps the relative insolubility of MnO₂ in aqueous electrolytes compared to m-DNB provides a partial explanation. The preceding polaro graphic data do show conventional behavior in that the magnitude of cathodic reduction current for the more soluble m-DNB depends on its

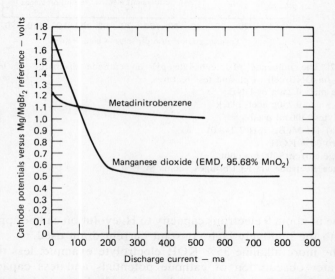

Figure 4.8 Experimental cathode polarization characteristics comparing metadinitroben zene and manganese dioxide. Cell and test features:
Depolarizer mixes
  1. m-DNB 0.20g + 0.10g acet. blk.
  2. MnO₂-electrolyte 0.20g + 0.10g acet. black
Mix Cross-sectional area
  a = 16.23 cm²
Electrolyte 350 ml, 1M MgBr₂
  pH 8.2
Reference electrode
  Mg in gelled electrolyte
  E = −1.20 volts vs. SCE
    ref
EMD-Electrolytic Manganese Dioxide

oncentration. On the unfavorable side of the ledger, however, this olubility is expected to, and does, result in wasteful reduction at the zinc r magnesium anodes during shelf storage. Highly colored intermediates n the magnesium surface are a common sight, particularly where nsufficient chromate inhibitors are used.

Stepwise Reduction of Intermediates Isolated and Identified after Galvanic Cell Discharge

Reference has already been made to the attempted analysis and identification of electrode products from experimental wet cells and from dry cell prototypes.

From the discharge products isolated and identified, it is inferred that there is a tendency for the initially possible alkaline condensations to be minimized and that reduction progresses via intermediates to eventually reach nine electrons per mol.

Very analogous to Haber's [3] stepwise chemical reduction of nitrobenene, Fig. 4.9 shows the complex, stepwise reduction of m-DNB in the mildly acidic to alkaline environments used in practical primary cells. As noted, a complex variety of paths were found depending on discharge rate and other conditions. The discharge capacities and the distribution of products formed were highly dependent on current drain, initial and progressive pH, operating potential, catalytic nature of the metal current collectors, and the presence or absence of oxygen.

The one predominant path in development, noted for the constant current tests, was a stepwise reduction to m-aminonitrosobenzene and m-aminophenythyroxamine which condenses to form 3, 3' diaminooxybenzene.

A number of typical products either stable enough to be isolated or those identified by spot tests as a transient intermediate specie are shown in Fig. 4.9.

While a mixture rather than pure isolated products was generally found in the analyses of cells where cathodic potentials progressively changed during discharge, the similarity and analogy to Haber's electrolytic cell nitrobenzene results is reasonably good. At lighter current drains and more acidic electrolyte reductions were more complete and the overall cell discharge efficiencies to useful cutoffs were higher approaching the theoretical maximum saturation limit of 12 electrons for the m-phenylene diamine product.

Reduction in the lower pH aqueous electrolytes generally were found to follow the lower end path to yield m-nitrophenylhydroxylamine, phenylamine, dihydroxyazo, and m-dihydroxylamine benzene, m-phenylendiamine (1, 3-diaminobenzene).

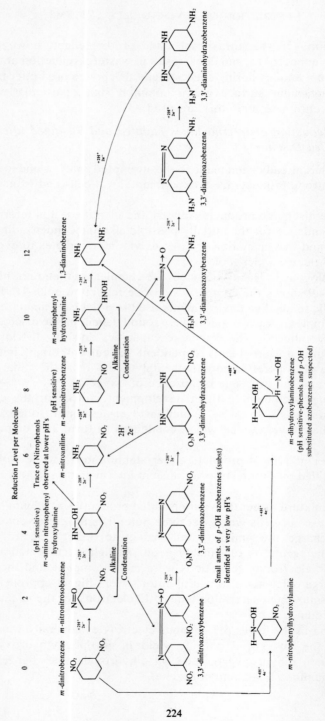

Figure 4.9 Experimentally indicated reduction paths and cathodic intermediates isolated or identified from practical cell discharges.

As illustrated in the following section and in Fig. 4.10, cathode service and coulometric limitations were often determined by the formation of a more stable and generally less soluble intermediate—usually accompanied by an increase in cell impedance and a resulting drop in operating voltage.

Subsequent reductions would continue at lower potentials and current drains since a fixed load resistor was usually used. Anodes with higher driving potentials could increase the extent of reduction and, of course, the service to a given cutoff voltage.

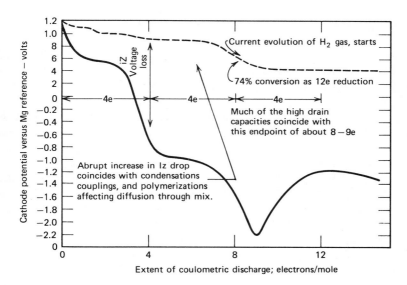

Figure 4.10 Constant current discharge behavior of m-DNB in experimental test cell illustrating the coincidence between reduction intermediate products and their coulometric boundaries. Cell and test features:

m-DNB, 0.20g; acet. blk. 0.10g
Electrolyte 350 ml 1M $MgBr_2$, pH = 7.7 initially
Apparent mix CD = 1500 ma/g
    Cd = 185 ma/cm^2 By Area
    - - - open circuit voltage;
    —— closed circuit voltage.

The sequential changes in the impedance—reflective dE/dt slopes at coulometric boundary values of 4, 8 and 12 electrons in the cathodic discharge curve above- -indicate kinetic or rate limitations experienced during progressive cathodic discharge. The predominant intermediate products isolated corresponded with the extent of coulometric reductions indicated. The E^{oc}_{cell} (10 msec) curve useful for kinetic and impedance characterizations, shows potentials measured during each 10 msec interruption of cell discharge current which was done in about 5-minute intervals.

*Coulometrically Indicative iz Voltage Drops Associated with Specific
Reduction Intermediates*

For the medium and higher drains an average eight to nine electron
capacity limit to the same cutoff voltage is representative, see Fig. 4.10.
Dividing up the total 12 electron reductions into three stages of four
electron each for product analyses, the first stage contained significant amounts of *m*-DNB, *m*-nitro phenylthydroxylamine, 3,3'-
dinitroazoxbenzene and the corresponding azo compound. The dinit-
roazoxy, azo, and hydrazo compounds along with *m*-nitroanilines were
found present as cathodic products of the second 4-electron reduction
stage. For the last four electron reduction step *m*-nitroaniline, *m*-
aminohydroxylamine, 3,3-diaminoazoxybenzene, some diamino, azo, and
hydrazo compounds, some *m*-phenylene diamine, plus some other un-
identified oxidation and coupling products were found. The 3, 3' diamino-
azoxybenzene predominated in the neutral to alkaline electrolyte
environments.

Figure 4.11 shows typical averaging discharge data for a considerable
number of developmental prototype cells using magnesium anodes and
depolarizer mixes containing *m*-DNB. A fixed load resistor rather than a
constant current was used here. Most of the cells for this curve employed
the cathodic envelope construction in order to maximize cross sectional
areas perpendicular to the mass and charge fluxes. The dependence of the
practical cell service and coulometric capacity to a useful 0.9 V endpoint
for a given load resistor (or discharge drain range) should be noted. An
almost linear curve is observed for coulometric capacities up to about
eight or nine electron limit corresponding to an "average product" of
diaminoazoxy compound. For lower cutoff voltage endpoints and at
lighter drains occasional higher capacities were achieved. Some of these
were traced to reduction of intermediates which were partially oxidized
by oxygen ingress into poorly sealed cells. Analytical results for isolated
product mixtures were approximately equivalent to the preceding analytical results on experimental test cells discharged at constant current. Thus,
conclusions reached from an integrated experimental study involving
polarographic, potentiostatic, galvanostatic, coulometric, and constant
load resistor discharges with both experimental half cells and complete
cell systems confirm that mixtures of reduction products rather than pure
products are the rule particularly for practical cells discharging at higher
drains and more drastically changing cathode potentials and electrolyte
pH values. For constant load discharges (fixed resistor) where cathode
potential and current progressively decrease, the coulometric capacity

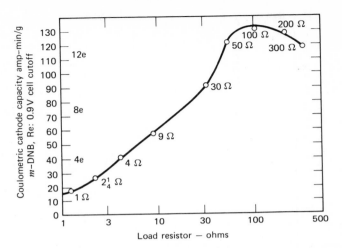

Figure 4.11 The relative effects of current or load resistor magnitude on m-DNB coulometric capacity for development $Mg/MgBr_2/m$-DNB primary cells discharged to 0.9 V cutoffs. Cell and test features:
$Mg/MgBr_2/m$-DNB development cell system
(continuous discharge)
$T = 25°C$

As indicated, decreasing the current drain (discharge rate) by using higher values of external load resistors with this cathode limiting developmental prototype cell yields higher coulometric capacities and efficiencies to the selected 0.9 V service endpoint. Discharge with resistors higher than about 25 ohms and discharge to between 8–9 electrons shows a departure from the linearity observed for the correspondingly higher current drains and lower resistor values. The data are taken from a variety of sources.

conforms to an average, of several discharge products, identified from isolation and analysis following cell discharge.

Hypothetical Cell Design Considerations

Finally, Fig. 4.12 illustrates some hypothetical stoichiometric design considerations for an overall $Mg/MgBr_2/m$-DNB cell system based on neutral to alkaline electrolyte and on an anticipated high drain limitation of nine electrons. For illustration of materials balance estimates of magnesium corrosion and physical wetting requirements are included.

ORGANIC ANODE CONSIDERATIONS

Complementing the cathode function, the organic anode acts as an electron source and the organic matter is oxidized at the electrode surface. However, degradation, formation of free radicals undergoing

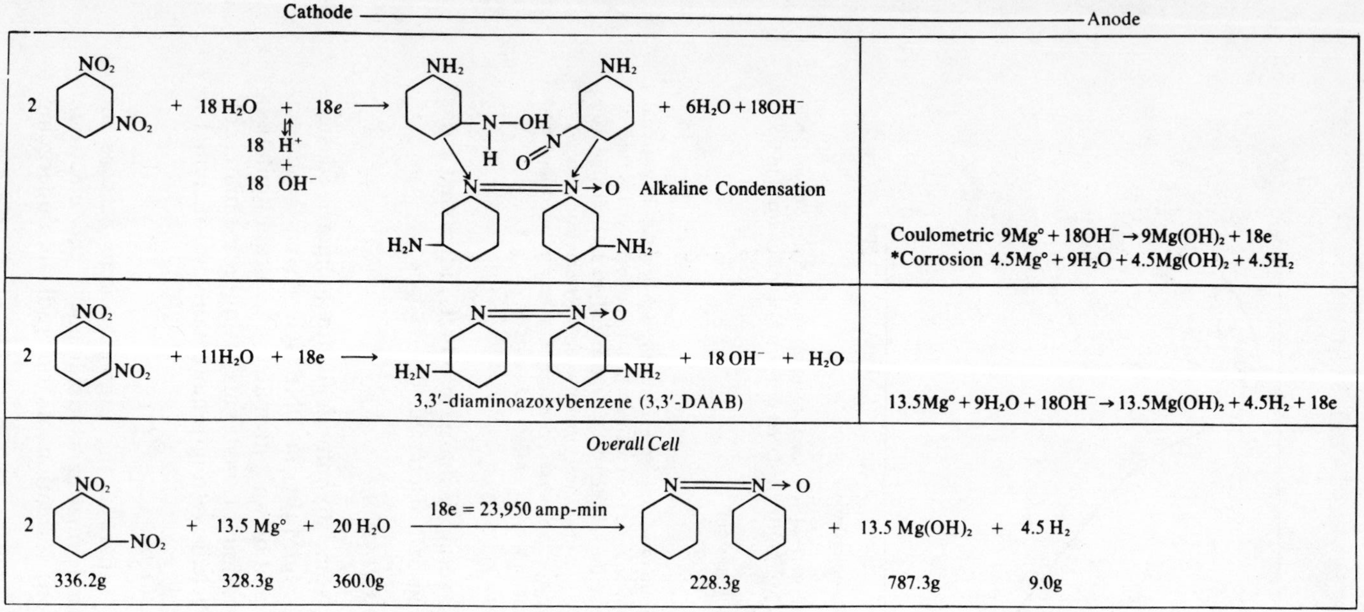

Cathode
Anode
18 H₂O
18e
18 H⁺
18 OH⁻
6H₂O + 18OH⁻
Alkaline Condensation
Coulometric 9Mg° + 18OH⁻ → 9Mg(OH)₂ + 18e
*Corrosion 4.5Mg° + 9H₂O + 4.5Mg(OH)₂ + 4.5H₂
11H₂O + 18e
18 OH⁻ + H₂O
3,3′-diaminoazoxybenzene (3,3′-DAAB)
13.5Mg° + 9H₂O + 18OH⁻ → 13.5Mg(OH)₂ + 4.5H₂ + 18e
Overall Cell
13.5 Mg° + 20 H₂O
18e = 23,950 amp-min
13.5 Mg(OH)₂ + 4.5 H₂
336.2g
328.3g
360.0g
228.3g
787.3g
9.0g

Figure 4.12 Hypothetical materials balance in the design of a $Mg/MgBr_2/m$-DNB cell.

Requirements for a 1/1 carbon/*m*-DNB Ratio Mix per Gram of *m*-DNB

Initial Mix

		Weight	%	
m-DNB		1.00g	12.00	
C, Acet. Blk.		1.00	12.00	
H_2O Coulometric	1.07			
To wet C, etc.	4.00**			
Total	5.07	5.07	60.78	76% of the mix is a
$MgBr_2 \cdot 6H_2O$	1.27	1.27	15.22	20% solution of
Totals		8.34g	100.00	$MgBr_2 \cdot 6H_2O$
Mg Anode		0.976g		

Final Products

		Weight	%
3,3′-DAAB		0.68g	5.52
C, Acet. Blk.		1.00	8.10
H_2O To wet C	4.00**		
To wet $Mg(OH)_2$	3.05***		
Total	7.05	7.05	57.13
$Mg(OH)_2$		2.34	18.96
$MgBr_2 \cdot 6H_2O$		1.27g	10.29
H_2 Gas 300 cm^3/g *m*-DNB at STP			
Totals		12.34g	100.00

subsequent chemical reaction, rupture of rings and various bonds, are all quite common in the practical anode experience, in fact, more so than for the cathode. Factors of interest here are similar to those listed for the cathode.

Numerous examples are known and discussed in other parts of this book concerning organic anodes used in fuel cells, and in some secondary or rechargeable cell systems.

Hydrocarbons, alcohols, substituted ammonia types, and of course hydrogen, represent typical examples of the classes of compounds finding use as anodes. One methanol-air cell(54) is used in signalling devices operated at about 91% efficiency based on the weight of methanol anodically converted to carbonate in a typical dehydrogenation step.

$$q_s = 18,071 \text{ coulombs/g methanol}$$
$$q_s = 301 \text{ amp-min/g}$$
$$q_s = 5.02 \text{ amp-hr/g}$$

In contrast with the situation in fuel cells, for many of the primary organic cells, metal anodes are still commonly featured and several prototype examples are discussed in this chapter.

RESERVE CELL WITH N-HALOGEN CATHODE

Discharge characteristics for several experimental reserve-activated cell systems using N-halogen and polyhalogen organic electrodes are shown in Fig. 4.13. Flat plate cell constructions and either aqueous 25 percent NH_4Cl or eutectic composition $CaCl_2$ electrolytes were used with the areas of anode and cathode being approximately 10.2 in.^{2}.

Since such materials are unstable when wet, they must be assembled in dry form and "activated" by introducing electrolyte only at time of discharge. The very high cathodic potentials of halogens are characteristic for the active halogen organic depolarizers.

For coulometric capacity estimates, whether one presumes the process

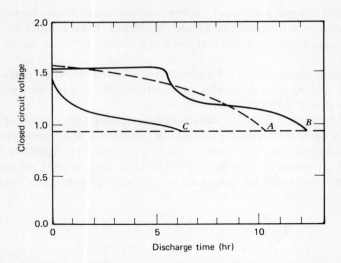

Figure 4.13 Comparison of Zn/N-halogen and polyhalogen depolarizer cells subjected to continuous 200 ma discharge. A = N-N′-dichlorodimethylhydantoin with 46.0% utilization to 0.9 V cutoff. B = Tetramethylammonium tetrachloroiodide with 78.9% utilization to 0.9 V. C = Electrolytic manganese dioxide with 17.8% utilization to 0.9 V.

order to be first the hydrolysis of the active halogen compounds to form inorganic halide species which is then followed by electrochemical charge transfer processes; or direct electrochemical mechanisms, the net stoichiometric coulometric capacities for the organic depolarizers shown in Fig. 4.13 might be represented as follows:

N-N′-dichlorodimethylhydantoin

$$
\begin{array}{c}
H_3C \quad Cl \\
| \qquad | \\
CH_3-C-N \\
\quad \quad \quad \backslash \\
\qquad \qquad C=O + 4H^+ + 4e^- \\
O=C-N \\
\quad \quad | \\
\quad \quad H
\end{array}
\quad 6432 \text{ amp-min}
\longrightarrow
\begin{array}{c}
H_3C \quad H \\
| \qquad | \\
CH_3-C-N \\
\quad \quad \quad \backslash \\
\qquad \qquad C=O \\
O=C-N \\
\quad \quad | \\
\quad \quad Cl
\end{array}
+ \ 2HCl
$$

MW 197 g/mol   Specific capacity $= \dfrac{6432}{197} = 32.65$ amp-min/g

Tetramethylammonium tetrachloroiodide

$$6432 \text{ amp-min}$$

$$(CH_3)_4NICl_4 + 4H^+ + 4e \rightarrow \underbrace{(CH_3)_3N + CH_3Cl}_{} + HI + 3HCl$$

$$MW = 343 \text{ g/mol} \qquad \downarrow$$
$$(CH_3)_4{}^+Cl^-$$

$$\text{Spec. cap.} = \frac{6432}{343} = 18.8 \text{ amp-min/g}$$

Thus, for the two above examples, the theoretical capacities for these cells becomes 32.65 and 18.8 amp-min, respectively. Of course, capacities for discharges to specified cell cutoff voltages usually fall short of the maximum theoretical capacities but can be expressed as a percentage of theoretical capacity and this should be related to the distribution and types of discharge intermediates formed.

As an illustrative example of cell assembly, 10.0 g of tetramethylammonium tetrachloroiodide were mixed with 8.0 g of acetylene or halo-black[2] and then stirred with sufficient 24 percent aqueous NH_4Cl solution to form a coherent mix. This cathodic depolarizer mix was then applied to both sides of $\frac{3}{16}$ in. thick porous carbon plate serving as a cathode current collector. Two amalgamated zinc powder anode plates were positioned on either side of the cathode mix assembly with a single sheet of Dynel separator intervening between each cathode and anode face. About 80 ml of the NH_4Cl electrolyte were added to the plastic case container holding the electrode assemblies.

SECONDARY AND HIGH DRAIN CELLS WITH ORGANIC CATHODES

Tetramethylammonium Tetrachloroiodide

High drain capabilities and secondary cell characteristics are attainable with certain classes of organic electrodes as illustrated in Fig. 4.14.

Weights and formulations used in cells of this type were essentially identical with those given for the previous example except that 19 g of acetylene black were used and the mix was compressed. These cells were cycled on charge and discharge at a 5 amp rate utilizing only about 37.2 percent of the theoretical 18.8 amp-min/g to maintain a desired higher operating voltage level.

[2]A commercially obtained product from Binney and Smith Co., 380 Madison Ave., New York 10017.

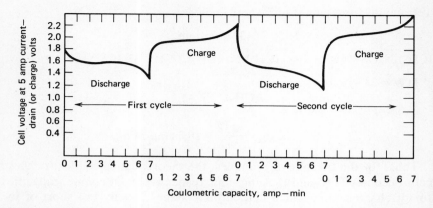

Figure 4.14 High drain secondary cell using a polyhalide organic depolarizer, $(CH_3)_4NICl_4$, cathode. Cycling at 5 amp current.

TETRACHLOROQUINONE (CHLORANIL) CATHODE

Another secondary type rechargeable organic cell system is represented by one example in Fig. 4.15 using tetrachloroquinone (chloranil) mix as the depolarizer and sheet zinc as anodes. Chloranil is an example of the quinonoid class of cathodic-anodic depolarizers and the more stable ring substituted chlorine atoms are not normally reduced.

Figure 4.15 shows that the Chloranil/$ZnCl_2$-NH_4Cl-H_2O/Zn cell had an initial working voltage of about 1.5 V, substantially higher than that shown by m-DNB in previous examples. At the drains used and for electrolyte pH's shown, this compound apparently was reduced in two steps as shown by the discharge curve. The pH of the electrolyte used in this example was 4.3. The cell was charged as shown, apparently at relatively good efficiency.

Alt and co-workers(50) have recently reported on the use of quinones as cathodic depolarizers in a low pH electrolyte of 3N H_2SO_4. They used chloranil successfully for a small rechargeable cell and report 50 cycles were obtained. They also emphasize that this cathode can support very heavy currents, up to 100 ma which corresponds to a discharge in 1.10 hour. Under such conditions the curve is not flat as it is at lighter drains. It is interesting to note that their tests indicated reasonably flat discharge curves even though the testing temperature was reduced to $-10°C$, although the total output dropped considerably from the level at room temperature.

The two-step discharge curve shown in Fig. 4.15 differs from the flat discharge curve given by Alt et al.(50). One suspected reason for the

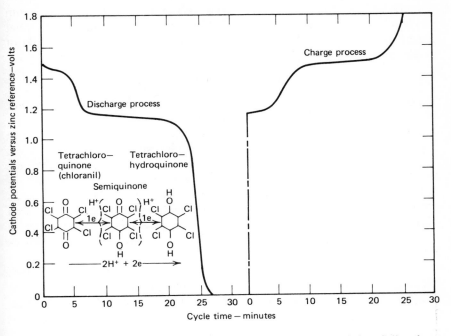

Figure 4.15 Tetrachloroquinone/Zn secondary cell cycling characteristics. Cell and test features:

Developmental cell using approximately 4 in. × 3½ in. plates of cathode mix and zinc separated by Dynel sheets
Cathode mix
   Tetrachloroquinone   0.20g
   Acet. Blk.   0.10g

Current = 0.100 amp continuous
   CD wt = 500 ma/g
   CD app area = 6.25 ma/cm²
Anode = sheet zinc 0.016 in. thick

Electrolyte: 28% $ZnCl_2$, 23% NH_4Cl,
             99% H_2O
   350 ml pH = 4.3
Potential of Zn ref = 0.82 V versus SHE

$$t_{cell}^{100\%} = \frac{Q}{i} = 13.08 \frac{AM}{g} \times \frac{0.20g}{0.10 \, amp} = 26.2 \, min$$

233

difference in behavior of the chloranil electrode is the difference in pH of the electrolytes. The two-step curve was obtained in an ammonium chloride-zinc chloride aqueous solution of pH 4.3 whereas the flat discharge curve was obtained in an electrolyte of 3N H_2SO_4 with a pH of about zero.

Poly-Benzofurazan Oxides and Azodicarbonamides

Work at American Cyanamide has lead to the development of newer primary and secondary organic cathode materials (30–35).

The poly benzofurazan oxides perform very well as primary cell depolarizer materials (30). Their small but finite solubility severely limits the shelf life of a battery, and its low drain rate applications. Some substituted polybenzofurazans, in particular, 5, 5' sulfonylbisbenzofurazan, 3, 3' dioxide (BBFOS), are water insoluble yet react as cathodic materials in dry cells (34). When coupled with a high surface area carbon black they yield a satisfactory cathode mix. Compared with MnO_2, BBFOS in $MnCl_2$ electrolyte gives long service although at a lower voltage in a similar cell construction as shown in Fig. 4.16.

Davis and Kraebel reported that azodicarbonamides perform as active rechargeable cathode materials in secondary batteries (30), as shown in Fig. 4.17. Substitution stabilizes the compound against hydrolysis for high

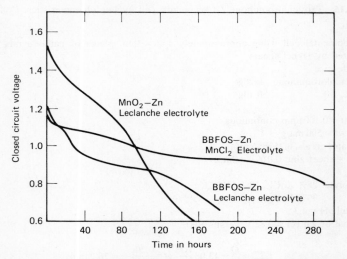

Figure 4.16 Discharge curves through 150 ohms of AA size cells comparing standard MnO_2 cathode cells with BBFOS organic depolarizer in two electrolytes (30a).

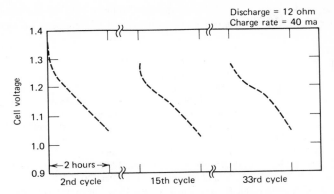

Figure 4.17   Discharge curves of rechargeable AA-size biurea-Zn cell after a 5-hour charge. The cell had a laminated cellophane separator (32a). Cathode composition was: Biurea 2.0g; Zinc oxide 0.6g; Carbon black 0.8g; Electrolyte (35% $ZnCl_2$, 20% NH_4Cl, 0.3% $HgCl_2$) 2.5 ml.

temperature storage in the charged state and also reduces the solubility in battery electrolyte. The m-butyl substitution appears to be the most satisfactory. Rechargeable cells can start with the discharged biurea product. Biurea is very stable on storage so that prolonged storage in the discharged condition is possible.

Some additional pertinent references and comments on practical cell systems using organic materials are reported by Murphy (72).

REFERENCES

1. Kolbe, B., *Annalen* **69**, 279 (1849).
2. *Gmelin's Handbook*, English translation, Vol. I (1848), Vol. III (1953), and in Moser, H., *Die Electroytischen Progresse der Organische Chemie* 1–13, Leipzig (1910).
3. Haber, F., *Z Elektrochem.*, **4**, 506 (1898); Haber, F. and Schmidt, C., *Z. Physik Chem.*, **32**, 271 (1900); Haber, F. and Russ, T., *Z. physik, Chem.*, **47**, 257 (1904); Haber, F., *Z. physik, Chem.*, **32**, 193 (1900); **32**, 271 (1900); *Z. angew. Chem.* 433 (1900); *Z. Elektrochem.* **7**, 304 (1900).
4. S. Swann, Jr., *Technique of Organic Chemistry*, Vol. II, 2nd ed., Ed. A. Weissberger, Interscience, New York (1956).
5. Fieser, L. F., *J. Am. Chem. Soc.*, **51**, 3101 (1929); *Organic Chemistry*, D.C. Heath, Boston (1944).
6. Brockman, C. J., *Electro-Organic Chemistry*, Wiley, New York, 1926.
7. Nicholson, I. E., *Nicholson's Journal*, **5**, 2 (1802).
8. Grove, W. R., *Ann. Chim. Phys.*, **58**, 202 (1893).
9. Gaugain: cf. Tommasi, D., *Traite des Piles Electriques*, Paris (1889).

10. Maiche: cf. Bottone, S. R., *Galvanic Batteries*, Whittaker, London (1902).
11. Braeuer, K. H. M., Intersociety Energy Conversion Engineering Conference, p. 525 (1969).
12. Bauer, W. C., U.S. Pat. 1,134,093 (April 6, 1915).
13. Arsem, W. C., U.S. Pat. 2,306,927 (Dec. 29, 1942).
14. Bloch, R., U.S. Pat. 2,566,114 (August 28, 1951).
15. Sargent, D. E., U.S. Pat. 2,554,447 (May 22, 1951).
16. Morehouse, C. K. and Glicksman, R., U.S. Pats. 2,836,644 (May 27, 1958); 2,836,645 (May 27, 1958); 2,855,452 (October 7, 1958); 2,880,122 (March 31, 1959); also Morehouse, C. K. and Glicksman, R., *J. Electrochem. Soc.* **105**: 299: **105**: 306 (1958); **108**: 303 (1961); **106**: 457 (1959); also Lozier, G., U.S. Pat. 3,060,155 (October, 1962); U.S. Naval Ord. Lab. NOLC Rept. 597 1. 8–17 (1964).
17. Morehouse, C. K. and Glicksman, R., U.S. Pat. 2,897,249 (July 28, 1959).
18. Dereska, J. S., Koehler, J. O., and Vinal, A. F., U.S. Pat. 3,057,760 (October 9, 1962).
19. Cohn, J. G. E., Ruetschi, P., and Chan, B. D., U.S. Pat. 3,080,444 (March 5, 1963).
20. Klopp, D. W., "The Use of *m*-DNB in Ammonia Batteries," U.S. Naval Ordance Laboratories NOLC-Rept. 559, Symposium on Ammonia Batteries, Berkely Calif. (January 23, 1962).
21. Davis, B., "Preliminary Experiments on a Microbial Fuel Cell," *Science* **137**: 615 (1962).
22. Elroy, W. D., *Cellular Physiology and Biochemistry*, Prentice Hall, Englewood Cliffs, N.J. (1961).
23. Fuel Cell Corporation, "Fuel Cell Progress, Status of Biocells." 4300 Goodfellow Blvd., St. Louis, Mo., the author (September, 1962).
24. Sullan, J. A. and Garrick, J. D., U.S. Bureau of Mines Information Circular 8003 (1961).
25. Sisler, F. D., "Electrical Energy from Biochemical Fuel Cells," *New Scientists* (Britain), **12**, 110–111 (October 12, 1961).
26. Rohrback, G. H., Scott, W. R., and Canfield, J. H., "Biochemical Fuel Cells, *Proc. 16th Annual Power Sources Conf.*, Atlantic City, (May 22–24, 1962).
27. Cohn, E. M., "Perspectives on Biochemical Electricity," NASA Office of Advanced Research and Technology, Washington, D.C., Paper presented before Society of Industrial Microbiology, Corvallis, Ore. (August 29, 1962).
28. Rohrback, G. H., U.S. Pat. 3,228,799 (January 11, 1966).
29. Pinkerton, R. D., U.S. Pat. 3,073,884 (January 15, 1963).
30. Davis, S. M., Kraebel, C. M., and Parent, R. A., U.S. Pat. 3,357,865 (December 12, 1967).
31. Davis, S. M., Kraebel, C. M., and Parent, R. A., U.S. Pat. 3,481,792 (December 2, 1969).
32. Kraebel, C. M., U.S. Pat. 3,594,231 (July 20, 1971).
33. Coleman, R. A., Parent, R. A., and Vorhies, J. D., U.S. Pat. 3,152,017 (October 6, 1964).
34. Hardy, W. B. and Parent, R. A., U.S. Pat. 3,163,561 (Dec. 29, 1964).
34a. Shaw, J. T., Vorhies, J. D., and Davis, S. M., U.S. Pat. 3,260,621 (July 12, 1966).
35. Davis, S. M., U.S. Pat. 3,438,813 (April 15, 1969).
36. Cassidy, H. G., Wegner, G., and Nakabayashi, N., U.S. Pat. 3,600,411 (August 17, 1971).
37. Foley, R. T. "Cells and Storage Batteries" Ed. J. Hladik *Physics of Electrolytes*, Vol. 2, (1972); Research on Electrochemical Energy Conversion Systems, R. T. Foley, D. H. Bomkamp, and C. D. Thompson DA-44-009-AMC-1386 (T) Interim Tech. Report May 1971, Nov. 1971 and Jan. 1972.
38. Baizer, Manuel M. Ed., *Organic Electrochemistry*, Marcel Dekker, New York (1973).

39. Parks, G. S. and Huffman, H. M., *Free Energies of Some Organic Compounds*, Chemical Catalog Co., New York (1932).
40. Bockris, J. O'M. and Conway, B. E., *Modern Aspects of Electrochemistry*, Plenum, New York (1971).
41. Prette, L. H., Ludwig, P., and Adams, R. N., *Anal. Chem.* **34**, 916 (1962); *J. Am. Chem. Soc.* **84**; 4112 (1963).
42. Clark, W. M., *Oxidation-Reduction Potentials of Organic Systems*, Williams and Wilkins, Baltimore (1960).
43. Latimer, W. M., *Oxidation Potentials*, Prentice-Hall, Englewood Cliffs, N.J. (1954).
44. Meites, L., "Controlled Potential Electrolysis," *Technique of Organic Chemistry*, Vol. I, Part IV, Ed A. Weissberger, Interscience, New York (1960), p. 3281–3335.
45. Kolthoff, I. M. and Lingane, J., *Polarography*, Vol. II, Interscience, New York (1962), p. 448.
46. March, J., *Advances in Organic Chemistry, Reactions, Mechanisms, and Structures*, McGraw Hill, New York (1961).
47. Hendrickson, J. B., Cram, D. J., and Hammond, G. S., *Organic Chemistry*, 3rd Ed., McGraw-Hill, New York (1970).
48. Fichter, F., *Organische Electrochemie*, Leipzig, Th Steinkopf (1942).
49. Rossman, J., *Trans. Electrochem. Soc.*, **84**, 121 (1943).
50. Alt, H., Binder, H., Kohling, A., and Sandstede, G., *J. Electrochem. Soc.* **118** 1950 (1971).
51. Elving, P. J., "Mechanisms of Organic Electrode Reactions," *Advances in Chemical Physics*, Vol. III, Interscience, New York (1961).
52. Allen, M. J., *Organic Electrode Processes*, Reinhold, New York (1958).
53. Kortum, G. and Bockris, J. O'M., *Textbook of Electrochemistry*, Elsevier, New York (1951).
54. Vielstich, W., *Batteries*, Ed D. H. Collins, Proc. 4th Int. Symp. Brighton (September, 1964), Pergamon Press, London (1965), p. 359.
55. Gutmann, F. and Lyons, L. E., *Organic Semiconductors*, John Wiley, New York (1967).
56. Maki, A. H. and Geske, D. H., *J. Chem. Phys.*, **33**, 825 (1960); **38**, 1999 (1963); **30**, 1356 (1959); *J. Am. Chem. Soc.*, **82**, 2671 (1966); **83**, 1852 (1961).
57. Conway, B. E., *Theory and Principles of Electrode Processes*, Ronald Press, New York (1965).
58. Ehlers, V. B. and Sease, J. W., *Anal. Chem.*, **31**, 16 (1959).
59. Schwabe, K., *Polarographie in Chemische Konstitution Organische Vergindung*, Akademie Verlag, Berlin (1947).
60. Pearson, J., *Trans. Faraday Soc.*, **44**, 683 (1943); **45**, 199 (1949).
61. Richter, G. H., *Textbook of Organic Chemistry*, Wiley, New York (1946).
62. Degering, F., *Organic Chemistry Outline Series*, 6th Ed., Barnes and Noble, New York (1966).
63. Groggins, P. H., *Unit Processes in Organic Synthesis*, Prentice-Hall, Englewood Cliffs N.J. (1944).
64. Adams, R. N., *Electrochemistry at Solid Electrodes*, Marcel Dekker, New York (1969).
65. Zuman, P., *Substituent Effects in Organic Polarography*, Plenum Press, New York (1967).
66. Shikata, M. and Tachi, I., *J. Chem. Soc. Japan* **53**, 834 (1934); *Czechoslav. Chem. Commun.*, **10**, 368 (1938).
67. *Oil, Paint, and Drug Reporter* **200**, No. 41 (*July 26*, 1971).

68. Jackson, G. W. and Dereska, J. S., *J. Electrochem. Soc.* **112**, 1218 (1965).
69. Endre, A. L. and Reilly, T. A., "Reduction Mechanisms of Aromatic Nitro Compounds," *Proc. 23rd Annual Power Sources Conf.* pp. 91–96 (1969).
70. Davis, S. M., Vorhies, J. D., and Schuroak, E. J., U.S. Pat. 3,468,708 (Sept. 23, 1969).
71. Haglind, J. B., *Adv. Chem. Ser.*, **78**, 24 (1948).
72. Murphy, J. J., *Chem. Tech.*, **1**, 487 (1971).
73. Moffitt, W. E., *Proc. 14th Annual Power Sources Conf.* (May 17–19 1960).

5.

Low-Temperature Aqueous Battery Systems[1]

J. W. PAULSON

One of the unique features of a primary battery is that it is readily portable and, as a result, it is often built into electrical equipment. Thus, some users may not hesitate to carry such equipment and its batteries into climates where the usual battery either fails to operate or provides substandard performance. The commonly available Leclanché cells do not furnish their customary good performance under conditions much below room temperature.

The need for low-temperature batteries has been primarily of a military nature, involving activities in polar climates. Other areas of use are in cold oceans and the upper atmosphere. There are also outdoor winter commercial applications such as road warning flashers, railroad lanterns, and sporting activities where special low-temperature batteries would be desirable. The major effort in providing low-temperature batteries has been sponsored by the U.S. Army with the objective of furnishing power to operate portable radio communications equipment in arctic temperatures.

GENERAL CONSIDERATIONS

Most military applications involve a relatively heavy drain, (i.e., 5–30 hour service) and a high cutoff point in the range of 1.0–1.2 V. These specifications can be adequately met at room temperature but become exceedingly difficult at temperatures such as $-40°$. The battery requirements for communication applications at low temperatures are described in military specifications(1). Examples are given later in this chapter.

[1]The text follows generally the outline originally prepared by the late Dr. W. S. Herbert of the Ray-O-Vac Division of ESB Inc.

PERFORMANCE OF AVAILABLE BATTERY SYSTEMS AT LOW
TEMPERATURES

The major part of the technical effort to produce a battery for operation at
low temperatures has been spent on various modifications of the
Leclanché dry cell, which, cf. Fig. 5.1(2), in its conventional form, suffers
serious loss in capacity with lowering of temperature, and becomes
practically inoperable below −20°C(−4°F). The open-circuit voltage is
affected little by temperature (about 0.0004 V/°C) thus giving no indication of substandard service.

The ampere-hour capacity at low temperature is reduced by:

1. Decreased chemical activity of the MnO_2 cathode;
2. Increased cell internal resistance due to an increase in electrolyte
   viscosity with decreasing temperatures and decreased ion mobility
   with decreasing temperature;
3. Reduced solubility of reaction products at electrode surfaces giving
   thicker passivating layers having high resistance to diffusion of reacting ions.

In spite of extensive studies on improving battery performance at low

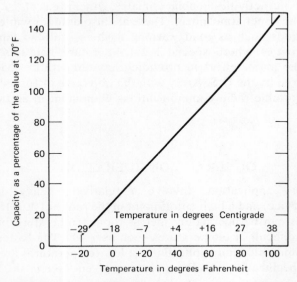

Figure 5.1 The effect of temperature on the capacity of conventional Leclanché dry cells.
Tests made on a $22\frac{1}{2}$ V battery unit of "F" size cells discharging through 1250 Ω to a cut-off
of 15 V.

temperatures, no particularly definitive work has been reported which has isolated any specific factors responsible for service loss beyond those given above. Most of the effort has been directed toward finding cell formulations that will operate, even to a limited extent, at $-40°C$.

Several different methods have been proposed for providing battery service at low temperatures. These can be broadly classified as thermal methods, wherein the battery is warmed by the application of external heat, and, as changes in composition and construction of the conventional Leclanché cell. Most of this chapter will be devoted to the latter, although various systems involving heating or insulation will be briefly described. Dry batteries which inherently operate well at low temperatures offer the preferred solution to the problem since the need for auxiliary heating devices is thus obviated. Such batteries are obtained by the use of nonfreezing electrolytes, very active depolarizers, and constructions that provide the lowest possible internal resistance.

THE MODIFIED LECLANCHÉ SYSTEM

LOW TEMPERATURE ELECTROLYTES

The selection of a low-temperature electrolyte solution involves the consideration of solubility–temperature data. When the $ZnCl_2$-NH_4Cl-H_2O electrolyte used in the conventional Leclanché cell is gradually cooled, NH_4Cl crystallizes out until the temperature reaches about $-20°C$, when the remaining solution freezes(3). This fact helps explain the reduced service at lower temperatures shown in Fig. 5.1. Obviously, to obtain improved performance at $-40°C$, electrolytes with lower freezing points are necessary. The eutectic point on the phase diagram is usually the starting point in developing a low-temperature electrolyte, since it represents a minimum-freezing-point composition. Figure 5.2 shows the solubility–temperature relationship(4) for aqueous solutions of NH_4Cl, $ZnCl_2$, $CaCl_2$, and $LiCl$. It is clear from these data that the solutions with the lowest freezing points for the four solutions are: for NH_4Cl $-15°C$ at 20 percent concentration; for $ZnCl_2$ $-62°C$ at 51 percent; for $CaCl_2$ $-50°C$ at 30 percent, and for $LiCl$ $-80°C$ at 25 percent. Although three of these salt solutions have adequately low freezing points experience has shown that a low freezing point alone is not the only criterion of a satisfactory electrolyte. Solution conductivities and compatibility with the electrode materials are also important. Other components may then be added to the eutectic composition to improve cell shelf life and increase conductivity.

Since most practical low-temperature battery electrolytes contain at

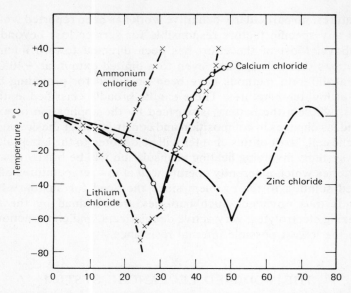

Figure 5.2   A comparison of the solubility-temperature relationships for aqueous solutions of the salts NH₄Cl, ZnCl₂, CaCl₂, and LiCl.

least four components, it is difficult to represent system behavior on one diagram. Hence, the establishment of practical compositions has been arrived at largely by empirical means involving actual battery testing. Most practical low temperature electrolytes were produced by additions of lithium chloride(5) or calcium chloride(3) to a selected composition containing zinc and ammonium chlorides and water. The exact proportion of the salts has been found to influence the cell performance. Ternary phase diagrams showing cell performances at the various composition points are useful in obtaining an indication of the general direction in which to move on the composition diagram in order to obtain the "best" battery performance from a particular system. Similarly, temperatures at which crystallization first appears can be shown on the ternary diagram, to maintain the electrolyte composition within a region where the freezing point is sufficiently low. Figure 5.3 shows an example of this type of comparison. Here the resistivity and solubility of a part of the system LiCl-ZnCl₂-H₂O with 5 percent NH₄Cl are shown at − 40°C(5). Figure 5.4 shows the temperatures when crystals first appear when solutions of CaCl₂-ZnCl₂-H₂O with 10 percent NH₄Cl are cooled(3). In a system as complex as this it is unlikely that the so-called "first crystal point" occurs at the same temperature as the freezing point.

Methylamine hydrochloride has been suggested(3) as an ingredient for

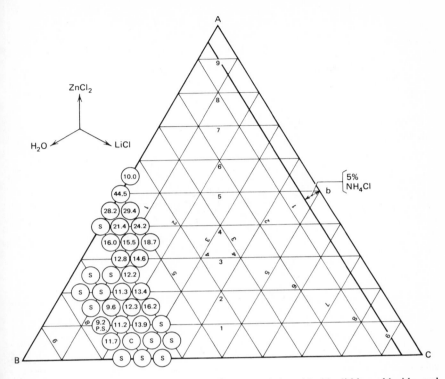

Figure 5.3 The solubility and resistivity of solutions of zinc chloride, lithium chloride, and water, all containing 5% ammonium chloride, at −40°C. The position of the center of each circle shows the electrolyte composition which it represents. The letters within the circle indicate the following: S, solution solidifies; PS, solution partly solidifies; C, solution crystallizes out in large part. The numbers within the circles indicate that the solutions remain liquid. The numbers also indicate relative resistances at −40°C (±2°C). These resistances are given in terms of a particular conductivity cell whose cell constant is 0.12^{-1}. To get resistivity in ohm-cms multiply by 8.3.

low-temperature electrolytes because of the low specific resistance of its solutions. Early efforts to take advantage of the properties of methylamine hydrochloride showed that the preferred composition was a solution containing 47 percent of methylamine hydrochloride, 3 percent zinc chloride, and 50 percent water. This solution had a freezing point of −45°C. Table 5.1 gives the resistance data for this solution at various temperatures (3).

The resistance values are quite comparable to those of a representative $ZnCl_2$-NH_4Cl-H_2O solution of the type normally used in dry cells through the temperature range of 20°C to −20°C. It was reasoned that such a

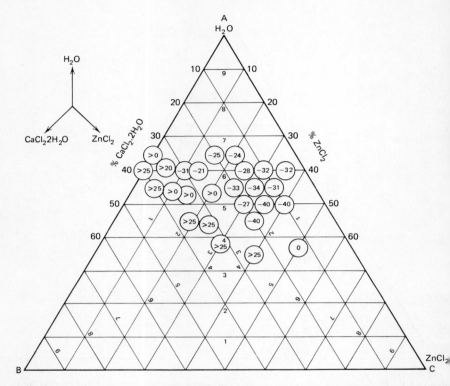

Figure 5.4 The calcium chloride, zinc chloride, water system containing 10% added ammonium chloride. The numbers on the chart give the temperature in degrees centigrade when crystals first appear.

TABLE 5.1. A COMPARISON OF THE RESISTIVITIES OF METHYLAMINE HYDROCHLORIDE AND AMMONIUM CHLORIDE–ZINC CHLORIDE SOLUTIONS AT LOW TEMPERATURES

Electrolyte	+ 20°C ohm-cm	0°C ohm-cm	− 20°C ohm-cm	− 30°C ohm-cm	− 40°C ohm-cm
μ solution[a]	3.50	5.18	8.84	12.6	19.0
As in NH₄Cl cell[b]	2.42	4.10	8.4	Frozen	

[a]The methylamine hydrochloride solution described in text.
[b]An NH₄Cl-ZnCl₂-H₂O solution of the type used in conventional dry cells.

TABLE 5.2. REPRESENTATIVE LOW-TEMPERATURE ELECTROLYTES

Type	LiCl "A"	LiCl "B"	$CaCl_2$	$CH_3NH_2 \cdot HCl$	$ZnCl_2$
References	5,7	5,7	7	3	7
% NH_4Cl	8.0	—	11.0	8.0	10.0
% $ZnCl_2$	12.0	5.0	19.0	2.8	47.5
% $CaCl_2 \cdot 2H_2O$	—	—	19.0	—	—
% LiCl	15.0	25.0	—	—	—
% $CH_3NH_2 \cdot HCl$	—	—	—	43.2	—
% H_2O	65.0	70.0	51.0	46.0	42.5
Crystal formation point, °C	$-40°$	$-54°$	$-37°$ ca.	$-45°$	$-54°$
Resistivity, Ohm, Cm, 20°C	5.0	6.8	5.2	3.50	10.0
$-20°C$	13.7	21.3	13.3	8.84	33.0
$-40°C$	38.4	58.8	33.3	19.0	143.0
Specific gravity 15°C/15°C	1.210	1.168	1.330	—	1.543

solution might have attractive low temperature possibilities. Actual trials with electrolytes containing methylamine hydrochloride showed substandard results. It appeared that this salt reacted with active types of MnO_2 and the cells did not possess adequate keeping quality. On these accounts it was dropped from further consideration.

The ultimate criterion for judging the merits of a particular low-temperature electrolyte is largely its performance in batteries tested under actual low-temperature conditions. Compositions of some various low-temperature electrolyte solutions, which have been developed over the years, are given in Table 5.2

A quantitative treatment of low-temperature cell mechanisms is lacking and the true role that electrolyte variables play in affecting performance is not clear. Conductivity per se is not fully adequate for comparing various electrolyte solutions. Experience has indicated that ammonium chloride in the electrolyte solution improves conductivity and shelf maintenance.

CATHODES

Next to the electrolyte, the cathode material or depolarizer is perhaps the most important factor responsible for good low-temperature performance. Early low-temperature dry cells had a nonfreezing electrolyte but used a relatively inactive natural MnO_2 ore alone or blended with an artificial MnO_2 of the hydrate or hydrous oxide type. More recently, electrolytic MnO_2 has been the preferred material because of its superior

TABLE 5.3. PROPERTIES OF VARIOUS KINDS OF MnO_2

Type MnO_2	Surface Area m^2/g	% MnO_2	MnO_2 Phases	Pore Volume cm^3/g
African Gold Coast (Ghana) ore	8–10	86	$80\%\gamma, 20\%\beta$	0.25
Artificial (hydrate)	15–25	77	Δ	0.44
Electrolytic	35–50	92	$95\%\gamma$	0.27

performance. A comparison of some of the pertinent properties is given in Table 5.3.

Since there is some variation possible in the manufacture of electrolytic MnO_2, some selection should be made to obtain the optimum material. A high surface area promotes good chemical reactivity and high content of γ-MnO_2 also seems to be beneficial. The capacity of the MnO_2 for absorption of electrolyte is also important since the best results are usually obtained with a relatively wet mix. The absorptivity of the conductive carbon black (usually acetylene black) in this respect is an even more important factor.

MnO_2 to carbon mix ratios in the range of 6:1 to 8:1 give the best results in cylindrical (round) cells. Mix ratios up to 20:1 are preferable for thin flat type cells(6). A light milling of the dry mix may be used to improve conductivity and MnO_2:carbon contacts. The wet mixes required for low-temperature cells are difficult to form or tamp into flat cakes or cylindrical cores. Most such mixes are hygroscopic so a low ambient relative humidity is desirable. The mixes for flat cells should be sufficiently strong and resilient so that they do not flow excessively under the pressure applied to the stack of cells. Typical mix formulas for both round and flat dry cells with lithium chloride electrolyte are given in Table 5.4.

TABLE 5.4. REPRESENTATIVE CATHODE MIX FORMULATIONS FOR LOW-TEMPERATURE CELLS

Material	Round Cell	Flat Cell
Electrolytic MnO_2—g	2,270	2,600
Acetylene Black—g	325	130
MnO_2/C ratio	7:1	20:1
Solid NH_4Cl—g	290	270
Lithium chloride electrolyte—ml	1,170	435

ANODES

Extruded or drawn zinc cans used for ordinary dry cells can be used in the Leclanché low-temperature units. As with room-temperature formulations, careful cleaning and proper amalgamation are essential to good shelf life. Most low-temperature electrolytes do not contain enough $ZnCl_2$ to get much benefit from the common ion (Zn^{+2}) effect. The methylamine hydrochloride electrolyte was abandoned largely because it caused severe corrosion, and early failure on shelf, due to perforation of the zinc cans.

STRUCTURAL FEATURES

The primary factor to be considered in the mechanical design of a battery for low-temperature operation is the apparent current density, the latter defined as the total current load divided by the apparent electrode area. The optimum configuration must be sought with respect to area and amount of active electrode materials to obtain maximum performance per unit battery volume(8), where a systematic, quantitative procedure for developing optimum designs is described.

In general, performance at low temperatures improves as the amount of electrolyte solution in the cell is increased. However, a practical limit for this variable is reached because of structural problems. This limit is around 0.5 cm^3 of electrolyte solution per gram of cathode material in the Leclanché-type cell.

In order to operate satisfactorily at low temperatures a maximum amount of electrolyte must be present. It is possible that not all the liquid added in the manufacture of the cell will be available. Presumably, as the electrolyte is cooled some hydrated salts may separate, leaving less liquid electrolyte for cell operation. However, when the cell is warmed up again, as during storage, the water present in the hydrates may be released. Under such conditions if an excess of electrolyte develops, over that amount which can be held immobilized in the cell, leakage may occur, causing corrosion of terminals and generally poor performance.

Low temperature battery constructions fall into two main classifications: (1) conventional round or cylindrical cells; (2) those employing the flat-cell type of unit cell.

ROUND CELLS

Conventional round cells, such as the N, A, D, and F sizes have been directly adapted to low temperature requirements simply by substituting low-temperature electrolyte systems for the conventional dry battery electrolytes.

In addition to the selection of a low freezing point electrolyte system, the type of separator used is important in affecting the performance. Of the two types, paste and supported film, the latter is more widely used. The paste type of separator consists of an electrolyte solution plus a gelling agent added to give an immobile layer between the anode and cathode of the cell. Cornstarch, wheat-flour, guar gum, methyl cellulose and numerous other natural and synthetic gelling agents have been used for this purpose. The principal objection to the use of the paste type of separator is the tendency for the paste to be soaked up by the mix core, resulting in discontinuities in the separator gel layer and poor contact to the zinc.

The more recent practice is to employ a paper separator having a dried coating of the gelling agent on one side. This coating then swells upon coming in contact with the cell electrolyte solution at the time of cell assembly. Methyl cellulose gels are quite stable at room temperature and below, retaining the electrolyte solution in the separator layer without excessive loss to the cathode mix. However, it is characteristic of methyl cellulose gels to become more fluid as the temperature is decreased. The gel then gradually soaks through the supporting paper and is absorbed by the cathode mix. To help alleviate this condition, an insolubilized layer of the synthetic gum is formed next to the paper to serve as a barrier against such migration(9).

FLAT CELLS

The main advantage in employing the flat cell type of construction is that the active electrode area can be greatly increased over that obtainable with round cells in a battery of the same fixed overall dimensions.

Several variations of the flat construction are possible. For example, the individual cell may consist simply of an anode and a cathode, or an anode with a cathode layer on each side, thus utilizing both sides of the metal anode. In any case the current density, along with the amounts of electrode materials, are essential parameters governing performance.

Parallel (A-section) and series (B-section) constructions employing flat cells are illustrated in Fig. 5.5. The cathode mix cakes are backed up by a strip of conductive plastic film which provides an electronic contact to the moist cathode material. A metal foil strip rests against the outer side of the conductive plastic to serve as a current collector at that point.

The multiple cell strip is contained in an outer plastic envelope and completely sealed off from the atmosphere. Sealing is accomplished either with adhesives or by heat-sealing of the plastic films. The metal anode is bonded to the plastic envelope by means of an adhesive. Due to the relatively large amounts of electrolyte solution normally used in

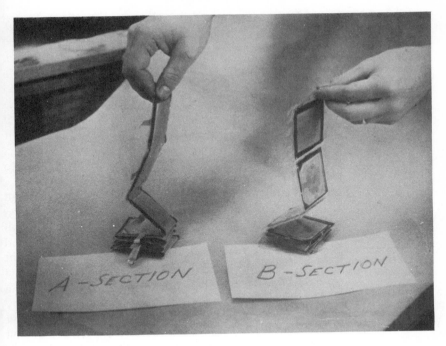

Figure 5.5 Parallel (A-section) and series (B-section) constructions of flat cell type.

low-temperature batteries, close inspection and control in the sealing operation are essential to prevent leakage during extended periods of battery storage.

SUCCESSFUL TYPES

In the Leclanché system the lithium chloride type of electrolyte (given as LiCl "A" in Table 5.2 and containing 15 percent LiCl, 12 percent $ZnCl_2$ and 8 percent NH_4Cl), has come into considerable usage in military low-temperature batteries, due to its consistently superior performance over other types. A comparison of typical performances obtained with this and the calcium chloride type is given in Table 5.5. The order of relative capacity may, however, be changed considerably at other low temperatures, that is, $-35°C$ and $-29°C$.

The choice and maintenance of a constant test temperature is important since variations of a few degrees in this range can cause large differences in capacity test results.

TABLE 5.5. TYPICAL PERFORMANCES OF LiCl AND CaCl$_2$ TYPES OF LOW TEMPERATURE ELECTROLYTES IN PASTED AND PAPER-LINED ROUND A SIZE CELLS AT $-40°C$

Storage time at 21°C	Cereal Paste Separator[a]		Methocel-Coated Paper Separator	
	LiCl	CaCl$_2$	LiCl	CaCl$_2$
0 Mos.	1.3 hr	0.7 hr	1.5 hr	1.2 hr
3 Mos.	0.7 hr	0.2 hr	1.4 hr	1.0 hr
6 Mos.	0.7 hr	—	1.4 hr	0.8 hr
12 Mos.	—	—	1.0 hr	0.5 hr
Electrolyte Composition:	LiCl: 15% LiCl, 12% ZnCl$_2$ 8% NH$_4$Cl 65% H$_2$O	CaCl$_2$: 19% CaCl$_2 \cdot$ 2H$_2$O. 19% ZnCl$_2$ 11% NH$_4$Cl, 51% H$_2$O		

Depolarizer mix = 7 parts electrolytic MnO$_2$/1 part acetylene carbon black.
Test: 52 ohm 18 min/20 min, 26 ohm 2 min/20 min, Continuous Hours to 1.10 V.

[a]The cereal paste separator contained 12% cornstarch plus wheat-flour.

It is quite clear from the data of Table 5.5 that the cells utilizing both pasted and Methocel-coated paper constructions give better performance with an electrolyte containing lithium chloride than with one made with calcium chloride. This conclusion is valid for both initial and delayed tests. With both electrolytes the cells with the Methocel-coated paper separator gave somewhat greater service than the pasted cells.

A comparison of performances between the flat and round cell constructions for a typical battery used under low temperature conditions is given in Table 5.6. This illustrates the marked superiority of the flat cell over the round cell in low temperature service.

Table 5.6 shows the results of a study utilizing an electrolyte containing LiCl-ZnCl$_2$-H$_2$O-NH$_4$Cl as applied to both round and flat cell packages for the BA-2270/U battery. It is quite clear from the data that the battery composed of flat cells gave from 4 to 4.5 times the service obtained from the battery composed of round cells.

Due to high electrolyte contents, battery leakage frequently occurs during storage, which obviates obtaining true delayed performance values. Where batteries are free of constructional defects, which may permit leakage and cause premature failure, service after six months at room temperature is about eighty percent of that obtained initially (0 mos.). Shelf life can be greatly extended by low-temperature storage which is readily available in the cold areas of anticipated use.

A considerable amount of effort has been spent to establish the

TABLE 5.6. FLAT CELL VERSUS ROUND CELL CONSTRUCTION IN THE BA-2270/U BATTERY(1)

	Hours of Service[a]					
	Flat Cell Construction			Round Cell Construction		
MnO_2:Carbon Mix Ratio	20:1			7:1		
Section	A	B_1	$B_1 + B_2$	A	B_1	$B_1 + B_2$
Electrode area, in.^{2}	77.0	3.5	3.5	53.0	2.2	1.1
Grams MnO_2	293	244	67	126	98	49
Performance at						
$-40°C$, hr	8	9	9	2	2	2
Hr/100 g MnO_2	2.7	3.7	13.4	1.6	2.0	4.1
Hr/cu. in. of battery						
volume	0.41	0.48	0.86	0.12	0.14	0.29

[a]BA-2270/U Test:
A section (1.5 V nominal) 2.5 ohm/18 min, 1.25 ohm/2 min repeating cycle to 1.10 V cutoff.
B_1 section (45 V nominal) 3,750 ohm continuous drain to 34 V cutoff.
B_1 plus B_2 section (90 V nominal) 2727 ohm/2 min out of 20 min, to 72 V cutoff.

optimum current density, mix ratios, electrolyte compositions and requirements for both round and flat cells designed for low-temperature operation. Figure 5.6, shows a cross section through a typical flat cell(6) used for this application. Figure 5.7, shows the relationship between the capacity at $-40°C$, in hours to 1.10 V, and the mix ratio for various anode areas in both types of cells(6). For given package-dimensions, such as one must consider for military applications, the flat cell has considerable flexibility and may be more attractive than the round cell. Thus, by making thinner cells, a larger total active area can be developed, using flat cells than can be attained with the round type. Data representation as shown in Fig. 5.7 facilitates this type of design.

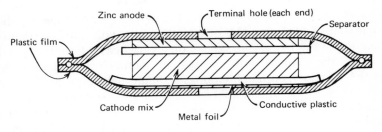

Figure 5.6   A cross section of a flat-type low-temperature cell.

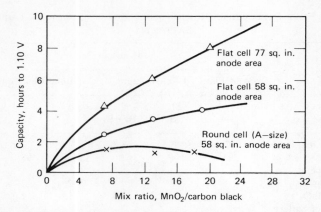

Figure 5.7 A comparison of the performance of flat and round low-temperature cells. Capacity versus mix ratio, BA2270/U A section, −40°C, flat cell mix cake thickness = 0.120 in.

OTHER LOW-TEMPERATURE SYSTEMS

Although most of the recent development work has been concerned with improvement of the modified Leclanché cell, other low-temperature systems have been considered.

ALKALINE MnO₂ CELL

The $Zn/KOH(aq)/MnO_2$ system has been reported to give better performance than the acid Leclanché system down to − 40°(12) as shown in Fig. 5.8. This system shows a marked advantage over the low-temperature Leclanché system throughout the range of resistances used.

Figure 5.9 shows the temperature-performance relationship for the $MnO_2/KOH/Zn$ and the $HgO/KOH/Zn$ systems at two current drains. Although this chart does not extend into the range of temperatures below zero it does show that the $MnO_2/KOH/Zn$ system outperforms the $HgO/KOH/Zn$ system at temperatures below 40° to 60°F. Table 5.7 shows a comparison of "D" size $MnO_2/KOH/Zn$ cells with a low-temperature modification of the Leclanché system on a very heavy drain test(13). It is clear that the $MnO_2/KOH/Zn$ system outperforms the modified Leclanché system. Too much emphasis should not be placed on such a comparison since in most military applications smaller currents are commonly used. Moreover the weight of the battery is an important consideration and the $MnO_2/KOH/Zn$ cells are considerably heavier than the modified Leclanché type.

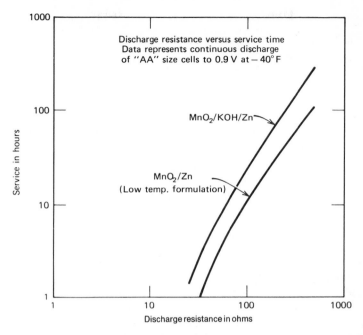

Figure 5.8 The relationship between discharge resistance and service time for "AA" size cells made in the Alkaline MnO₂–Zn and low-temperature Leclanché systems. Cells discharged to 0.9 V at −40°C.

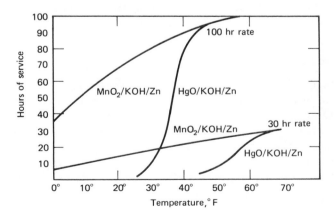

Figure 5.9 Effect of temperature on service of "AA" size cells at the 100-hr and 30-hr rates to 0.9 V.

TABLE 5.7. A COMPARISON OF THE SERVICE OF ALKALINE AND
LOW-TEMPERATURE LECLANCHÉ "D" SIZE CELLS

Continuous service to 0.8 V at 0.5 Ampere Starting Drain

	Service (minutes)		Capacity (amp-hr)	
Temperature	Alkaline E-95	Leclanché 950	Alkaline E-95	Leclanché 950
70°F	1,200	154	6.93	0.89
32°F	835	100	4.45	0.58
0°F	275	50	1.39	0.26
20°F	162	9	0.82	0.04

Mercuric Oxide-Zn System

The RM(Zn/KOH/HgO) cell, described in Chapter 5, Vol. I, has a very good capacity/volume ratio but its performance is seriously reduced below about 40°F(5°C). Considerable early improvement was obtained by using a more conductive electrolyte (41 KOH, 1.5 ZnO, 100 water), but, this caused a decrease in the shelf life. Performance at medium low temperatures has been substantially improved by deep corrugation of the zinc foil anode to increase its active area. Efficiency of both anode and cathode in the Zn/KOH (aq)/HgO system were found to increase with increasing porosity(10), particularly depending on the number of pores greater than 100 microns in diameter. In this case specific surface area of the porous mass decreased with increasing pore size, which suggests that ionic diffusion in the bulk electrolyte phase is the controlling step limiting cell performance.

Silver Oxide-Zn System

In studies of the performance of the $Zn/KOH(aq)/Ag_2O$ system(11), passivation of the zinc anode is attributed to dehydration of $Zn(OH)_2$ to form ZnO at low temperatures. In an important paper, Shepherd and Langelan(11a) show that the use of zinc electrodes of very high surface area can support surprisingly large currents at $-40°C$. This work should provide the basis for the improvement of the low temperature performance of all cells using a zinc electrode in a KOH electrolyte.

Zn:O₂ System

The air-depolarized $Zn:O_2$ battery, (cf. Chap. 8, Vol. I), when supplied with KOH electrolyte, will deliver its rated capacity at temperatures as low as 0°F (− 18°C). Its applicability is limited by its commercial availability only as large-size (600–3000 amp-hr) wet cells.

MAGNESIUM CELL

The magnesium dry cell (Leclanché type) potentially should show good low-temperature performance with the use of a $MgBr_2$-LiBr aqueous electrolyte. However, the delayed action of the magnesium anode is increased at low temperatures, thus restricting its use in some applications. The more active electrolyte and wetter mix required tend to reduce the shelf life. On medium-heavy drains, the wasteful corrosion of the magnesium has a beneficial heating effect.

RESERVE SYSTEMS

The water activated Mg/CuCl and Mg/AgCl systems (14) (cf. Chap. 7) have important uses at low temperatures. These batteries (14) are self heating due partially to excessive wasteful corrosion of the magnesium (already referred to) in the $MgCl_2$ electrolyte which is formed. This heat can be useful in many applications; for example, properly insulated batteries in radiosonde or pilot balloon usage can thus operate at high altitudes where the temperature reaches − 65°F (− 55°C). Similarly, such batteries can be used in seas at about 32°F (0°C) in sonobuoy applications. The Boron Trifluoride battery (14a) is an example of another reserve system. In the boron trifluoride cell activation is accomplished by coordination of BF_3 gas with the water of crystallization in $Ba(OH)_2 \cdot 8H_2O$ to yield fluoboric acid electrolyte. Any one of a number of anodes, Al, Cd, Fe, Mg, Mn, Pb, Zn may be used with MnO_2 or PbO_2 cathodes in cells with this electrolyte. A cell, Pb/BF_3-$Ba(OH)_2 \cdot 8H_2O/PbO_2$, is described as capable of current densities up to 320 ma/cm^2.

Operation as low as − 65°F is reported although the exothermic nature of the activation reaction between BF_3 and $Ba(OH)_2 \cdot 8H_2O$ does raise the battery temperature above its initial value.

NONAQUEOUS SYSTEMS

The relatively new gas-activated batteries should have favorable low-temperature properties. These include the ammonia vapor activated (AVA) battery species. Typical of the former is the $Mg/KSCN$-$NH_3/HgSO_4$ system (opencircuit voltage = 2.4) described by Spindler (16) who discharged the cell at temperatures as low as − 100°F with successful results. In this battery all ingredients, except the liquid ammonia, are dry. Only when ammonia gas is allowed to contact the KSCN does liquid electrolyte form and the cell go into operation.

SYSTEMS WITH ORGANIC ELECTROLYTES

A battery that operates as low as $-100°F$ ($-73°C$) is the Ca/acetonitrile and salts/AgCl system(15). The Ca is used as thin flakes compacted with a binder into a sheet. Acetonitrile or proprionitrile are solvents which effect good ionization of several suitable salts. These cells perform well at very low temperatures, but probably are not competitive with the usual low temperature cells at more normal low temperatures such as $-40°$. Figure 5.10 shows some representative curves obtained with this system. In some cases, depolarizers other than AgCl can effectively be used.

Depolarizers such as PbO_2, can also be used effectively with organic electrolytes. One such cell(15a) using an electrolyte containing 20 percent $AlCl_3$ dissolved in a mixture of equal parts of propionitrile and n-butyronitrile with an anode of strontium and a cathode of PbO_2 gave an open circuit voltage of 3.6 at room temperature and 2.55 at $-90°C$.

A COMPARISON OF LOW-TEMPERATURE SYSTEMS

A comparison of the performance(17) of a number of battery systems throughout the temperature range of $+20°$ to $-60°C$ is shown in Fig. 5.11. The data are presented as watt-hours per gram, thus taking into considera-

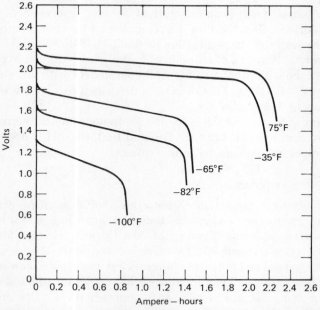

Figure 5.10 Discharge curves of "arctic" test cells at various temperatures. Cell weight—126 g; cell electrolyte—LT-2; discharge current—100 ma (5 ma/in.²).

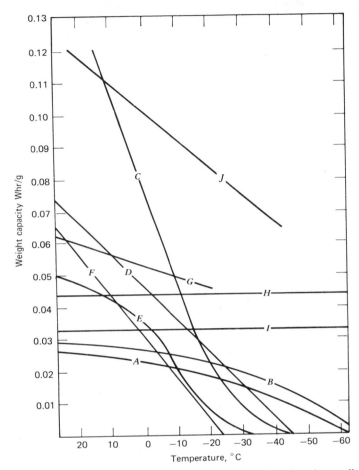

Figure 5.11 Temperature dependence of the weight capacities of various cells. (A) Pb-H₂SO₄-PbO₂(2.3a), (B) Cd-KOH-Ni(OH)₃(0.4a), (C) Zn-KOH-Ag₂O(3a), (D) Zn-NH₄Cl-MnO₂(LT)(0.1a), (E) Zn-KOH-HgO (0.10a), (F) Zn-KOH-MnO₂(10 hr rate), (G) Cd-KOH-Ag₂O(6a), (H) Mg-KSCN in NH₃-S(0.4a), (I) Mg-MgCl₂-Cu₂Cl₂(0.23a), (J) Mg-Mg(ClO₄)₂-HgO(2.0a).

tion the weights of the various units tested. Two systems identified as H and I show horizontal lines across the chart thus indicating no effect of decreasing temperature. System H is the ammonia-activated cell, that is, Mg-KSCN in NH₃-S, a system similar to an example previously described. System I is a reserve cell of the water-activated variety, that is, Mg-MgCl₂-CuCl₂, which develops considerable heat due to wasteful corrosion during its operation. With the exception of these two systems

all others show a sharply decreasing output with decreasing temperature
Too much emphasis should not be placed of the performance of system
which has the highest output over the range studied. This system show
unsatisfactory shelf-storage characteristics, and cannot be seriousl
considered at this time.

OTHER APPROACHES TO THE PROBLEM

The important military need for operation of primary batteries at lov
temperatures has led to other ideas than the development of inherentl
operable low temperature batteries. Four examples are given below.

THERMAL BATTERIES

Thermal batteries, for example, operate rather independently of th
ambient temperature and thus are suitable for low-temperature use. Initia
heating is usually provided by ignition of a non-gassing chemical mixtur
by an electrically or frictionally operated match. Since temperatures i
the range of 400°–550°C are attained, the outside temperature is not to
important. There are several cell systems employed, one of the mor
common being $Mg/LiCl \cdot KCl$ eutectic/Fe_2O_3. Such batteries are used onl
for special applications since they cannot readily be tested prior to us
and are relatively expensive.

INSULATION

A heat-retaining mass (e.g., metal) can be preheated and contained in th
battery to maintain battery temperature during operation in a lov
temperature environment. Actually, the electrolyte solvent, particularly
water, is as good as most metals in this respect on an equal volume basi
as found by comparing the product specific heat times density fo
different materials.

A general survey of the effectiveness of thermally insulating conven
tional low temperature primary batteries has been made(17). Severa
hours of useful operating life, sufficient for certain military operations ca
be obtained by this means. The rates of cooling both BA-23 and BA-270/L
batteries protected with Styrofoam insulation is shown in Fig. 5.12. It i
clear that an alternative method would be desirable.

BODY HEAT

One of the earliest attempts involved the use of body heat. Thus, in one
case, the N size cells of a B battery were enveloped in a fabric or rubbe
vest which had to be worn close to the body, the effectiveness depending
on the amounts of under- and outer-garments for heat transfer and
insulation. This expedient is limited to small size cells because of the bulk

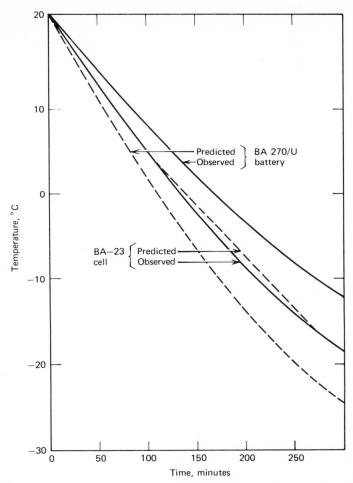

Figure 5.12 Comparison of predicted and observed cool-down for a battery and a cell. 1°C Initial battery temperature, −53°C Ambient temperature, 2.54 cm Styrofoam insulation.

actor. They also are much more expensive than batteries of more conventional constructions.

HEATING ELEMENTS

Another approach to improving the low temperature performance of regular batteries has been the use of a heating element or heat cartridge. The latter is usually composed of mixtures of chemical powders which are ignited by means of a mechanical or electrical fuse. The burning of

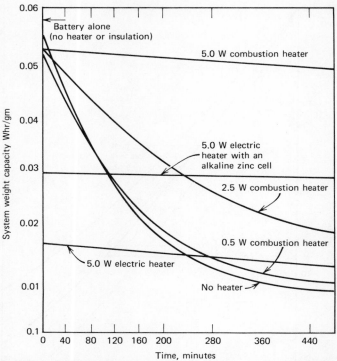

Figure 5.13 Effect of different heaters on the performance of insulated batteries. 21°C Initial temperature, −53°C Ambient temperature, 2.54 cm Styrofoam insulation, BA-270/U Size, weight, and cool-down, low-temperature type Leclanché cell performance.

fuels such as alcohols or light gasoline on activated charcoal has also been considered. Since the size of the battery for each operating equipment cannot usually be changed, space for the heating device must be provided inside the battery. This can be accomplished by using smaller cells or by the use of fewer magnesium dry cells of the same size, since the latter have a higher operating voltage (ca. 0.3 V) level. Other space must also be provided for means for heat distribution and a heat sink. This is usually done by some type of metal fins or grids. The heat cartridges themselves can attain temperatures in the range 300°–500°C. Satisfactory low-temperature operation, for about three to five hours, can be obtained and it is possible to insert new cartridges for several more such periods of operation.

Figure 5.13 shows how combustion heaters of various capacities will increase the electrical output of both a low-temperature Leclanché battery and an alkaline zinc cell. It is clear that a 5.0 W combustion heater is almost enough to maintain the room temperature output for a period of over seven hours even though the package was exposed to a temperature of −53°C.

If cold or frozen batteries are required for rather immediate use, it is possible to bring them to operating temperature range by passing an ac current through them. The details of such ac heating have been given by Vinal(1). This method takes advantage of the IR drop present in all batteries. The passage of small alternating currents through such units does not discharge them but does raise their temperature.

REFERENCES

1. MIL-B 18 Military Specifications.
1a. Vinal, G. W., *Primary Batteries*, John Wiley, New York (1950).
2. "Electrical Characteristics of Dry Cells and Batteries (Leclanché Type)" Letter Circular LC 965, U.S. Dept of Commerce, National Bureau of Standards, Washington, D.C. (Nov. 15, 1949).
3. Otto, E., Morehouse, C. K., and Vinal, G. W., "Low Temperature Dry Cells" *Trans. Electrochem. Soc.*, **90**, 419 (1946).
4. Seidell, A. and Linke, W. F., *Solubilities of Inorganic and Metal Compounds*, Am. Chem. Soc., Washington, D.C. (1965).
5. Wilke, M. E., "Dry Cell Designed to Operate at − 40°C, Containing Lithium Chloride in the Electrolyte," *Trans. Electrochem. Soc.*, **90**: 433 (1946). Id, U.S. Pat. 2,403,571 (July 9, 1946).
6. Paulson, J. W., "Low Temperature Dry Cell," U.S. Pat. 3,060,256 (Oct. 23, 1962).
7. The Ray-O-Vac Battery Co. (now a part of E.S.B. Co.), "Industrial Preparedness Study on Dry Batteries, Low-Temperature Type," U.S. Army Signal Corps Contract No. DA-36-039-SC-30280 (1953) (unclassified).
8. Paulson, J. W., "Low Temperature Dry Batteries," *Proc. 12th Annual Battery Res. Develop. Conf.*, U.S. Army Signal Corps, p. 85 (1958).
9. Cahoon, N. C., "Primary Galvanic Cell," U.S. Pat. 2,534,336 (Dec. 19, 1950).
10. Przybyla, F. and Kelly, F. J., "Low Temperature Performance of the Zinc-Mercuric Oxide System," *Proc. 6th Int. Symp. Power Sources*, 1968, Pergamon, Brighton, Eng. (1970).
11. Goodkin, J. and Solomon, F., "A Zinc–Silver Oxide Cell for Extreme Temperature Application," *Proc. 4th Int. Symp. Batteries*, 1964, Pergamon, Brighton (1965).
11a. Shepherd, C. M. and Langelan, H. C., "High Rate Battery Electrodes," *J. Electrochem. Soc.*, **114**, 8 (1967).
12. Daniel, A. F., Murphy, J. J., and Hovendon, J. M., "Zinc-Alkaline MnO_2 Dry Cells." *Proc. 3rd Int. Symp. Batteries*, 1962, Pergamon Press, Bournemouth, Eng. (1963).
13. Winger, J., 18th Annual Proc. Power Sources Conf. (1964); Cf. also Vol. I, p. 260 ff.
14. Adams, B. N., U.S. Pat. 2,322,210.
14a. Evans, G. E., Proc. 11th Ann. Batt. R & D Conference, Army Signal Corps (1957).
15. Solomon, F., "A-100°F Non-aqueous Battery," *Proc. 12th Annual Battery Res. Devel. Conf.*, U.S. Army Signal Corps. p. 94 (1958).
15a. Horowitz, C. and Breitner, T., U.S. Pat. 3,098,770, July 23, 1963.
16. Spindler, W. C., "Ammonia Batteries," *Proc. Advances in Battery Technology Symp.*, Electrochem. Soc., AD-624, 768 (4 Dec. 1965).
17. Horne, R. and Richardson, D., "Low Temperature Operation of Batteries," *Proc. 18th Annual Battery Res. Devel. Conf.*, U.S. Army Signal Corps (1964).

6.

Thermal Batteries[1]

C. W. JENNINGS

hermal batteries containing fused-salt electrolytes are special-purpose ystems, for which aqueous-electrolytes are not adapted, which fulfill nique military and aerospace applications. They are reserve sources of ower in which the solid electrolyte is nonconducting at ambient tempera- ires. The battery is activated by melting the electrolyte, thus making the itter conductive. In the conventional thermal battery, activation, requir- ıg from a fraction of a second to several seconds, is achieved by igniting yrotechnic heat sources within the battery. Lifetime is usually measured ı minutes and operating temperature is between 400° and 600°C. Other /pes, developed mainly for space applications, must operate for hours or ven days in a high-temperature environment. Both types become inoper- ble on cooling, when the electrolyte solidifies.

The primary advantages of thermal batteries are rapid and reliable ctivation, virtually no deterioration on storage, moderately high dis- harge rates for short times after activation, and operability over a wide mbient-temperature range of −54° to 75°C. They are ideal for military nd aerospace systems and for alarm and sensing applications because ıey require no maintenance, are compact and mechanically rugged, and re operable in any position. Some units are designed to withstand high ıock and spin forces.

In this report, published and unpublished information on current and rojected electrochemical cell systems for thermal batteries is reviewed. ıcluded are the basic types of cell designs; requirements and characteris- cs of the anode, electrolyte, and depolarizer; investigations of single ells; descriptions of long-life cells and heat-sensing batteries; and a iscussion of the factors affecting the operation of a few of the cell ystems used in production batteries. The details of battery construction

[1]This work supported by the U.S. Atomic Energy Commission.

263

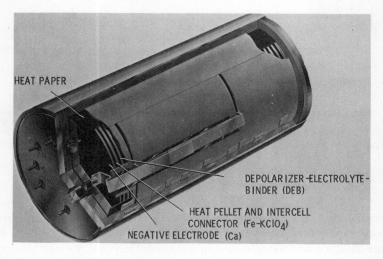

Figure 6.1 Thermal battery with pellet-type cells.

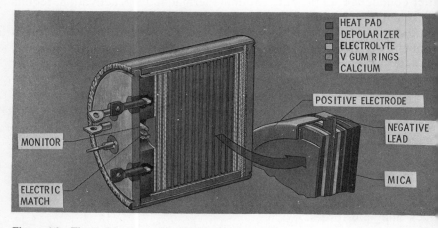

Figure 6.2 Thermal battery with closed-cup-type cells.

and activation are omitted, since this information is frequently propriet ary or unavailable.

CELL DESIGN

General characteristics of thermal battery cells have been discussed b Goodrich and Evans(1), McKee(2), Goldsmith and Smith(3), Jasinski(4)

nd Reddy(5), and special types have been described by various
uthors(6–30). Although many fused-salt cells have been studied with the
bjective of achieving maximum energy density, few have actually
eached production stage. Energy density has been increased more often
y improved assembly techniques and better thermal regulation than by
hanges in cell electrochemistry. Each new application usually requires
nodification of the battery design, and smaller weight and volume for the
ame energy output. Because of the many diverse applications, there are
robably more battery designs and fewer of each type produced than for
nost primary batteries.

Typical conventional thermal batteries are shown in Figs. 6.1 and 6.2.
The cells of Fig. 6.1 are of the open-pellet type which contain an anode of
olled sheet calcium, a pelletized mixture of LiCl-KCl eutectic electro-
yte, SiO_2 binder, and $CaCrO_4$ depolarizer, together with a heat pellet
ontaining iron powder and a small amount of $KClO_4$. The excess iron
erves as the cathode collector and intercell connector. Peak voltage is
etween 2.5 and 2.8 V, depending on the discharge rate. Cell rate to a
.1-V cutoff ranges from 5 minutes at 155 ma/cm^2 to 23 minutes at
5.5 ma/cm^2. Cell weight is 13.2 g, yielding 46.7 Whr/kg or 21.2 Whr/lb.

The closed-cup cell design, illustrated in Fig. 6.2, has been in use for
everal years and is described by Goldsmith and Smith(3). Separate
lectrolyte and depolarizer wafers are located between a cathodic en-
elope and a central anode. The cell, Ca/LiCl, KCl/CaCrO$_4$/Ni, has an
pen-circuit voltage of 2.8 V. A typical discharge provides 50 ma/cm^2 for
our minutes between a peak voltage of 2.5 V and a cutoff voltage of
.0 V.

The cell reaction is postulated as

$$3Ca + 2CaCrO_4 + 6LiCl \rightarrow 3CaCl_2 + Cr_2O_3 \cdot 2CaO + 3Li_2O, \qquad (1)$$

hough the exact composition of the mixed oxide produced at the cathode
as not been determined.

Although various specific features have been reported for production
atteries(3)(10)(11)(26)(29)(30), the cell designs differ primarily in the
nethod of heating and in the arrangement of electrolyte and depolarizer.
Heat is furnished by a cylindrical block of pyrotechnic material enclosing
he cell stack. Alternatively, pyrotechnic heat pads or pellets are in-
erspersed between the cells and ignited by a separate pyrotechnic strip
laced along the cell stack. Since most cells operate efficiently only within
 limited temperature range, the pyrotechnic material and the battery
nsulation are designed to bring and maintain the cells within this range
or both the lowest and the highest ambient temperatures expected.

There are three general techniques for incorporating the depolarizer. It

may be combined directly with electrolyte and binder to form a pellet o
disc, combined with binder to form a separate pad, or applied as a fuse
coating on the cathode. Separate electrolyte-binder pellets or discs are
used with the pad or cathode coating. Glass or ceramic fiber, cloth, o
powder serves as binder.

The unactivated batteries are essentially inert over several years o
storage when properly constructed, thoroughly dried, and sealed t
exclude moisture. There is little, if any, self-discharge and cell resistanc
is several orders of magnitude higher than in the activated state. For
short activated life, the electrochemical system need not be ther
modynamically stable in the ordinary sense. Deterioration, such a
wasteful corrosion of the anode, must be slow compared with th
discharge rate.

ANODE MATERIALS

Anodes are usually made of calcium or magnesium. Lithium is elec
trochemically more attractive because of its high oxidation potential, lov
equivalent weight, and low polarization is fused-salt electrolytes, but it
low melting point (180°C) and high oxidation rate in air cause anode
confinement and processing problems which render it less desirable fo
general battery use. Calcium and magnesium can be readily obtained an
can be handled without excessive oxidation in reasonably controlle
environments. Calcium, because of its high oxidation potential, is prefer
red for applications involving minimum battery weight and volume, shor
life, and high discharge rate, but magnesium[2] is more attractive wher
long life, stability, and economy are desired.

Calcium must be stored and handled under conditions of low humidity
(usually less than 5 percent RH), and cells must be hermetically sealed t
reduce $Ca(OH)_2$ formation. The anodes are stamped from sheet stock o
bimetallic strips. Stamped discs are mechanically bonded to iron or nicke
discs for support. Bimetallic strips are made by vapor-depositing calciun
onto iron or nickel. The sheet stock and bimetallic strips contain 92 to 9?
percent and 98 to 99 percent calcium, respectively.

Upon cell activation, calcium undergoes a reaction with lithium ion ir
lithium-containing electrolytes to form a calcium-lithium alloy,

$$2Ca + 2Li^+ \rightarrow Ca^{++} + CaLi_2. \qquad (2$$

This alloy melts at 230°C, forming a liquid interface between the anode

---

[2]Magnesium anodes are made primarily from sheet stock; however, compressed mag
nesium powder is used in one type of pellet cell(30).

nd the electrolyte.[3] Extrusion of the alloy across the cell edge to the athode can form short circuits when more than a surface film of the alloy s present and the cell is under pressure. Gas formation at the anode vould aid in this process. Internal shorting has also been observed.

Lithium salts are used in all production batteries, and the formation of he calcium-lithium alloy is believed to be essential to efficient operation. 'he alloy must be confined or its rate of formation regulated if shorts are o be eliminated or minimized. In one design, mica rings around the alcium discs serve as barriers to alloy flow. Clark(32) treated the surface f calcium electrodes with a solution of glacial acetic acid in acetone to educe alloy formation and electrical noise during discharge. The rate of lloy formation has been reduced by changing the amount and composi- ion of the electrolyte. Oxidizing species such as chromate in the lectrolyte provide dimensional stability by forming a film on the anode nhibiting the alloy reaction. However, a balance must be achieved etween inhibition of alloy formation and excessive reduction of the lectrochemical discharge rate.

VOLTAGE MEASUREMENTS

t is well recognized that the lack of a satisfactory reference electrode, vith an emf accepted by the scientific community, handicaps the predic- ion of voltages of cells made with molten salts. Plambech(33) presented a omprehensive review of potentials measured by a number of workers in his field. In one table data are given versus a platinum wire immersed in a ilute solution of $PtCl_2$ in the molten solvent, LiCl-KCl at 450°C thus cting as a true reference electrode. In another table the data obtained vith a metallic silver electrode immersed in a dilute solution of AgCl in nolten NaCl-KCl at 700°–900°C are presented. Unfortunately, these ables do not include data for calcium which is used extensively in hermal batteries. Such a situation is not an oversight but is presumably lue to the difficulties of obtaining satisfactory potential values with this netal.

Hamer and co-workers have reported theoretical potentials calculated rom the available data for a large number of metals in fused hlorides(34), bromides and iodides(35), and for systems including an xide(36). The open circuit voltages obtained on actual experimental cells re usually within a few tenths of a volt of those calculated from tables

---

[3]Danly and Walker(31) found no ternary mixture of LiCl, KCl, and $CaCl_2$ that did not orm the alloy. They estimated the equilibrium constant for alloy formation from eutectic .iCl-KCl at 500°C to be 25.

such as those referred to above provided the appropriate fused salt typ
and the temperature are selected properly. Thus, Plambech(33) gives th
measured potential of lithium at 450°C as 3.304 V while Hamer calculate
it to be 3.385 V.

Nissen(37) measured the potential of $CaLi_2$ against a Ag/AgCl (0.1 m
LiCl-KCl reference electrode in eutectic LiCl-KCl. He obtained 2.3(
2.35, and 2.40 V at 400°, 500°, and 600°C, respectively. The calciur
potential was reduced to -0.5 V by adding 1.1 mole percent $CaCrO_4$ to
large volume of the same electrolyte at 490°C. As the temperature wa
increased above 500°C, the potential again approached that of $CaLi_2$. Thi
passivation below 500°C was attributed to an anodic film. Selis an
McGinnis(38) also observed this calcium passivation in a NaCl-KC
K_2CrO_4 electrolyte, but not in LiCl-KCl-K_2CrO_4.

Magnesium anodes do not require lithium ion electrolytes; howevei
the latter are desirable since they form low-melting solutions witl
magnesium reaction products, thus avoiding film formation and makin,
high discharge rates possible.

Lithium and its alloys are polarized to a negligible extent in LiCl an
LiCl-KCl melts(39–41), indeed, a Li/LiCl electrode was not polarized
even at 40 amp/cm^2(39). The absence of activation polarization at higl
current densities is not surprising in view of the high exchange current
reported by Laitinen and co-workers for nonpassivating metal/metal io
electrodes(42)(43). In addition to the disadvantages of the lithium elec
trode for thermal battery applications already mentioned, the solubility o
lithium metal in some electrolytes poses a self-discharge problem. Allo
electrodes have shown promise, although production usage has not bee
reported.

Giner and Holleck investigated the performance of an aluminum anod
in AlCl-NaCl-KCl melts, with a chlorine-carbon cathode(44). Steady
state current densities of up to 50 ma/cm^2 were reported for the aluminun
anode at 150°C. Passivation occurred when the melt was saturated witl
$AlCl_3$, and a salt layer formed on the anode. Polarization was lowere
when the temperature was increased or the melt composition wa
changed to permit greater $AlCl_3$ solubility. Polarization is predominantl
ohmic with this electrode and electrolyte.

The anodic character of lithium hydride in eutectic LiCl-KCl wa
studied by Indig and Snyder(12). They reported high anode efficiencie
and current densities up to 1.55 amp/cm^2 without appreciable polariza
tion. A pellet cell, Ag/LiH/LiCl, KCl/$CaCrO_4$/Ag, gave an open-circui
voltage of 2.15 V and a flash current of 3.1 amp/cm^2. The system is les
attractive for battery usage since hydrogen is evolved at the anode.

ELECTROLYTES

'he electrolyte serves as both a conducting medium and a solvent for the
lectrode reaction products and depolarizer. It should have good electri-
al and thermal conductivity, a melting point between 150° and 450°C, and
hemical stability for at least 250°C above the melting point, and it should
e compatible with both the anode and cathode materials, a requirement
1at eliminates many low melting salts that are otherwise satisfactory.
'he solubilities of the depolarizer and electrochemical reaction products
1 the electrolyte are important to cell operation and often limit its use.
hase diagrams should be consulted, or these solubilities should be
etermined experimentally, before any cell system is seriously considered
or battery use. Alkali and alkaline earth chlorides, fluorides, bromides,
itrates, sulfates, and hydroxides have been investigated as electrolytes.
utectics of LiCl (45 wt percent)-KCl (55 wt percent), melting point
53°C, and LiBr (54 wt percent)-KBr (46 wt percent), melting point 322°C,
re most commonly used in production batteries.

The cell should be designed to contain electrolyte in sufficient quantity
o ensure low internal cell resistance, without, however, causing exuda-
on during activation. Glass cloth, inert powders, and fibers are fre-
uently used to reduce fluidity and to permit handling in strip or pellet
orm during battery fabrication, as well as to retain the electrolyte during
ctivation. Large volumes of electrolyte are not required, since fused-salt
onductivities are high. Specific conductances range from 0.1 to 10 ohm^{-1}
m^{-1}. Nonaqueous and aqueous solutions seldom exceed 0.01 and 1 ohm^{-1}
m^{-1}, respectively. Diffusion constants for ions in LiCl-KCl eutectic at
50°C are higher than for similar ions in aqueous solutions(4). Exchange
urrents for metal/metal ion couples are many orders of magnitude higher
1an in aqueous solutions(42)(43).

The solubilities of the depolarizer and electrode reaction products are
unctions of the cell temperature and the extent of discharge. The IR drop
cross a cell is usually increased by these materials. Depolarizer sol-
bilities in the LiCl-KCl eutectic range from very low for copper and iron
xides to very high for V_2O_5. The solubility of $CaCrO_4$ in this eutectic is 10
/t percent at 350°C and 34 wt percent at 600°C. The phase diagram of
1e system LiCl-KCl-$CaCrO_4$ determined by Clark, Blucher, and
ioldsmith(45) shows a ternary eutectic at 342°C for 41 wt percent LiCl,
0 wt percent KCl, and 9 wt percent $CaCrO_4$. The same authors
etermined the electrical conductance for a variety of concentrations for
his system(46). Specific conductance for the LiCl-KCl eutectic saturated
/ith $CaCrO_4$ is between 0.4 and 0.5 ohm^{-1} cm^{-1} from 350° to 600°C; the

LiCl-KCl eutectic conductance is 1.1 to 2.4 ohm^{-1} cm^{-1} for these tempera‑
tures.

Blucher and Goldsmith(47) presented phase diagrams and reporte
conductivities for the ternary systems LiBr-KBr-K$_2$CrO$_4$ and LiCl-KC‑
K$_2$Cr$_2$O$_7$. Specific conductivity was increased by a factor of 2 at mos
compositions by increasing the temperature from 400° to 550°C. The LiC‑
KCl-K$_2$Cr$_2$O$_7$ system conductivity decreases with increasing K$_2$Cr$_2$O$_7$ a
constant KCl concentration. There is no unique composition for max
imum conductivity at all temperatures. The optimum compositions cor
tain more LiCl at higher temperatures. The selection of an electrolyt
composition for a given discharge requirement is based on the operatin
temperature, the depolarizer requirement, the desirability of a hig
lithium concentration, and the specific conductance.

Relatively dry assembly conditions (less than 5 percent RH) and
sealed battery design are required when the electrolyte contains lithiu‑
salts; however, the tolerance limits for moisture in the electrolyte hav
not been determined. Cells of Ca/LiCl, KCl, CaCrO$_4$/Fe which have bee
extensively vacuum dried develop lower outputs than those that are nc
dried. Cells exposed to higher humidities are characterized by more allo
formation and noisier electrical discharge. Gas generated from th
reaction of moisture with calcium increases electrolyte and alloy exuda
tion. An excessive moisture content is indicated by a cell potential a
ambient temperatures when measured with a low-impedance voltmete‑
Electrolyte pads or pellets are usually stored in closed containers, an
production batteries are usually vacuum dried at 60°C for 14 to 20 hour
prior to sealing.

DEPOLARIZERS

Depolarizer requirements include reproducible reduction potential, lo‑
internal resistance, low polarization, and suitable solubility in the ce
electrolyte. Chemical stability is a special requirement which excludes ga
formation over the entire temperature range of cell operation.

More than 70 distinct chemical compounds have been evaluated a
depolarizers. Those most frequently used—either singly or i
combination—in thermal batteries are the chromates of potassium, ca‑
cium, lead, sodium, zinc, and copper, and oxides of tungsten, vanadium
molybdenum, iron, copper, and silver. Silver and lead chloride and othe
soluble heavy metal salts although attractive depolarizers, are undesirabl
for most applications because they tend to undergo a displacemen
reaction at the anode and form a tree or electronic short through th
electrolyte to the cathode. Wider applications for such salts may resul
when a suitable separator is devised to prevent this transport. Chemica

urity, particle size, surface area, crystalline structure, and oil adsorption dex are properties that have been measured for depolarizer powders. The depolarizer, its choice determined primarily by the end use of the ell, is usually incorporated in the electrolyte for low-current drain cells. , greater amount, in pad or pellet form, is used in high current-drain cells. ischarge characteristics vary with the physical state as well as the omposition of the depolarizer.

Laitinen and Liu(48) tabulated standard potentials for electrode sysms in the LiCl-KCl eutectic at 450°C. They used Pt/1M Pt(II) as the ference electrode. A similar tabulation for metals in LiCl-KCl—10 ercent K₂CrO₄ was reported by McKee(2). It is difficult to determine the andard potentials for many of the battery depolarizers, because the oltage drifts with time. Depolarizers are often ranked by comparing the eak voltages generated at a common current density in cells with Mg or a anodes and LiCl-KCl or LiBr-KBr electrolytes.

SINGLE-CELL INVESTIGATIONS

lany different fused-salt systems have been investigated for thermal atteries, some to achieve higher energy density or greater life in onventional applications, others to meet high-temperature or sensing pplications. In this section, the cells are classified by the depolarizer, as is the principal component differentiating the system. Ranking of cells nd depolarizers is difficult because each application has a different ctivation and discharge requirement. Most of the systems were investiated in test tubes or as single cells at constant temperature. Such testing nds to give optimum performance and frequently does not translate irectly to multicell battery operation. Performance can vary with elecolyte volume and amount and method of incorporating the depolarizer to the cell. Temperature varies with the location of a cell in an activated attery, and with the duration of the discharge. Predicting battery erformance from single-cell tests is especially difficult for cell systems ith a narrow operating temperature range. Battery output may be gulated more by a few cold cells than by many operating at optimum mperature.

Goldsmith and Smith(3) compared different depolarizers in the cell ystem, Ca/LiCl, KCl/D/Ni, where D was CuO, CaCrO₄, Fe₂O₃, WO₃, and ₂O₅. Both closed-cup and two-layer pellet-type single cells were disharged at constant current (between heating platens held at 2.7 kg/cm²) nd a constant temperature between 550° and 650°C. The energy output as a function of the type of cell, the temperature, and the current ensity. The pellet cell was superior to the closed-cup cell in most

272 THERMAL BATTERIES

TABLE 6.1. PELLET-CELL DISCHARGES

Depolarizer	Wt%[a]	Peak Voltage	Life[b]
CuO	95	1.98	58
CaCrO$_4$	85	2.20	111
WO$_3$	89	2.22	67
Fe$_2$O$_3$	75	2.20	58
V$_2$O$_5$	62.5	2.80	44

[a]Weight % of depolarizer in the cathode layer of the two-layer pellet.
[b]Life to 80% of peak voltage at 155 ma/cm^2 and 600°C, given in seconds.

discharges. Pellet-cell discharges at 155 ma/cm^2 and 600°C are compared in Table 6.1. CuO had a longer life in closed-cup cells than in pellet cell but the other depolarizers had shorter lives. Although the general conclusions of this work are valid, it should be noted that, greater energy outputs have been obtained for some of these systems in other investigations conducted at lower temperatures.

IRON OXIDES

Selis, McGinnis, McKee, and Smith(10) studied closed-cup cells with the system Mg/LiCl, KCl/FeO$_x$/Ni. A typical discharge yielded 1.6 V at 60 ma/cm^2 for 120 seconds at 420°C. They found essentially all the magnesium oxidized during the discharge to be in the form of MgCl. Water-insoluble mixed oxides of iron and lithium were found at the cathode. Energy output was greater for cells with excess electrolyte than for those with just sufficient electrolyte to provide conductivity. Test tube cells with a large excess of electrolyte ran for 30 minutes at 75 ma/cm with a plateau voltage of 1.4 V. Oil adsorption index of the iron powder provided an index for the amount of electrolyte. With equivalent amount of electrolyte, cell discharge varied widely with the type of iron oxide. The authors suggested that the oxide be selected to meet the particular discharge requirement, a γ-Fe$_2$O$_3$ or Fe$_3$O$_4$ being recommended for high power output and an α Fe$_2$O$_3$ for a flat discharge at lower voltage. The cell voltage was increased or decreased by incorporating small amounts of TiO$_2$ or SiO$_2$, respectively, into the iron oxide. A high-voltage peak was obtained by blending the iron oxide with carbon. An intermediate behavior was obtained by blending the oxides.

VANADIUM (V) OXIDE

Very little information on the discharge characteristics of V$_2$O$_5$ cells has been published, even though this depolarizer is used in many production

,atteries. In some cell designs the V_2O_5 is applied to the cathode collector metal with boric oxide(29) or aluminum phosphate to form a fused-glass oating(20). In others, it is combined with the LiCl-KCl eutectic and ;aolin, or glass binder, in pellet form. The coated cathode collector may ,r may not be used with the pellet. A separate disc, or pellet, composed of ,inder and LiCl-KCl eutectic electrolyte is used in all cell designs.

Laitinen and Rhodes(49) investigated the reduction of V_2O_5 in a molten LiCl-KCl eutectic solution by voltammetry, chronopotentiometry, and :onstant-current electrolysis. Their results indicate that V_2O_5 is reduced o a mixed lithium-vanadium bronze on the electrode surface, at a ,otential which is a function of the V_2O_5 concentration. Chemical analysis ,f the lithium vanadate, which was formed in the potential region, 0.00 to).02 V versus Pt(II) 1M/Pt reference electrode, gave a compound with a ;toichiometry of $Li_2O \cdot 2V_2O_4 \cdot 4V_2O_5$. These lithium vanadium bronzes liffer from those obtained in the thermal decomposition of Li_2CO_3 and V_2O_5. Lithium vanadates of limited solubility are also formed during electrochemical reduction. V_2O_5 is very soluble in a LiCl-KCl melt and, at higher concentrations, it reacts with the melt to form Cl_2 and some type of insoluble reduced vanadium compound. The V_2O_5 must be added carefully to the melt to avoid high local concentrations.

One would expect a lithium vanadium bronze to be formed at the cathode during discharge in a thermal battery cell. This bronze would probably include small and variable amounts of MgO or CaO, depending on the anode used and the rate of discharge.

CHROMATES

Chromate is one of the most popular depolarizers for production batteries. Discharge characteristics of cells with $CaCrO_4$ have been mentioned and will be discussed in more detail in a subsequent section. Laitinen and Propp(50) have shown that the electrochemical reduction product from K_2CrO_4 in a LiCl-KCl eutectic containing dissolved $MgCl_2$ is a single unstoichiometric compound of formula $Li_x Mg_y CrO_4$, where $x + 2y = 5$. The values of x and y depend on current density, temperature, and molar ratio of Mg(II) to Cr(VI) dissolved in the melt. Typical values of x range between 0.3 and 0.5. Hanck(51) observed that the reduction of chromate in the presence of Zn(II) was shifted from -1.0 to -0.5 V versus Pt(II)/Pt reference electrode. Analysis of the deposit indicated a $LiZn_2CrO_4$ composition. The stoichiometry was not affected by the electrolysis conditions. Popov and Laitinen(52) found the reduction product from a LiCl-KCl eutectic melt containing $NiCl_2$ to depend upon the temperature, the deposit at 500°C approaching the composition $LiNi_2CrO_4$. Current densities in the range 1–10 ma/cm^2 are not significant factors in determining the chemical composition of the electrode deposit.

Laitinen and Bankert(53) concluded from chronopotentiometric measurements of K_2CrO_4 in molten LiCl-KCl eutectic that the primary
reduction product is the ion CrO_4^{-5}, which decomposes to yield CrO_3^{-3} and
O^{-2}. At sufficiently high surface concentrations of CrO_4^{-5}, solid Li_5CrO_4 is
deposited. The reduction potential is shifted by $+0.8$ and $+0.6$ V in the
presence of $MgCl_2$ and $CaCl_2$, respectively. The insoluble reduction
products formed in the presence of $MgCl_2$ and $CaCl_2$ proved to be
extraordinarily resistant to all forms of aqueous acid attack. They were
also resistant to electrochemical reoxidation. The insoluble product
containing calcium was similar to that obtained by short-circuiting the cell
$Ca/LiCl-KCl-K_2CrO_4/Ni$. The composition may be represented approximately by $2Li_2O \cdot CaO \cdot Cr_2O_3$ or $Li_4Ca(CrO_3)_2$.

One may conclude from the above discussion that cathodic reduction
products are mixed oxides of varying stoichiometry. They are usually
insoluble and form a film or layer on the cathode. These products must
transmit charge since electrochemical reduction at the cathode continues
without excessive polarization. Semiconductor behavior is suggested, but
no quantitative conductivity measurements have been related to cell
behavior. It is of interest that potassium was not found in the reduction
products, while calcium, magnesium, zinc, and nickel ions play potential
determining roles in the reduction.

The influence of the collector metal on the cathodic reaction is
uncertain. Selis and McGinnis attributed the cathodic potential in the
system, $Mg/LiCl-KCl-K_2CrO_4/Ni$, to nickel oxides formed by the oxidation of nickel by chromate ion(9). This interpretation has been
questioned(53), since the direct reduction of chromate was not considered. The collector metal can be oxidized by the depolarizer; however
the effect of this reaction on the reduction potential has not been
quantitatively demonstrated. Nickel, iron, stainless steel, and silver are
used frequently as collector metals.

COPPER OXIDE

Thaller(16) discharged cells of $Mg/LiCl$, $KCl/CuO_x/Cu$ with excess
electrolyte over a period of six days at 425°C. Anode efficiency was 72
percent. Other cells, discharged over shorter periods of time, had anode
efficiencies greater than 90 percent. He concluded that slightly soluble
cuprous oxide was reduced to metallic copper at the cathode, and
magnesium was oxidized to soluble magnesium ion at the anode. Insoluble magnesium oxide was found in the region between the cathode and
anode. The cell potential was 1.48 V, which is in agreement with that
calculated for the reaction,

$$2Cu^{+1} + Mg \rightarrow 2Cu^0 + Mg^{+2} \tag{3}$$

Both MgO and Li_2O were found in the cathode region.

Earlier cells were limited by self-discharge. Copper ions diffused from the vicinity of the cathode to the magnesium anode, where they reacted to form submicron particles of copper. The latter, suspended in the electrolyte, agglomerated to form a short circuit between the anode and the cell case or the cathode. This type of self-discharge was characterized by erratic voltage and current fluctuations. Woven glass separators prevented internal shorts for certain cell configurations(54). For others, (the rate-controlling step in the self-discharge process was the diffusion of copper ions from the cathode to the anode) the rate of self-discharge was reduced considerably by increasing the electrode spacing and by adding CsCl to the electrolyte.

In a subsequent investigation, Thaller observed a large drop in open-circuit potential (~ 0.25 V) beyond 50 percent discharge(55) which he attributed to the formation of a Li_2O-CuO complex that reduced the activity of the cuprous ion by about two orders of magnitude. The complex could be formed by adding Li_2O to a LiCl-KCl electrolyte, whereas the addition of $MgCl_2$ retarded its formation but caused a much higher self-discharge rate. Potentiometric titrations with Li_2O and $MgCl_2$ allowed more accurate characterization of this interaction. Thaller suggests that magnesium ions move to the catholyte, where they are precipitated as magnesium oxide during the initial stages of discharge. During the latter stages of discharge, lithium ions predominate in the charge neutralization process, creating conditions favorable for the formation of the cuprous complex.

ANTIMONY AND COPPER OXIDE

Senderoff, Klopp, and Kronenberg(56, 57) investigated single cells of Ca/LiCl-KCl/CuO or Sb_2O_3/Ag. CuO and Sb_2O_3 were selected for their thermal stability, low solubility in the electrolyte, and low reactivity with the electrolyte. The electrolyte or separator layer contained a compressed mixture of eutectic LiCl-KCl and powdered ZrO_2, MgO, Al_2O_3, or kaolin. In some cells with CuO, an iron powder was added to the separator layer to reduce any diffusing copper ions and prevent a bridge, or short, from forming between the calcium and cathode. The cathode layer of CuO cells contained copper metal, powdered CuO, MgO, and LiCl-KCl eutectic. MgO was added to obtain mechanical consistency within the layer. Mixtures of Sb_2O_3 and LiCl-KCl eutectic, with and without graphite, were used in the cathode layer of Sb_2O_3 cells. A wt percent of 45 or more Sb_2O_3 was needed unless an excess of graphite was used. The copper or graphite in the cathode layer increases conductivity during the initial phases of the discharge. As the discharge progresses, the conductivity is increased by the metallic copper or antimony produced in the cathodic reaction reaction.

CuO cells with this construction had open-circuit voltages from 2.2 to 2.5 V. A typical single-cell discharge at 600°C through a 2-ohm load produced 120 ma/cm² at a peak voltage of 1.84 V. Discharge life to 80 percent of peak voltage was seven minutes. For the Sb_2O_3 cell 113 ma/cm² and 1.80 V were typical. Discharge life was 27 minutes, and open-circuit voltage was 2.2 to 2.4 V.

NITRATES

Electrolytes containing nitrates melt at lower temperatures than most fused-salt systems; however, gas can form from corrosion and decomposition reactions at high cell temperatures. Increasing the nitrate concentration lowers both the activating temperature and the maximum operating temperature, thus, shifting the operating temperature range downward.

The cell, $Mg/LiNO_3$, KCl/Ag, was investigated(58) between 300° and 500°C with electrolyte compositions of 0 to 70 mole percent KCl. With less than 10 mole percent KCl, open-circuit voltage decreased with time and a film formed on the magnesium electrode. With greater than 20 mole percent KCl, steady open-circuit voltages of 1.4 to 1.6 V were obtained. Optimum discharge characteristics were obtained at 50 mole percent KCl. In cells with excess electrolyte of this composition, voltage was 1.4 V at 4 ma/cm² and 0.9 V at 100 ma/cm².

The cathode reduction mechanism is uncertain. Open-circuit voltage is lower and less polarization is observed at a Ag or Ni cathode than at a Pt or graphite cathode. This difference is attributed to a reaction between nitrate and Ag or Ni, since both metals were etched in the $LiNO_3$-KCl electrolyte. The oxidation of silver has been questioned(5) since polarographic studies(59) have shown that nitrate does not react directly with silver in $NaNO_3$-KNO_3 melts. Melts containing chloride were not included in the later studies.

Test tube cells with the system, $Mg/LiCl$, KCl, $LiNO_3/Ag$, where $LiNO_3$ is from 1 to 10 mole percent in an LiCl-KCl eutectic, have provided attractive discharge rates. Batteries using this system would have to be restricted to applications in which the maximum cell temperature could be regulated to avoid the possibility of gas formation.

SULFATES AND PHOSPHATES

It has been reported(23, 57) that at the higher temperatures at which nitrates decompose, the oxygenated anions of Groups IIIA-VIA, such as sulfates, phosphates, carbonates, silicates, aluminosilicates and borates become reactive and act as depolarizers. In general their reaction rates below 450°C are so slow as to render them essentially inert, but at about 600°C and above many of them become effective depolarizers. The phosphates and sulfates appeared to be most effective in single cell tests(23), and one of the better cells,

Ni, Ca|KCl-LiCl|KPO₃-LiPO₃ (with 1.5% NaF), Ni

was discharged at 600°C for one hour at about 2 V and 100 ma/cm². Its open circuit voltage was 3.22 V. Severe materials problems for containers, insulators, and separators were encountered at the high temperatures required.

FLUORIDES

Root and Sutula(60) examined high-oxidation state fluoride compounds in an LiF-NaF-KF eutectic electrolyte (melting point 458°C) at 500°C. The study was directed toward achieving a cathode with a high-energy density through the formation of cathodic reduction products which did not limit cell performance. Open-circuit voltages for cells Mg/LiF, NaF, KF/D/Pt, where D is a fluoride depolarizer, are: (1) FeF_3, 2.3 V; (2) CrF_3, 2.3 V; (3) VF_4, 2.4 V; (4) MnF_3, 2.5 V; (5) K_2MnF_6, 2.5 V; (6) CuF_2, 2.6 V; (7) K_3CuF_6, 2.8 V; and (8) AgF_2, 3.2 V. Only cells employing K_3CuF_6, AgF_2, and VF_4 showed any appreciable voltage under load. In three-layer pressed-pellet cells (Mg layer, LiF-NaF-KF layer, and fluoride depolarizer layer) only K_3CuF_6 showed promising behavior at 500°C. This cell furnished 21 ma/cm² for 12 minutes to 80 percent of the peak voltage of 2.2 V. At 155 ma/cm² peak voltage was 1.7 V and discharge life was four minutes.

Several of the materials which exhibited no performance under load were retested, with a glass frit separator and a conductive powder mixed with the cathode material. Only MnF_3 and a silver powder performed well, delivering 50 ma/cm² at 2.8 V and 1.2 amp/cm² at 0.5 V. Thermogravimetric studies showed that silver powder does not react with MnF_3 at 500°C; therefore, the cathodic reaction was concluded to be the reduction of MnF_3.

SILVER OXIDE

The system, Zn/NaOH, KOH/Ag₂O/Ni, was investigated by Doan(61) in an effort to approach the characteristics of aqueous silver-oxide: zinc cells with their excellent capacity, high drain, and voltage control, and to find an electrolyte system that would melt at a much lower temperature than the LiCl-KCl system.

Experimental cells never approached their aqueous counterparts. Typical performance of a 5- × 5-cm cell was 100 seconds of discharge to 80

percent of peak voltage at 700 ma/cm² and 250°C. Open-circuit voltage
ranged from 1.86 V at 200°C to 2.05 V at 300°C. Activation was slow
Difficulties encountered with the cell were (1) decomposition of the Ag₂C
above 200°C and (2) interaction of zinc with the electrolyte, both reaction
yielding gas which could expel electrolyte from the cell and increase cel
resistance. To modify these difficulties, CaO was added to the electrolyte
to reduce electrolyte fluidity, and HgO was added to reduce gassing
Electrolyte was dried at 500°C to reduce moisture and to ensure inertnes
during storage. Since safe operating temperature range was only 200° t
275°C, ambient temperature spread permitted for this cell was less than
the usual −54° to 74°C.

SILVER CHLORIDE, POTASSIUM CHROMATE

Selis, Wondowski, and Justus(13) working with cells that were no
restricted to the small anode-to-cathode distances necessary for rapid cel
activation, obtained high power outputs for the system, Ca/LiCl, KCl
AgCl, K₂CrO₄/Ag. In test tube cells with a large excess of electrolyte
1.5 amp/cm² was obtained for over 10 minutes at 2.6 V and 450°C. The
electrolyte composition in mole percent was 33.3 KCl, 50.1 LiCl, 10 AgCl
and 6.6 K₂CrO₄.

The high output was related to the high rate of depolarization of AgCl a
this temperature and the continuous oxidation of Ag by the chromate
present. Lithium ion in the electrolyte promotes the formation of lithium
calcium alloy, preventing silver or oxide film from covering the calciun
surface and polarizing the electrode. The chromate is present in the
electrolyte in such concentration that its reaction with the anode physi
cally stabilizes the lithium-calcium alloy, enabling it to be contained
within the anode area.

A rectangular can cell, 8.9 × 8.9 × 1.6 cm yielded, at 450°C, a current o
100 amp (0.86 amp/cm²) at 1.7 V. In another cell, effective volume 118 cm
operated at a load of 30 amp (0.26 amp/cm²), 8 W-min/cm³ of energy wa
obtained to 1.9 V (20 percent of peak voltage). Additional volume would
be needed for the heating mechanism, thermal insulation, and leads. Thi
type of cell contained, roughly, 10 times the electrolyte volume of norma
thermal battery cells.

SILVER CHLORIDE IN FUSED ORGANIC ELECTROLYTE

Wallace and Bruins(62) found molten acetamide with dissolved KCl to be
the best of several fused organic electrolytes investigated, several o
which, though thermally stable and good conductors, were eliminated
from consideration because they reacted spontaneously with chromate
or silver ion depolarizer. The cell, Zn/2.5 wt percent KCl

cetamide/AgCl/Ag, provided the best discharge characteristics of several examined. It furnished 2 ma/cm^2 at 1.0 V for 360 minutes at 100°C. Termination was caused by cathodic polarization. Voltage regulation and uniform discharge were a function of the porosity of the silver chloride electrode. Cell resistance and polarization decreased directly with temperature from 90° to 140°C. Cells of this type have the advantage of low-temperature activation; however, as with other low-temperature activating cells, their temperature range of operation is limited. This discourages their use for general thermal-battery applications.

LONG-LIFE CELLS

Long-life, low-current-drain cells have been investigated for applications where activation time is not a problem and a continuous source of heat is available. Thaller(22) has described conventional fused-salt galvanic cells while those with ceramic and ceramic/fused-salt electrolytes have been reported by other investigators(63–68).

As a result of the observation that two metals separated by a porcelain enamel are able to produce electric power when heated in air, a cell was developed(63) with an aluminum alloy (20Mn, 4Li, 4Si) on one side of a lithium borosilicate enamel and a porous coating of silver on the other side. It produced an open-circuit voltage of 2.62 V at 650°C. Lithium or aluminum was assumed to oxidize at the anode, while oxygen was reduced at the porous silver cathode. Although this cell is not a fused-salt cell, it is included to illustrate possible solid-separator capability. Ionic conductivity in the enamel was sufficient to produce an initial output of 0.1 ma/cm^2 at 2.3 V, which decayed to 2.5 ma/cm^2 and 0.65 V in 80 minutes at 650°C. Increasing the operating temperature of the same cell to 700°C and redischarging at the same load increased the output to 6.3 ma/cm^2 and 1.6 V; dropping in 90 minutes to 2.9 ma/cm^2 and 0.75 V. Such cells were proposed for reentry and other aerospace applications where large electrode areas and high temperatures (450° to 700°C) would be available.

Lozier(64) proposed the cell-type, alkali metal/ceramic separator/fused-salt alkali chloride electrolyte/metal ion—metal for high-energy-density applications. Hamby, Steller, and Chase(65) investigated separators for such cells, using Li-Sn and Na-Sn alloy anodes and NiCl$_2$-Ni and Cl$_2$-C cathodes. Maximum continuous anodic current density above 1 V was 42 ma/cm^2, obtained with a cell of a Na-Sn anode, a sodium aluminosilicate membrane 1 mm thick, NaCl-CaCl$_2$-NiCl$_2$ electrolyte and a nickel cathode operating at 730°C. Subcasky(66, 67) investigated a similar cell with Li-Mg anode, zeolite separator, and a molten CuCl, Cu cathode. Open-circuit voltage was 3.29 V. Cell voltage under

6.7 ma/cm^2 load was 1.6 V at 427°C. When a porous glass separator wa·
used, a voltage of 2.1 V was obtained at the same current drain. Lifetim·
was measured in hours. High operating temperatures and problems suc♭
as the membrane cracking, or being attacked by the alkali metal, limit th·
effectiveness of cells of this type for conventional thermal batter♩
applications. Further improvement in conductivity and mechanical stabil
ity of the membranes would be required to make them attractive.

Plizga, Arrance, and Berger(68) investigated single cells with Li and M♩
anodes, NiCl$_2$ and CuO cathodes, an excess of LiCl-KCl eutectic electro
lyte and an Astroset inorganic separator of undisclosed composition·
Relatively flat discharge curves were obtained at 5 to 7 ma/cm^2 with a▪
Mg anode, CuO or NiCl$_2$ cathode, and a 0.25-cm-thick separator. Voltag·
decreased from 1.72 to 1.58 V in 10 hours at 425°C for the CuO cell. In th·
discharge interval from 10 to 60 hours, with the NiCl$_2$ cell at 425°C
voltage decreased from 1.70 to 1.47 V. The separator was characterize·
by a resistance area product of 1.39 ohm-cm^2. No mention was made o·
separator deterioration.

HEAT-SENSING THERMAL BATTERY

Levy(27, 69) investigated various low-temperature-activating cell sys
tems for use in heat-sensing applications such as fire detection. A battery
was desired that would not only detect a temperature rise above a given
threshold value but would also furnish the power necessary for ·
communication or alarm system.

Systems of Ca/LiCl, KCl, LiNO$_3$/MoO$_3$/Ni and Ca/LiCl, KCl
LiNO$_3$/Ag$_2$CrO$_4$/Ni activated at 163° to 232°C, but high internal impedance
at these temperatures limited the power output. It was recognized that the
temperature at activation would have to be 50 or more degrees above the
melting point of the electrolyte for adequate power production; therefore
a heat source was sought which would autoactivate above a predeter·
mined temperature and furnish heat to bring the electrochemical system
to the desired temperature range. Slow heating of pyrotechnic powders
used in conventional thermal batteries did not cause their ignition in this
temperature range. The heat generator developed specifically for this
application makes use of the exothermic heat furnished by the alloying of
lithium metal with a lead-tin solder. Li and a 60 Sn-40 Pb solder melt at
approximately 186°C. When a mixture of the two reaches this tempera·
ture, the liquids interact, forming Li$_7$Sn$_2$ and Li$_7$Pb$_2$, and instantaneously
release a large (up to 230 cal/g) pulse of thermal energy. This heat is

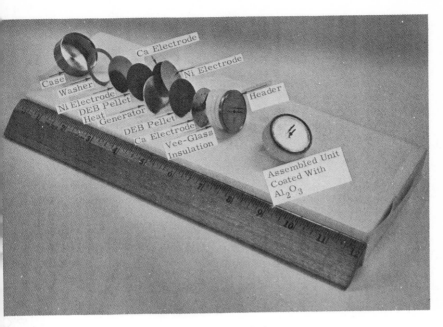

Figure 6.3 Heat-sensing thermal battery, exploded view.

sufficient to bring pellet cells of the type Ca/LiCl, KCl/Ag₂CrO₄/Ni to the required operating temperature range.

Batteries consisting of two pellet cells and a heat-generator pellet, as shown in Fig. 6.3, have furnished 22.5 amp (2.8 amp/cm²) at 2.2 V for four seconds. The pellet cell is made up of a calcium disc with nickel backing; a pellet of LiCl, KCl, Ag₂CrO₄ and molecular sieve powder binder; and a nickel disc cathode collector. The heat pellet contains a lithium disc sandwiched between solder discs. Battery volume is 4.8 cm³ (0.75 in.³) and weight is less than 45 g (0.1 lb). Half the volume is insulation.

The temperature of activation may be varied by additional battery insulation if one is able to approximate the external heating rate. Electrical output may be varied to meet different requirements by changing the area and cell arrangement or by using a different electrochemical system. Storage is not expected to be a problem, since hermetically sealed heat pellets have been stored at 93°C for six months without detrimental effect. Use of this battery is not limited to the detection of fires. Any other source of heat will cause the battery to activate, regardless of the heating rate.

BATTERY OPERATION

Many cell systems are used in production batteries and a few of the more familiar systems which have been used by Sandia Laboratories are discussed in more detail to illustrate both the capabilities and limitations inherent in their operation.[1] This coverage, admittedly is not complete for example, V_2O_5 cells are not included and chromate is the only depolarizer discussed. However, factors influencing the behavior of these should also influence other systems.

HIGH-VOLTAGE BATTERIES

One of the oldest systems used for high-voltage batteries is Ca/LiBr, KBr K_2CrO_4/Ni. The electrolyte pad contains a depolarizer and a glass-cloth binder. Activated life is from 30 to 200 seconds at current drains of 1 to 10 ma/cm^2. Percentage of available capacity expended electrochemically is of the order of 2 percent. Consumption of active cell materials by direct reaction of chromate with calcium, and high cell impedance from cell cooling, limit this efficiency.

A noneutectic electrolyte composition is chosen so that both solid and liquid phases are present over most of the expected operating temperature range for the cell; that is, 325° to 575°C. The eutectic melts at 312°C and contains, in wt percent, 30KBr, 60LiBr, 10K_2CrO_4. A typical electrolyte composition is 40KBr, 40LiBr, 20K_2CrO_4. The solid phase restricts the transport of chromate to the calcium, thus reducing the consumption of active materials by direct reaction. The amount of liquid need only be sufficient to assure cell conductivity and diffusion of chromate to the cathode.

In actual tests, more than 95 percent of the chromate capacity and 65 to 70 percent of the calcium capacity were lost when cells were maintained on open circuit for 420 seconds at 560°C. (The cells contained an excess of calcium over chromate capacity.) The same cells at 332°C for the same time interval lost less than 5 percent of their chromate capacity. Chromate transport to the anode is much higher in the liquid state, above 550°C, than in the liquid-solid temperature region. The effect of temperature on life to 80 percent of peak voltage is shown in Table 6.2. Current drain was 0.3 to 0.5 ma/cm^2. Open-circuit voltage at 580°C decreased from 2.90 to 2.15 V in five minutes; whereas, at 320°C, the same voltage decrease occurred in 370 minutes.

High-voltage battery production now utilizes a new design, in which miniature cells are stacked in electrically insulating but thermally con-

[1]The information reported here has been extracted from unpublished reports of several authors at Sandia Laboratories.

TABLE 6.2. LOW-CURRENT THERMAL CELLS DISCHARGED AT 0.5 MA/CM²

	Temperature (°C)				
	300	350	400	450	500
Peak voltage (volts)	1.30	2.70	2.80	2.68	2.69
Life to 2.15 V (seconds)	—	2210	1218	970	540
K_2CrO_4 remaining (mg/cell)	14.4	1.26	0.19	0.27	0.20
Percent K_2CrO_4 remaining (assuming original content of 14.4 mg/cell)	100	8.7	1.3	1.9	1.4

ducting beryllium oxide tubes. They are heated by a pyrotechnic ($Fe/KClO_4$) source which surrounds the tube. The cell stack consists of alternating bimetallic discs and DEB[2] pellets. Bimetallic discs are stamped from calcium evaporated on thin iron sheet. The DEB pellet contains $CaCrO_4$ depolarizer, LiCl-KCl eutectic electrolyte, and SiO_2 binder. A special assembly fixture is used so that the bimetallic disc always has the same side up. Each cell is about 0.05 cm thick and 0.32 cm in diameter. A number of cell stacks are connected in series to achieve the desired high voltage. A battery with a volume of 16.1 cm³ and a weight of 41 g can deliver 3 ma at 1000 V for 40 seconds. Prior to this design, the smallest available high-voltage thermal battery was capable of 3 ma at 500 V for 60 seconds, weighed 227 g, and occupied 108 cm³.

LOW-VOLTAGE BATTERIES

The closed-cup design was for many years the principal one for heavy-current-drain applications. The depolarizer—most commonly $CaCrO_4$, WO_3, or Fe_2O_3—is initially separated from the electrolyte. Eutectic LiCl-KCl is the prevailing electrolyte, although other compositions have been used for special purposes. The operating temperature range is 425° to 600°C. Current drains up to 200 ma/cm² have been obtained with a cell voltage of 2 or more volts. Output of a typical cell is 14 Whr/kg or 37.5 Whr/dm³.

Various cell parameters for the system, Ca/LiCl, KCl/CaCrO₄/Ni, were studied in this laboratory by utilizing single-cell discharges. At an optimum discharge temperature of 475°C, activated life[3] increased directly with the quantity of electrolyte, as shown in Fig. 6.4. With a limited amount of electrolyte, activated life varied according to the source of the

[2]The term DEB is used to represent depolarizer-electrolyte-binder combinations.
[3]Activated life is defined as the discharge time to 80 percent of peak voltage.

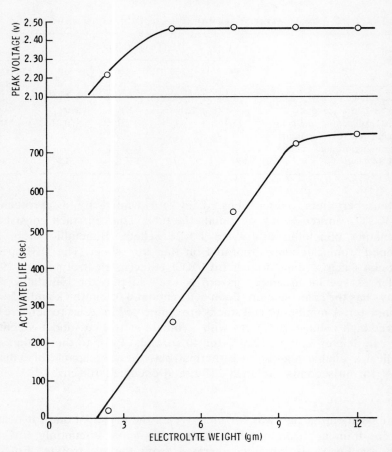

Figure 6.4 Activated life and peak voltage versus weight of electrolyte (cells discharged at 40 ma/cm² and 475°C).

$CaCrO_4$ used in the cell. With an excess of electrolyte, activated life was longer and the $CaCrO_4$ differences disappeared or were minor. Electrolyte leakage and Li-Ca alloy formation limit the amount of electrolyte. Predrying the $CaCrO_4$ pad for one or more hours at 500°C reduced electrolyte leakage.

Cell components change in weight and chemical composition as a result of activation. These changes were measured for constant-temperature discharges, using the cell assembly shown in Fig. 6.5. Cells were disassembled after discharge while still hot into anode, electrolyte, and cathode sections. Above 500°C separation was difficult and often not

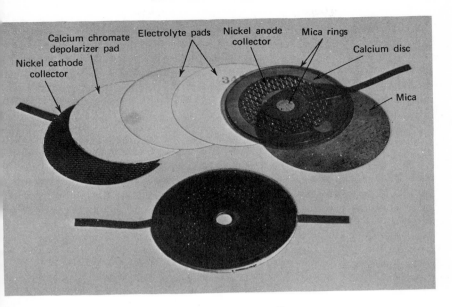

Figure 6.5   Cell assembly used for material-balance studies.

complete. At 425° and 475°C weight measurements showed that 35 to 50 percent of the electrolyte was taken up by the cathode and 25 to 35 percent by the anode. Cell resistance was calculated to be 0.002 to 0.005 ohm for the electrolyte remaining in the electrolyte pad. Since the quantity of electrolyte absorbed by the cathode was relatively constant, a decrease in the quantity of cell electrolyte reduced the amount of electrolyte available for the anode reaction.

Material analyses were run on cells discharged at 50 ma/cm² and 475°C to a cutoff voltage of 2.1 V. For discharge times of six to seven minutes, approximately 13 percent of the $CaCrO_4$ and 10 to 11 percent of the calcium were consumed electrochemically. From 62 to 94 percent of available chromate and 86 to 96 percent of the calcium were accounted for by cell reaction and chemical analyses of discharged electrodes. Material accountability was greater at the lower operating temperature.

Cells discharged at 375° and 425°C could be redischarged without alteration at temperatures of 475° and 550°C. Activated life for the second discharge was often more than that obtained for the first. Several anode and cathode sections of discharged cells were reassembled and discharged with new cathode and anode sections and new electrolyte. In one series of discharges at 475°C, with an average current drain of 30 ma/cm², one anode was used for seven discharges and one cathode for six. Electrochemical utilization of $CaCrO_4$ to a 2.1-V cutoff was 50 percent. Anode sections of discharged cells that were used in an unaltered condition with new cathode and new electrolyte had only 50 percent of the life of a new cell. When the calcium surface was scraped free of electrolyte prior to the second discharge, cell life was 80 to 100 percent of normal life.

An estimate of the ohmic voltage drop in the cells was made by interrupting the current for short intervals during discharge. Since the ohmic voltage drop ceased immediately upon removal of the load, its magnitude could be estimated by the straightline portion of the recovery curve obtained on a high-speed voltage recorder. Also included in this measurement was the activation polarization; its contribution was believed to be small in comparison to the ohmic term. This assumption is based on studies of Piontelli and co-workers(70)(71) and on the high exchange currents for electrode reactions in fused salts(42). The magnitude of the straightline portion of the curve increased with a decrease in amount of cell electrolyte, a decrease in operating temperature, and the extent or period of the discharge. These conclusions were confirmed by Clark and Zaffery(72) using an instrument developed to provide a continuous measurement of internal resistance of the cell during discharge.

Individual electrode polarization effects were measured by means of a reference electrode and a high-impedance voltmeter. The reference electrode consisted of a platinum wire immersed in a 0.1-molal solution of K_2PtCl_4 in eutectic LiCl-KCl. It was contained in a thin-walled glass capillary tube.[4] By combining this reference electrode with current interruptions it was possible to determine which portion of the cell, anode or cathode, contributed most to the cell ohmic voltage drop. Comparisons were made at three points in the progress of discharges at 1-ohm load (40 ma/cm^2) for two weights of cell electrolyte. The results are given in Table 6.3.

The major portion of the ohmic voltage term, which is a considerable fraction of the cell voltage decrease from open-circuit voltage, is attributed to the anode. This ohmic voltage change is a function of the temperature and amount of electrolyte. Above 475°C less of the total voltage drop is attributed to ohmic polarization.

For these discharges we have seen that there are adequate amounts of electrode materials remaining at cutoff voltage and that there is sufficient

[4]Panzer(73) has discussed reference electrodes for measurements in pellet cells and recommends a Mg/MgO electrode.

TABLE 6.3. CELL DISCHARGE WITH REFERENCE ELECTRODE

Temperature (°C)	Electrolyte (gm/cell)	Time of Measurement (min)	Cell Voltage (volt)	Voltage Change from OCV[a] (volt)	Cell Ohmic Voltage Drop (volt)	Anode to Referenced Electrode Ohmic Voltage Drop (volt)	Cell Life to 2.1 V (sec)
425	4.7[b]	1	2.45	0.20	0.12	0.05	250
		5	2.00	0.65	0.42	0.26	
		10	1.33	1.32	0.96	0.72	
475	4.7[b]	1	2.47	0.18	0.11	0.05	340
		5	2.15	0.50	0.30	0.23	
		10	1.83	0.82	0.52	0.43	
475	7.0[c]	1	2.45	0.20	0.10	0.03	590
		5	2.35	0.30	0.11	0.05	
		10	2.09	0.56	0.23	0.13	

[a] Open-circuit voltage for this cell was 2.65 V.
[b] Two electrolyte pads.
[c] Three electrolyte pads.

electrolyte for cell conductivity, yet the useful life of the battery has ended. One may logically ask, What went wrong? Why are we utilizing but a small fraction of the theoretical energy density of the system? Two factors in addition to alloy shorting may be postulated which limit cell life at any given temperature. One, which is more prevalent at high temperature, is the direct reaction of chromate with calcium, forming products on the surface of the calcium. In some cases, there is tentative evidence for products building up from the calcium to the cathode, thereby causing a high-resistance electronic path or internal short. As the cell temperature increases, solubility and transport of $CaCrO_4$ in the electrolyte increase and more direct reaction occurs.

The other factor is the buildup of reaction products at the electrode making the interface between the active electrode material and electrolyte resistant to the passage of ions. At the calcium electrode, two reactions occur during discharge: (1) the displacement reaction between calcium and lithium chloride,

$$2Ca + 2LiCl \rightarrow CaCl_2 + CaLi_2 \tag{4}$$

and (2) the electrochemical reaction,

$$Ca + 2Cl^- \rightarrow CaCl_2 + 2e^- \tag{5}$$

Both reactions produce $CaCl_2$ in the vicinity of the calcium electrode, producing a concentration polarization effect and lower cell voltage. A more important effect of the increase in concentration of $CaCl_2$ next to the anode is the precipitation of solid, from the three-component system, $LiCl-KCl-CaCl_2$, on the calcium surface to impede the movement of charge through this interface and account for the large ohmic voltage drop indicated above. Measurements on cells of this type by Doan(74) at Eagle Picher Industries by means of an alternating current circuit have confirmed the large ohmic contribution to cell polarization. As the temperature is lowered or quantity of electrolyte at the anode is decreased, precipitation of solid and formation of a film or crust occur at an earlier point in the discharge and shortens cell life. Additional life of cells redischarged at higher temperatures after discharge at 375° to 400°C is attributed to the dissolution of this film. The temperature at which solid first appears and the fraction of solid present at any given temperature can be obtained from the phase diagram(75)(76) for $LiCl-KCl-CaCl_2$. Neglecting the reaction between Ca and $CaCrO_4$, precipitation of solid is calculated to occur within five minutes for a discharge of $50 \, ma/cm^2$ at 425°C. The fact that solid salt can form at the calcium electrode lends support to the film or barrier postulate. There is no indication that magnesium undergoes a similar displacement reaction with LiCl, but polarization of the magnesium electrode was found to be appreciable at 375°C to 400°C.

Recent studies of magnesium anodes in $MgCl_2$-LiCl-KCl melts by Thaller(77) show the polarization characteristics of the magnesium anode to be very dependent on the operating temperature, magnesium ion concentration, and the current density. He suggests that, in electrolytes containing magnesium ions, the mass transport processes are not always able to dissipate the magnesium ions resulting from anodic discharge, and the anolyte freezes adjacent to the anode. In an operating battery this situation could result in premature failure. To overcome this difficulty, he suggests that more electrolyte per unit of cell capacity or adjustment of the mass transport characteristics of the cell be used to allow better mixing of the anolyte and catholyte. The role of the electrolyte as not only an ionic conductor but also a sink for reaction products needs to be emphasized in cell design.

Pellet cells (Fig. 6.1) are supplanting the closed-cup type for low-voltage/high-drain applications, being more efficient in terms of output per unit weight and volume. Cells with the system, Ca/LiCl, KCl, $CaCrO_4$, SiO_2/Fe($KClO_4$), weigh between 13 and 14 g. The cells are stacked and maintained under a pressure of about 13 kg/cm^2 to ensure good material contact within the cells and low resistance contact between the cells. For a typical single-cell discharge at 60 ma/cm^2 and 525°C, voltage decreases from 2.6 to 2.1 V in 300 seconds. For such a discharge approximately 8 percent of the Ca and 11 percent of the $CaCrO_4$ react electrochemically. In cells maintained on open circuit for 300 seconds at 525°C, 15 percent of the chromate reacted with the calcium. At 600°C, 25 percent reacted. The amount of chromate reacting was determined from chemical analyses of activated and unactivated cells. As with other cell designs, activated life is not limited by consumption of electrode materials but by the rise in internal resistance of the cell. This rise occurs primarily from the buildup of insoluble reaction products at the electrodes. These products are formed by both electrochemical and direct reactions.

The effect of temperature and cell pressure on single cells discharged between heating platens is shown in Fig. 6.6. Temperature is an important factor in the operation of these cells. If temperature is too low, salts freeze from the electrolyte; if too high, the corrosion reactions accelerate, generating additional heat and reaction products. Temperature regulation is a perennial problem in thermal battery design. McKee(2) pointed out its significance in closed-cup cells several years ago in his statistical study of cell variables. It is even more critical for pellet cells with their more restricted operating temperature range, and is especially true for those using a minimum amount of electrolyte to provide a mechanically rugged

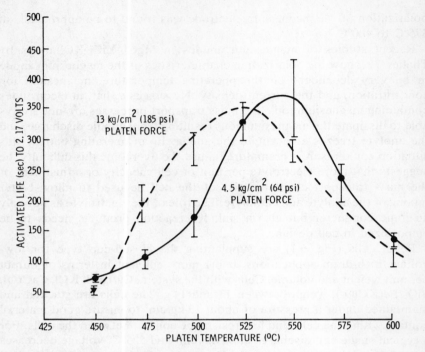

Figure 6.6 Effect of temperature on activated life of single cells tested between heating platens (five pellet cells of Ca/LiCl, KCl, CaCrO$_4$/Fe were discharged at 60 ma/cm^2 to 2.17 V at each temperature and platen force).

cell. An efficient ionic separator would be one means for increasing the operating temperature range of a cell.

Caution should again be emphasized regarding single-cell tests at constant temperature. However useful they are for screening and evaluating cell designs and materials, they do not always simulate the behavior of a cell in a battery. Ambient temperature prior to activation, heat generated from the heat source and cell reactions, and heat taken up by the battery components, including the insulation and container, determine the cell temperature, which usually decreases with discharge time for low temperature activation, but can increase for high-temperature activation. It is seldom that a cell in a battery operates at constant temperature.

The ideal thermal battery design should include a regulator for the thermal environment so that the optimum temperature range, as determined from single-cell tests, is maintained throughout the discharge, regardless of the temperature before activation.

SUMMARY

Thermal batteries, which are used primarily for military and aerospace applications, have unique storage, activation, and operational capabilities. The most common electrochemical systems contain calcium or magnesium anodes; LiCl-KCl electrolyte; and vanadium, iron, or copper oxide, or chromate depolarizer. Unless they are operated at optimum temperature and current density, only a very small fraction of the available energy density is utilized in a discharge. Through better regulation of the thermal environment within the battery and more efficient utilization of the electrolyte, the output of these systems could be increased considerably. If an efficient ionic conducting separator could be developed, operating temperature range could be increased and soluble metallic ion depolarizers could be used.

The fire-sensing battery is an interesting innovation which should extend the use of these batteries into the commercial market.

REFERENCES

1. Goodrich, R. B. and Evans, R. C., *J. Electrochem. Soc.*, **99**, 207 (1952).
2. McKee, E. S., *Proc. Annual Battery Res. Dev. Conf.*, **10**, 26 (1956).
3. Goldsmith, H. and Smith, J. T., *Electrochem. Tech.*, **6**, (1–2), 16 (1968).
4. Jasinski, R., *High Energy Batteries*, Plenum Press, New York, 96–123 (1967).
5. Reddy, T. B., *Electrochem. Tech.*, **1**, (11–12), 343 (1963).
6. Hamer, W. J. and Schrodt, J. P., *J. Am. Chem. Soc.*, **71**, 2347 (1949).
7. Rubin, B. and Malmberg, M. S., *Nat. Bur. of Stds. Report*, 3081, Washington, D.C. (1954).
8. Doan, D. J., *Proc. Ann. Batt. Res. Dev. Conf.*, **11**, 64 (1957).
9. Selis, S. M. and McGinnies, L. P., *J. Electrochem. Soc.*, **106**, 900 (1959).
10. Selis, S. M., McGinnis, L. P., McKee, E. S., and Smith, J. T., *J. Electrochem. Soc.*, **110**, 469 (1963).
11. Ubbelohde, R. and Rogers, S. E., *Brit. Pat.* 865,810 (April 19, 1961).
12. Indig, M. E. and Snyder, R. N., *J. Electrochem. Soc.*, **109**, 757 (1962).
13. Selis, S. M., Wondowski, J. P., and Justus, R. F., *J. Electrochem. Soc.*, **111**, 6 (1964).
14. Jerome, J. C., U.S. Pat. 3,194,686 (July 13, 1965).
15. Kurtzweil, T. J., Kronenberg, M. L., and Hansen, R. E., U.S. Pat. 3,201,278 (August 17, 1965).
16. Thaller, L. H., *J. Electrochem. Soc.*, **113**, (4) 309 (1966).
17. Klopp, E. M., Senderoff, S., and Hansen, R. E., U.S. Pat. 3,258,365 (June 28, 1966).
18. Nee, C., Pathe, C., Charnay, C., Clans, J., and Leroy, M., Fr. Pat. 1,502,006 (November 18, 1967).
19. Wallace R. A. and Bruins, P. F., *J. Electrochem. Soc.*, **114**, (3) 212 (1967).
20. Zellhoefer, G. F., U.S. Pat. 3,311,503 (March 28, 1967).
21. Zanner, J. H. and Zellhoefer, G. F., U.S. Pat. 3,345,214 (October 3, 1967).
22. Thaller, L. H., *J. Electrochem. Soc.*, **115**, (2), 116 (1968).
23. Senderoff, S. and Klopp, E. M., U.S. Pat. 3,361,596 (January 2, 1968).
24. Panzer, R. E., U.S. Pat. 3,367,800 (February 6, 1968).

25. Yonemaru, I., U.S. Pat. 3,404,041 (October 1968).
26. Osborne, F. H. and Hull, C. J., U.S. Pat. 3,421,941 (January 14, 1969).
27. Levy, S. C., U.S. Pat. 3,425,872 (February 4, 1969).
28. Jost, E. M., Ger. Pat. 1,291,394 (March 27, 1969).
29. Mead, R. T., *Proc. Ann. Batt. Res. Dev. Conf.*, **23**, 137 (1969).
30. Nielsen, N. C., *Proc. Ann. Batt. Res. Dev. Conf.*, **23**, 140 (1969).
31. Danly, D. and Walker, R. D., U. of Fla., unpublished report (1956).
32. Clark, R. P., U.S. Pat. 3,527,615 (September 8, 1970).
33. Plambech, J. A., *J. Chem. Eng. Data*, **12**, 77 (1967).
34. Hamer, W. J., Malmberg, M. S., and Rubin, B., *J. Electrochem. Soc.*, **103**, 8 (1956).
35. Hamer, W. J., Malmberg, M. S., and Rubin, B., *J. Electrochem. Soc.*, **112**, 750 (1965).
36. Hamer, W. J., *J. Electroanalytical Chem.*, **10**, 140 (1965).
37. Nissen, D. A., unpublished report (1969).
38. Selis, S. M. and McGinnis, L. P., *J. Electrochem. Soc.*, **108**, 191 (1961).
39. Swinkels, D. A. J., *J. Electrochem. Soc.*, **113**, 6 (1966).
40. Behl, W. K. and Beals, D. L., *U.S. Army Electronics Command Report*, ECOM-3166 (August 1969) (AD694368).
41. Rightmire, R. A. and Jones, A. L. *Proc. Ann. Batt. Res. Dev. Conf.*, **21**, 42 (1967).
42. Laitinen, H. A., Tischer, R. P., and Roe, D. K., *J. Electrochem. Soc.*, **107**, 555 (1960).
43. Laithinen, H. A. and Gaur, H., *Anal. Chem. Acta*, **18**, 1 (1958).
44. Giner, J. and Holleck, G. L., Tyco Laboratories NASA-CR-1541 (March 1970).
45. Clark, R. P., Blucher, R. L., and Goldsmith, H. J., *J. Chem. and Engr. Data*, **14**, 465 (1969).
46. Clark, R. P., Goldsmith, H. J., and Blucher, R. L., *J. Chem. and Engr. Data*, **15**, 277 (1969).
47. Blucher, R. L. and Goldsmith, H. J., *Electro Technology*, **77**, 109 (1966).
48. Laitinen, H. and Liu, C., *J. Am. Chem. Soc.*, **80**, 1015 (1958).
49. Laitinen, H. A. and Rhodes, D. R., *J. Electrochem. Soc.*, **109**, 413 (1962).
50. Laitinen, H. A. and Propp, J. H., *Anal. Chem.* **41**, 645 (1969).
51. Hanck, K. W., PhD thesis, Univ. of Illinois (1969).
52. Popov, B. and Laitinen, H. A., *J. Electrochem. Soc: Electrochemical Science*, **117**, 482 (1970).
53. Laitinen, H. A. and Bankert, R. D., *Anal. Chem.* **39**, 1790 (1967).
54. Thaller, L. H., *J. Electrochem. Soc.*, **115**, 116 (1968).
55. Thaller, L. H., NASA-TN-D-5731 (April 1970).
56. Senderoff, S., Klopp, E. M., and Kronenberg, M. L., Union Carbide Corp. Final Report No. NORD 18240 (June 1962) (AD277533).
57. Ibid (September 15, 1962), Part II.
58. Jennings, C. W., *J. Electrochem. Soc.*, **103**, 531 (1956).
59. Swofford, H. and Laitinen, H., *J. Electrochem. Soc.*, **110**, 814 (1963).
60. Root, C. B. and Sutula, R. A., *Proc. Ann. Batt. Res. and Dev. Conf.*, **23**, 100 (1969).
61. Doan, D. J., *Proc. Ann. Batt. Res. Dev. Conf.*, **11**, 64 (1957).
62. Wallace, R. A. and Bruins, P., *J. Electrochem. Soc.*, **114**, 209 (1967).
63. Westinghouse Electric Corp., Report ASD-TDR 62-397, Aerospace Electrical Div., Lima, Ohio (1962).
64. Lozier, G. S., *Proc. Ann. Batt. Res. Dev. Conf.*, **15**, 80 (1961).
65. Hamby, D. C., Steller, B. W., and Chase, J. B., *J. Electrochem. Soc.*, **111**, 998 (1964).
66. Subcasky, W. J., NASA Report No. NASA-CR-54731, 182 (1965).
67. Subcasky, W. J., Place, T. M., Parker-Jones, H. A., and Anderson, W. G., NASA Report No. NASA-CR-54404 (1965).

68. Plizga, M. J., Arrance, F. C., and Berger, C. Douglas Missile and Space Division Report No. 4219 (January 1967).
69. Levy, S. C., Sandia Laboratories Report No. SC-RR-69-697 (October 1969).
70. Piontelli, R. and Sternheim, G., *J. Chem. Phys.*, **23**, 1971 (1955).
71. Piontelli, R., Sternheim, G., and Francini, M., *J. Chem. Phys.*, **24**, 1113 (1956).
72. Clark, R. P. and Zaffery, E. D., *Rev. Sci. Inst.*, **38**, 492 (1967).
73. Panzer, R. E., *Electrochem. Tech.*, **2**, 10 (1964).
74. Doan, D. J., unpublished Eagle Picher report (1959).
75. Plyushchev, V. E., and Korolev, F. V., *Zhur. Neorg Khim*, **1**, 1013 (1956).
76. Seineutsova, D. V., *Zhur. Noerg Khim*, **12**, 1650 (1967).
77. Thaller, L. H., NASA-TN-D-5657 (February 1970).

7.

Water-Activated Batteries
Silver Chloride—Magnesium and
Cuprous Chloride—Magnesium Systems

D. J. DOAN

In the early days of dry battery manufacture the keeping qualities of the commercial product were unsatisfactory. Thus, between 1910 and 1930, a reserve Leclanché cell was marketed to a limited extent. This unit was manufactured completely dry with all the salts contained in the cathode mix. A carbon electrode with a central hole through its length and radiating holes at various depths provided the means for distributing water which was added when activation of the unit was desired. This cell was used in remote areas of the world where fresh supplies of conventional batteries were not readily available. This construction is no longer made and is now only of historical interest. Today, conventional batteries are widely available in the marketplace and customers usually have no difficulty in purchasing the types and sizes that are needed.

As demands grow for an increase in the energy output from primary battery systems, the battery designer looks toward the use of more reactive components than those utilized in conventional systems. Many attractive combinations of electrodes and special electrolytes would not possess the acceptable storage quality required for normal use due to undesirable side reactions. By combining all the dry elements of a cell in one package and the electrolyte in a second package, no deterioration of these separate materials takes place. When operation is desired the electrolyte is added to the dry part of the unit to "activate" the cell. This reserve cell can be stored in the inactive state often for many years and then placed in operation in a very short time when needed.

GENERAL CONSIDERATIONS

In the design of a reserve cell it is possible to select any active component of the cell for packaging separate from the remainder of the components. In practice, it is usually the electrolyte that is kept separate from the dry ingredients until activation is undertaken. The field of reserve cells may be divided into the following groups depending on the type of activating fluid or gas that is employed.

1. Water-activated cells
   (a) Those activated with freshwater
   (b) Those activated with seawater
2. Cells activated by special electrolytes
   (a) The activating fluid is supplied
   (b) The activating fluid is prepared from salts supplied and water which may or may not be furnished
3. Cells activated by the addition of a reactive gas.

Cells of this type were described in Volume I, Chapter 10, entitled "Low-Temperature Nonaqueous Cells."

A list of specific systems is shown in Table 7.1.

Although some of the systems shown in Table 7.1 were developed to meet commercial requirements, the greater part of the effort was spent to develop special cells for the U.S. Army during and following World War II. Here a group of new factors was introduced, including high current density, low-temperature performance, a long shelf life in the reserve state of 10 years or more, and compatibility with many military requirements. The first named group, those utilizing Cu_2Cl_2 and $AgCl$ and activated with water or seawater, comprise the subject of this chapter.

TABLE 7.1. A LIST OF RESERVE BATTERY SYSTEMS

Group	Battery System	Activating Electrolyte	Electrolyte Supplied in Form of
1	Mg/Cu_2Cl_2	Water or seawater	Furnished at the activation site
1	$Mg/AgCl$	Water or seawater	Furnished at the activation site
2	Mg/Air	18% NaCl solution	Furnished at the activation site
2	Zn/Air	NaOH or KOH solution	Water added to cell
2	Zn/CuO	NaOH solution	Flake NaOH plus water
2	$Zn/Silver$ Oxide	KOH solution	Prepared electrolyte
2	Mg/HgO	KOH solution	Prepared electrolyte
2	Mg/HgO	$Mg(ClO_4)_2$ solution	Prepared electrolyte
2	Pb/PbO_2	$HClO_4$ solution	Prepared electrolyte
3	$Mg/AgCl$ (KSCN)	NH_3 gas	NH_3 in ampul under pressure
3	$Mg/HgSO_4$(KSCN)	NH_3 gas	NH_3 in ampul under pressure
3	Mg/m-DNB(NH_4SCN)	NH_3 gas	NH_3 in ampul under pressure

Theoretically, any system using a solid solute in water as an electrolyte may be utilized as a reserve type with water activation. However, the silver chloride-magnesium and cuprous chloride-magnesium systems are the only ones that have had extensive development for appreciable civilian and military application. Both systems may be activated by seawater within a few seconds, but use of fresh water requires sufficient time for the introduction of soluble ions from the electrodes to produce electrolyte conduction before satisfactory functioning is obtained. As a consequence of this, more applications using seawater than fresh have been made for both systems. Where the use permits, the important advantage of these batteries over other reserve types is that no electrolyte need be carried within the battery package prior to activation. Thus, a weight saving results.

Both systems have been known for many years, the silver chloride type was mostly developed between 1943 and 1952 while the cuprous chloride development took place later, after 1949. Several reviews have been published of the silver chloride(1)(2)(3)(4), of the cuprous chloride(5), and both types(6)(7). Much of the battery effort has been published and reports of the early work have now become available(8)(45). Developments have been sponsored in Britain(9)(10) as well as in the United States by the Navy and Signal Corps(11)(12)(13)(14)(15). As evidenced from the patent literature, there exists a wide international interest in both systems(16)(17)(18).

DESCRIPTION OF THE BATTERY SYSTEM

The anode or negative electrode of both systems is, of course, the magnesium, while the cathode is the chloride in each case. The electrochemical reactions along with standard potentials(19)(20) are:

$$Mg \rightarrow Mg^{++} + 2e \qquad E° = 2.37$$

$$2AgCl + 2e \rightarrow 2Cl^- + 2Ag \qquad E° = 0.222$$

$$2CuCl + 2e \rightarrow 2Cl^- + 2Cu \qquad E° = 0.137$$

The standard cell potentials are respectively, 2.592 V and 2.507 V, but these, even when corrected for temperature and concentration differences, are never realized. The observed open circuit voltages in sea water are 1.7–1.9 for AgCl and 1.64 for CuCl-Mg cells. The difference between standard and actual voltages is believed to lie in the anode.

Discharge of the magnesium anode in aqueous electrolyte results in hydrogen production as a result of the wasteful corrosion reactions. The mechanism of this formation has been investigated by workers concerned with Leclanché type magnesium cells. (21 to 25 incl.) Also see Chapter 2.

The construction of a typical water activated battery is shown in Figure 7.1. Cuprous chloride cells produce more hydrogen than those with silver chloride, but both require cell design to permit escape of the gas uniformly, so that cell impedance and voltage will be predictable and reproducible. Some facility is provided either to use the gas to circulate electrolyte or to release the gas without appreciably exhausting the electrolyte during the useful life of the cells. Films of magnesium hydroxide develop on the anode surface through hydrolysis of the magnesium chloride during discharge. A magnesium alloy containing aluminum and zinc is used to prevent an adherent dense film building up so the anode polarization is reproducible during discharge.

Silver chloride batteries are stored in lightproof wrapping to prevent catalytic decomposition. Manufacture, however, is performed in lighted

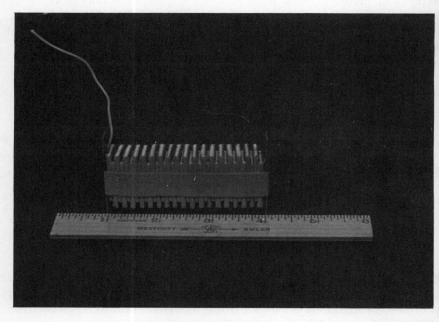

Figure 7.1   The construction of a typical water-activated battery. Courtesy of Eagle-Picher Industries, Inc.

rooms with no appreciable loss of either capacity or voltage from this exposure. Both types are usually stored with desiccant to reduce moisture exposure to the plates. Prolonged moisture exposure affects anode surface by producing a film of magnesium hydroxide on the anode that increases the activation time. Coulometric capacity is not ordinarily lowered since the cells are normally built with a large excess of negative capacity. Moisture results in a loss in capacity of the cuprous chloride positive through the hydrolysis reaction and the oxidation-reduction reactions:

$$2CuCl \rightarrow Cu + CuCl_2$$

$$2CuCl \rightarrow 2Cu + Cl_2$$

There is a belief that oxygen from exposure to air may result in similar chemical change since a few of the cuprous chloride patents(17) have been taken out incorporating reducing agents into the cathode plate mix to improve stand characteristics. As far as is known, no controlled data or results have appeared to show the efficacy of this. Stand characteristics above 110°F are not good since the rates of these degrading reactions increase with increasing temperature. During discharge, increasing the temperature decreases ionic resistance of the electrolyte which normally results in a higher cell voltage (decreased polarization).

ENERGY OUTPUT

The most attractive feature of the water-activated battery is its energy yield which is higher than that of some other battery systems. This feature, of course, is particularly attractive in military applications where both the space and weight of the battery must be kept at a minimum.

Table 7.2 shows a comparison of the energy yield, expressed as watt-hours output per pound of battery weight and per cubic inch of battery volume(28)(44). For comparison similar data for four conventional systems are included. It should be recognized that all battery systems provide their highest energy output under optimum conditions of current drain, discharge schedule and temperature. The data offered in Table 7.2 present maximum values. Since the output of any battery decreases as the current drain is increased, the technologist should expect to obtain reduced output on heavy drains and at temperatures below normal. Battery size is also a factor since the proportion of active ingredients to packaging materials varies somewhat with the size of the package. Thus the smaller batteries usually contain a smaller proportion of active ingredients than the larger units.

ACTIVATION OF RESERVE BATTERIES

Many battery systems have relatively poor stand characteristics after the electrolyte has been added to the plate structure. This defect has created a need for various procedures to introduce the electrolyte just before use The "manually activated" silver zinc battery, saving the weight of the electrolyte container, is one of the highest energy density (watt-hour per pound) units presently known. The outer case is made removable to provide access to the individual cells each of which carry a filler plug Hypodermic syringes are usually used to introduce the electrolyte into the cells.

Water activated silver chloride and cuprous chloride-magnesium bat teries either have provision for the "dunk" filling by immersion or ready manual pouring fill for either sea or fresh water which serve as the electrolyte.

A group known as "automatically activated" batteries comprise those that carry the electrolyte in a separate chamber and expel the electrolyte into the cells on firing an electrically activated match or a mechanically fired primer cap. A "double base" propellant (nitrocellulose-nitroglycerine) gas generator is usually used to displace the electrolyte Three concepts have been used. A long coil contains the electrolyte between diaphragms with the gas generator on one end and the battery manifold on the other. Figure 7.2 shows diagramatically how this method operates.

A collapsible piston type bellows contains the electrolyte with the generator functioning to compress the bellows cup and rupture a diaphragm with a lance. Similar to this is the type using a large ellipsoidal

TABLE 7.2. A COMPARISON OF THE ENERGY YIELDS OF BATTERY SYSTEMS

System	Open Circuit Voltage	Operating Voltage	Watt-Hours Output	
			per lb	per in.^{3}
Reserve Types (28)				
AgCl-Mg	1.9	1.3–1.6	75	7.5
AgO-Zn	1.86	1.3–1.5	50	3.5
CuCl-Mg	1.65	1.1–1.3	35	1.3
PbO$_2$-Pb	2.1	1.6–1.9	20	2.5
Conventional Types				
Leclanché	1.65	1.0–1.2	35	1.7
HgO-Zn	1.37	1.0–1.3	52	8.8
Alk. MnO$_2$-Zn	1.65	1.0–1.4	45	2.7
Ag$_2$O-O	1.6	1.3–1.5	50	1.4–4.0

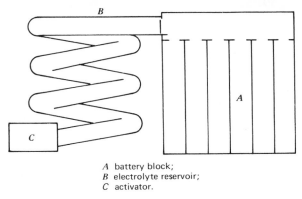

A battery block;
B electrolyte reservoir;
C activator.

Figure 7.2 One method used for the automatic activation of a reserve battery. (Furnished through the courtesy of Eagle-Picher Industries, Inc.).

reservoir having a metal diaphragm which is moved by the gas generator gases. Controlling the operating temperature at the time of activation and operation is performed in two ways. If external electrical power is available before the need for activation, heater windings are provided with thermostats to warm the electrolyte. A typical example of this type would be an aircraft missile which has aircraft power available before release and activation. If external power is not available, a chemically heated exchanger is provided to warm the electrolyte during passage on activation.

ELECTROLYTE

The electrolyte is one of the most important parts of the cell in that, in addition to dissolving the anode reaction products, it furnishes the ionic path between the two electrodes. Thus, its conductivity becomes an important consideration and one that may limit cell operation. In practice, both fresh and seawater activation may be considered.

Freshwater is a term that usually denotes any potable water, which, despite its low mineral content, is sufficiently conductive to permit electrochemical activity. When freshwater is used to activate a reserve battery and the system immediately placed on drain, the working voltage may be somewhat lower than normal at the start. However, as the discharge proceeds the magnesium chloride produced by the operation of the battery dissolves and soon produces an effective electrolyte. There have been suggestions that some electrolyte salt be incorporated in the battery to reduce the effect of this factor. Both cations and anions are introduced into the electrolyte during discharge with the systems.

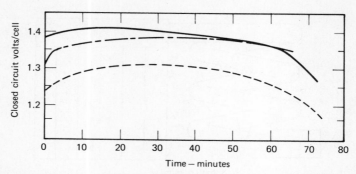

Figure 7.3 Typical discharge curves for AgCl/Mg batteries discharged under various conditions of temperature and salinity. ———— Stored at +20°C. Discharged in 3.6% NaCl at 30°C. ————— Stored at −20°C. Discharged in 3.6% NaCl at 0°C. ——————— Stored at −20°C. Discharged in 1.5% NaCl at 0°C.

When seawater is used to activate the reserve cell the situation is more favorable in that the initial stage of the discharge is usually at a somewhat higher working voltage than when freshwater is employed. Nevertheless, the development of magnesium chloride in the electrolyte, resulting from cell operation, is still an important factor in producing the high conductivity electrolyte so necessary for the efficient operation of the system.

Figure 7.3 shows typical discharge curves for AgCl:Mg batteries discharged under various conditions of temperature and salinity. The early part of the discharge curve is lower than the middle portion, reflecting the result of the increase in electrolyte conductivity that is attributed to the magnesium chloride generated by the operation of the cell. As would be expected, the battery operates more efficiently at higher than at lower temperatures.

In an effort to explain the increase in working voltage during the early part of the discharge, Faletti(36) has reported on the specific conductivity of aqueous solutions of magnesium chloride in a range of concentrations and at temperatures from 5° to 75°C. It is clear from Fig. 7.4, in which this data is presented graphically, that an increase from 2.5 to 25 g/l in magnesium chloride results in an increase in specific conductivity of an order of magnitude. Increasing the temperature also increases the conductivity, an effect used to good advantage in reserve cell applications, as will be discussed later in this chapter.

Cuprous chloride has an advantage over silver chloride because it gradually hydrolyses to produce hydrochloric acid, which in turn, dissolves magnesium metal, to produce more magnesium chloride helping to increase its concentration and the electrolyte conductivity. Some soluble cupric chloride may also be present to some extent, either as an initial

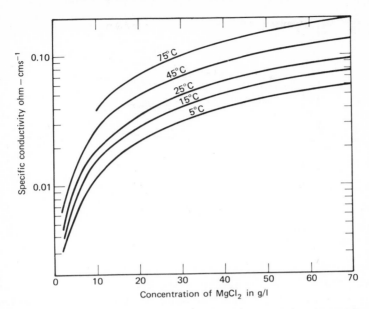

Figure 7.4 The conductivity of aqueous MgCl₂ solutions at various temperatures(36).

impurity or as an oxidation product formed during aging, and by reaction may also raise the conductivity of the electrolyte.

To decrease the time required for the voltage of the battery to reach the desired level a number of methods have been developed. Most important, perhaps, is the incorporation of a fusible link(28) as a short circuit across the battery terminals. This link places an additional load on the system, increasing the initial rate of magnesium chloride generation and also tending to heat up the battery. After a short time the fusible link melts and the extra load on the system is thereby removed, thus allowing the battery to achieve equilibrium with the normal load on the system.

Although seawater is the favored activation medium as explained earlier, once the battery is activated the seawater charge in the unit rapidly dissolves any magnesium chloride produced by the battery reactions. Thus the battery electrolyte becomes one containing both NaCl and MgCl₂. Faletti and Gackstetter have reported(34) the specific conductivity data on solutions containing both salts at a number of concentration levels and at several temperatures. A part of their data is shown graphically in Fig. 7.5. From this chart it is clear that the addition of MgCl₂ increases the specific conductivity of an NaCl solution at all temperatures in the range of 5° to 75°C. Thus this data helps to explain the increase in

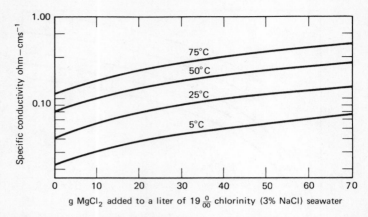

Figure 7.5 Specific conductivity of simulated sea water containing various additions of MgCl₂ at a variety of temperatures (34).

working voltage as the operation of the cell supplies more and more MgCl₂. Presumably the increased amounts of MgCl₂ increase the electrolyte conductivity and so reduce the IR losses in the electrolyte.

A practical example of the effect of accumulation of MgCl₂ in the battery electrolyte is shown in Fig. 7.3 taken from data presented by Strauss(28). The data emphasize that a much lower discharge curve is obtained when a low salinity (1.5 percent NaCl) is used at low temperatures. Obviously an electrolyte of higher salinity and a higher operating temperature result in a higher battery voltage and an increased output.

CATHODES

The type of silver chloride cathode plates to be used depends on capacity duration of discharge and current density of the particular application for which it is intended. For low-rate, long-duration discharge, a plate may be made by rolling a billet made by casting fused silver chloride. Sheets of about 0.025 in. (0.062 cm) thick are cut to size and small holes punched through for improving electrolyte circulation and electronic conduction after development. An electronic conductor or collector of silver is produced on the exposed silver chloride surface by treatment with a photographic developer. This conducting film is sufficiently porous to permit reduction of the underlying silver chloride during cell discharge. Another type of plate, more porous and better suited for high current-density operation, is produced by electrochemically anodizing pure silver screen or sheet in a sodium chloride electrolyte. To develop the electronic conduction of such plates, a short high current reversal (discharge) is

iven the plates before removal from the anodizing bath. Similar forma-
ion of silver chloride is given silver foil for low capacity plates by
reating with a sodium hypo-chlorite solution.

The cathode of cuprous chloride is commonly made by first producing a
·aste of the powder with a polystyrene-base binder and organic
olvent(15). Specific amounts are extruded onto an electronic collector of
opper screen and, with proper proportioning of binder and solvent a
·late of sufficient porosity, mechanical strength and active cathode
urface for satisfactory overall operation is obtained. Powdered graphite
·r carbon have been used in the mixture to decrease plate resistance.
Expanded copper screen has been coated by successive dipping in fused
uprous chloride and quickly removing until the desired quantity is
eposited. Addition of $HgCl_2$ to the cuprous chloride cathode active
naterial results in galvanically precipitated metallic mercury on the anode
fter gradually going into solution in the electrolyte and diffusing to the
node. The results of these additions and the use of magnesium contain-
ig lead or mercury on performance are given under battery weight(27) to
·ield the specification discharge shown in Table 7.3. It is quite clear from
he data presented in Table 7.3 that an alloy of magnesium and mercury
·as the most effective of the group shown since the battery weight for a
iven output was the least.

A number of alternative cathodic depolarizers have been considered in
·lace of copper and silver chlorides. Potassium persulfate(3) has been
escribed as a cathodic depolarizer in such a cell and the claim is made
hat its use results in a 90 percent improvement in cell output. However,
his type of cathode has not found favor with the military market.

TABLE 7.3. REPRESENTATIVE COMBINATIONS OF CATHODES AND
ANODES USED IN WATER-ACTIVATED CELLS

Anode Material[a]	Cathode Material	Battery Weight (grams)
AZ-31B (Dow)	100% Cu_2Cl_2	698
AZ-31B (Dow)	95% Cu_2Cl_2, 4% $HgCl_2$, 2% C	680
3.1% Lead (MEL)	94% Cu_2Cl_2, 4% $HgCl_2$, 2% C	615
3.2% Mercury (Dow)	94% Cu_2Cl_2, 4% $HgCl_2$, 2% C	568

[a]This item shows the magnesium alloy designation.

ANODES

'he anode used in both systems is rolled magnesium sheet, 0.010 in. to
.020 in. (0.025–0.050 cms) thick. The silver chloride type are an alloy
AZ-31-B, AZ-61-B, etc.(42)] of aluminum and zinc with magnesium

which, during discharge, reduces the tendency to form adherent films o
magnesium hydroxide by hydrolysis and thus reduces the ohmic polariza
tion at the anode surface. The cuprous chloride cell anode is usually
commercially pure magnesium or an alloy containing 1 percent mangan
ese. Electrical connections are made to the anode mechanically (rivets) o
by using inert gas arc welding. Other compositions of sheet magnesiun
alloys have been recently (1967–1968) made available in thicknesse
permissible for these applications. Both lead and mercury up to about
percent have been alloyed with the magnesium and rolled(27). A group o
anode alloys is shown in Table 7.3.

The anode in the battery reacts electrochemically when current is take
from the cell according to the equation

$$Mg + 2Cl^- \rightarrow MgCl_2 + 2e$$

The product, $MgCl_2$, dissolves in the electrolyte as described earlier i
this chapter. However, the anode also reacts chemically with the water i
the electrolyte in a wasteful corrosion reaction as follows:

$$Mg + 2H_2O \rightarrow Mg(OH)_2 + H_2 \uparrow$$

The products of this reaction may interfere with the further operation o
the unit. Therefore, provisions are made to permit the hydrogen gas t
move through the electrolyte away from the anode. As mentioned above
the choice of anode alloy helps materially in preventing the formation o
adherent deposits of $Mg(OH)_2$ on the anode surface and the resultan
polarization.

SEPARATORS

Several unique types of separation have been used in the silver chlorid
type, taking advantage of the very low solubility of silver and silve
chloride and the consequent low diffusion and galvanic precipitation o
silver metal on the anode. Also, due to the desirability and possibility o
using a circulated electrolyte, very open types of separation are feasible
Glass heads have been pressed into the surface of the solid silver chlorid
in such a manner that a uniform spacing results when the magnesiun
sheet rests on the opposite side of the beads. Nylon monofilament, 0.01
to 0.020 in. (0.030–0.050 cms) in diameter has been wrapped around th
silver chloride to produce the same effect.

Separation for the cuprous chloride type is nonwoven cotton clot
which provides retention of the electrolyte and sufficient restriction fo
the diffusion of copper ions for practical operation. The tendenc
ultimately is for the anode to precipitate the copper as metal, whic
subsequently grows through the separator to the cuprous chloride posi

ive to short the plates. Proper spacing and separator density in relation to desired capacity and current density provide control of this effect to permit a reliable cell structure. Usual procedure is to sew the cotton separator strips to the copper screen before deposition of the cuprous chloride paste. The magnesium is welded to the copper to provide intercell connection and ready assembly in batteries.

DEVELOPMENT OF HEAT IN THE ACTIVATED BATTERY

As long as the battery is stored in the inactive state there is no heat developed. However, one of the unique features of this system is that when water is added to activate either system a series of reactions occur resulting in the generation of heat and raising the battery temperature. This is a particularly useful feature enabling operation at temperatures that otherwise would be far too low. These reactions are listed below and will be considered separately.

1. The wasteful corrosion of the magnesium anode.
2. The effects of intercell leakage currents.
3. The possibility of Cu^{++} plating out on the anode.
4. The use of a shorting fuse.

THE WASTEFUL CORROSION OF THE MAGNESIUM ANODE

When an electrolyte contacts the surface of a magnesium anode, a wasteful corrosion reaction occurs which may be expressed in the overall reaction

$$Mg + 2H_2O \rightarrow Mg(OH)_2 + H_2 \uparrow$$

This is, of course, an electrochemical reaction. The magnesium anode is believed to have both anodic and cathodic areas on its surface. In this reaction the metal dissolves at the anodic area while hydrogen is liberated at the cathodic area. The appropriate equations are

$$H_2O \rightarrow OH^- + H^+$$

$$Mg \text{ (anodic area)} + 2OH^- \rightarrow Mg(OH)_2 + 2e^-$$

$$Mg \text{ (cathodic area)} + 2e^- + 2H^+ \rightarrow H_2 \uparrow$$

These reactions produce 75.89 kcal of heat for every 24.31 g of magnesium so consumed. In commercial magnesium dry cells inhibitors, such as chromates, are added to reduce the extent of the wasteful corrosion. However, in water-activated cells, designed for radiosonde use in the upper atmosphere where low temperatures are encountered, no inhibitors are used. Thus the above reaction can operate without restriction to keep the battery warm and operating even though the air temperature may approach − 65°F (18°C).

THE EFFECTS OF INTERCELL LEAKAGE CURRENTS

When a single quantity of electrolyte is poured into the top of a multicell battery, where the cells are connected in series, at some point in time this electrolyte contacts all or several cells. Under these conditions parasitic currents develop.

Two types of intercell discharge or leakage that lower plate capacity and voltage are present. Most applications include series-connected single cells to increase the voltage available. There is an additional electrolyte path between the series connected (shorted) positive and negative plates of adjacent cells. There ordinarily, is a path between plates separated in the series at a voltage greater than that required for decomposition of the electrolyte. Both mechanisms tend to discharge the plates and are dependent principally on the ohmic resistance and number of respective parallel electrolytic paths. The paths taken by these leakage currents are diagramatically shown in Fig. 7.6(28). It is obvious that these leakage currents be reduced to a minimum so the output of the battery is not seriously decreased. These may be minimized by design control. Thus the edges of the electrodes exposed to a common electrolyte may be covered with insulating tape or lacquer to increase the length and, thus the

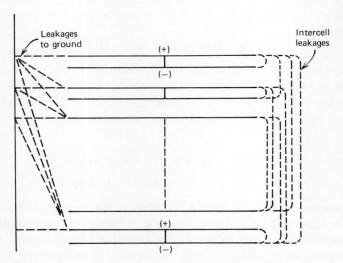

Figure 7.6 Leakage current paths in multicell batteries.

esistance, of the leakage path. Furthermore barriers of various types nay be placed at selected points to help reduce the magnitude of the eakage currents.

The leakage currents are usually most noticeable when the battery is being activated and as such are beneficial in raising the battery temperature. As the excess electrolyte becomes absorbed in the separator and electrodes there is less liquid to cause leakage currents and they become ess important.

THE POSSIBILITY OF THE CUPRIC ION PLATING ON THE ANODE

Although water-activated cells are manufactured with cuprous chloride or silver chloride there is some possibility that the former depolarizer salt may become oxidized to the more soluble cupric chloride. When such a battery is activated there is the possibility that a part of the cupric copper may plate out on the magnesium anode by displacement. If this occurs a metallic couple exists and wasteful corrosion of the anode can occur at an increased rate, and, like the wasteful corrosion reaction discussed above, it will generate heat. In some early forms of water-activated cuprous chloride cells a duplex electrode was made by plating one side of the magnesium anode with metallic copper. Obviously, any electrolyte that adhered to the edges of such an electrode would cause the reaction discussed above.

Fortunately this reaction may not be as serious as might be imagined since the magnesium anode becomes coated with a layer of $Mg(OH)_2$, which may act as an insulating layer and reduce the intensity of this parasitic problem.

THE USE OF A SHORTING FUSE

As mentioned in an earlier section, a fusible link is often connected across the terminals of the reserve battery. When the unit is activated the full current capacity of the cell flows through this fusible link and shortly thereafter it melts, opening the shorting circuit mechanism, and exposing the battery to its normal load. However, during the time period between activation and the melting of the fusible link, the battery acts like a self powered electrical heater with the battery temperature rising as a result.

CONSTRUCTION AND OPERATION

Both systems are built into batteries using various sheet and tubular plastic shapes to provide the least cost construction compatible with the specification requirements determined by the particular special purpose application. That is, each application clearly determines the lowest cost

TABLE 7.4. APPLICATIONS FOR BATTERIES OF THE
TWO SYSTEMS

Silver Chloride	Cuprous Chloride
Torpedo power supply	Radiosonde
Sonobuoy (antisubmarine)	Weather balloons
	Emergency power supply
Submarine masher ignition	Submarine sonobuoy
Air–sea rescue beacon	Sea surface lighting unit

system possible and the type construction to be used to obtain a uni
satisfactorily meeting the specifications set forth by the user. Essentially
all batteries are "special purpose" and are not designed for a wide numbe
of outlets. The applications, however, for these batteries overlap and ar
listed in Table 7.4. To determine specifically whether a certain applicatio
can be met with these type batteries involves specific calculation, design
and fabrication of several unit blocks which are discharged under the bes
and worst environment conditions to determine if the specification i
realistic. Design criteria for safety factor and reliability are not estab
lished and are usually subject for determination or negotiation with th
user. Table 7.5 lists organizations manufacturing batteries of these types
They have supplied unclassified performance information which has bee
used to prepare the summary information.

These type batteries at times are specified to supply multiple voltage
either isolated or tapped from a main power section. Connections are
made by simply soldering leads to connector plugs either mounted flexibl
on the leads or mounted as an integral part of the battery housing. Two
types of electrolyte activation are used. The "dunk" type is made fo
filling by immersion. Circulation is achieved as the anode gas "pumps"
the fresh or seawater through the unit. The other type activation is t

TABLE 7.5. ORGANIZATIONS THAT MANUFAC-
TURE EITHER OF THESE TYPES OF BATTERIES

Burgess Battery Company, Freeport, Ill.
Eagle-Picher Company, Joplin, Mo.
Eureka Williams Company, Bloomington, Ill.
General Electric Company, Pittsfield, Mass.
Peerless Roll Leaf Company, Union City, N.J.
Three Point One Four Corp., Yonkers, N.Y.
Yardney Electric Corp., New York, N.Y.

manually fill the unit with the appropriate water. These latter types are designed to contain the same water during discharge and, of course, do not expel the water electrolyte by the gas formation. Usual packing includes a protection from atmospheric moisture by sealing a desiccant into the package.

Figure 7.7 shows a sectional view of a silver chloride magnesium

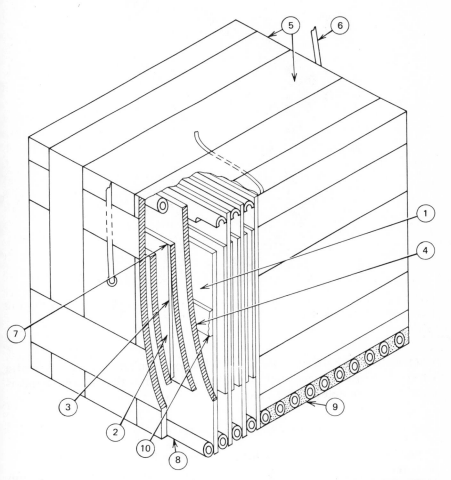

Figure 7.7 The construction of a silver chloride magnesium battery. (1) Positive Plate, fused silver chloride; (2) Negative Plate, magnesium (AZ 61); (3) Separator, $\frac{1}{16}$ in. polyurethane tubing; (4) Spacer, Fiberglass epoxy; (5) Covering, Fiberglass tape; (6) Wire, stranded copper; (7) Adhesive, epoxy cement; (8) Activation tubes, polyurethane; (9) Case potting, epoxy; (10) Intercell connectors, silver foil. Courtesy of Eagle-Picher Industries, Inc.

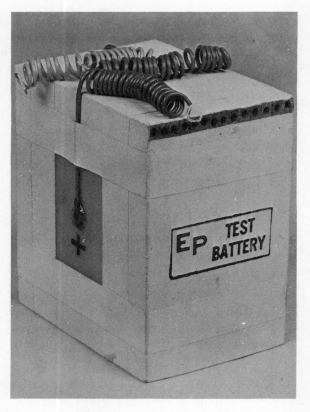

Figure 7.8 The external appearance of a silver chloride magnesium battery. Courtesy of Eagle-Picher Industries, Inc.

water-activated battery while Fig. 7.9 shows a similar cutaway view of a cuprous chloride magnesium unit. Figure 7.8 shows the external appearance of a completed silver chloride magnesium battery.

BATTERY CHARACTERISTICS

Typical discharge voltages exhibit a desirable flat characteristic at both high and low rates using either system. The silver chloride type is about 1.50 V per cell as compared with 1.40 for cuprous chloride. A typical Signal Corps battery characteristic is shown in Table 7.6 which lists the specification information as well as performance using either system. A more complete review of the overall field may be obtained through the

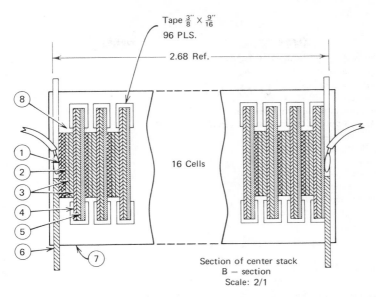

Figure 7.9 A sectional view of a cuprous chloride magnesium battery. (1) Copper foil; (2) Cuprous chloride and cotton gauze; (3) Cotton Webril (Kendall Mills); (4) Paper separator; (5) Magnesium; (6) Formica case; (7) Varnish-coated paper; (8) Void (for electrolyte). Courtesy of Eagle-Picher Industries, Inc.

specifications(26) which are available. Generally, batteries are available with voltages ranging from 1.5 to several hundred volts, currents from a few milliamperes to several hundred amperes and discharge rates from a few minutes to several hours. Thirty-five watt-hours per pound of activated battery may be accomplished with ideal requirements and optimum design.

Figure 7.10 shows the external appearance of a number of types of water activated batteries. Table 7.7 shows the dimensions and capacities of some of these units.

TORPEDO BATTERIES

The silver chloride magnesium reserve battery system has been used to power a torpedo. In this case the battery utilizes seawater as an electrolyte and a more or less continuous supply of the latter flows through the battery when the torpedo is in operation. Because of the military nature of the application many details of the battery construction are understandably classified and very little descriptive matter on the

TABLE   7.6. DATA   FOR   THE   BA-316/AM   BUILT   USING   SILVER   AND
                            CUPROUS CHLORIDE

BA-316/AM Specification (Abstract)
Size—$2\frac{1}{2}$ in. × $3\frac{1}{4}$ in. × 3.75 in. (max.)
Weight (activated)—450 g (max.)
Activation—75°F (24°C)
Activation time—20 minutes (max.)
Voltage, nominal, maximum, minimum:
    A Section—7.0; 7.5; 5.5
    B Section—120; 120; 100
Loads:
    A Section—19 ohms
    B Section—30,000 ohms
Capacity after activation at 75°:
    at −58°F—90 minutes
    at 75°F—100 minutes

BA-316/AM Actual Performance

	AgCl Type	CuCl Type
Activated Weight, grams	257	383
Peak Voltage, A Section	5.5	5.5
Peak Voltage, B Section	115.5	115.0
Life (−58°F) A Section min	149	149
Life (−58°F) B Section min	173	146
Watt-hours/pound	10.1	7.5
Watt-hours/cubic inch	0.263	0.206

battery performance has been published. However, a series of papers (29)(34)(35)(36)(38)(39) has covered various phases of the battery problems and we are indebted to these and to the specifications (43) for the limited data presented herewith.

In contrast to the batteries described earlier in this chapter, which are packages of a few cubic inches in volume and up to 18 lb in weight as shown in Table 7.5, the torpedo battery has a dry weight of about 486 lb. It delivers a voltage of between 295 and 180 on a test load for approximately 10 minutes. It is fed with seawater at an initial rate of 120 gal/min which may decrease to a lesser rate as the end of battery life approaches. From the fact that each battery contains about 4100 Troy oz of silver it is an expensive battery. (At $1.60 per oz this material alone costs $6550 per battery.) At the 1975 price of silver of $4.50 per Troy oz. the cost would be $18,450.

It is clear from the above brief comments that this torpedo battery is a unique electrochemical power plant and among the largest primary battery units constructed in modern times.

Figure 7.10 A variety of water-activated battery types. Courtesy of Eagle-Picher Industries, Inc.

Figure 7.11 shows the construction of an experimental unit used to study the characteristics of silver chloride magnesium cells for this application. Like most of the water-activated batteries, it consists of parallel plate electrodes separated by an electrolyte space, which, in this case is the path through which the seawater circulates.

Figure 7.12 shows a typical discharge curve obtained on a properly functioning stack of 10 experimental cells. The accumulation of by-

TABLE 7.7. SPECIAL WATER-ACTIVATED BATTERIES

Battery No.	Nominal Voltage (volts)	Dish. Current	Life	Size (inches)	Weight (pounds)
396	14	0.5 amp	40 min	$\frac{7}{8} \times 2 \times 2\frac{3}{16}$	0.18
395	1	1 amp	7 sec	$1\frac{1}{4}D \times \frac{3}{4}$	0.04
393	A+ 5.5	150 ma	9 hr	$5\frac{5}{8} \times 5 \times 6$	2.75
	B+ 95	20 ma			
385	A+ 6	150 ma	12 hr	$5\frac{1}{2} \times 4\frac{5}{8} \times 4\frac{5}{8}$	3.37
	B+ 120	21 ma			
373	14	9 amp	15 min	$2\frac{3}{4}D \times 3\frac{3}{4}$	1.0
375	6.6	7.5 amp	8 hr	$7\frac{3}{8}ID \times 9\frac{1}{8}OD \times 8$	11
362	A+ 6.5	40 ma			
	A+ 6.5	500 ma	48 hr	$6 \times 8 \times 17\frac{1}{2}$	18
	B+ 250	30 ma			
360	A+ 6	300 ma	2 hr	$5 \times 4 \times 2\frac{3}{8}$	1.0
356	B+ 95	20 ma			
	A+ 6.6	3 amp	$2\frac{1}{2}$ hr	$3\frac{1}{8} \times 3\frac{1}{8} \times 11$	4.5
	B+ 140	60 ma			
352	6	4 amp	2 hr	$1\frac{5}{8} \times 2\frac{7}{8} \times 7$	1.6
338	6	1.3 amp	4 hr	$1\frac{3}{8} \times 2\frac{7}{8} \times 7$	1.5

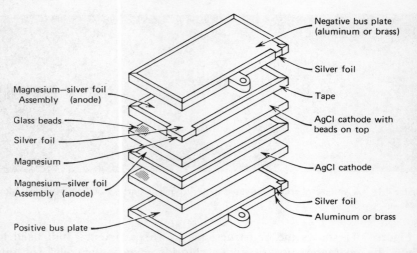

Figure 7.11 The construction of an experimental silver chloride magnesium battery for high currents. Duplex battery construction (two-cell).

product $Mg(OH)_2$ occasionally adheres to the anode and collects in the interelectrode spaces, to such an extent that it is not removed by the stream of electrolyte, and may reduce the working voltage and shorten battery life(39). The development of such deposits has been related to the treatment of the anode alloy, AZ-61, and the presence of $Mg_{17}Al_{12}$. Appropriate heat treatment schedules for the anode alloy maintain the

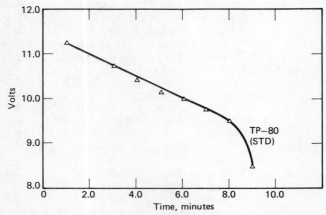

Figure 7.12 The discharge curve of the high current battery shown in Fig. 7.11. Ten cell batteries, 23.4 IN^2 anodes, 22.85 IN^2 cathodes, 44.0 ampere load, seawater 3-0, 25 ±0.5° inlet temperature, 0.0205 in AGCL processed by Gould-National Batteries Lot 2.

ɔulk of the aluminum in solid solution in the alloy and so prevent the ɪbove undesirable condition.

REFERENCES

1. Mullen, J. B. and Howard, P. L., *Trans. Electrochem. Soc.*, **90** 529 (1946).
2. Vinal, G. W., *Primary Batteries*, John Wiley, New York, (1950), pp. 274–279.
3. Blake, I. C., *J. Electrochem. Soc.*, **99** 202c (1952).
4. Permik, J., *Rev. Gen. Elec.*, **62** 294 (1953).
5. Pucher, L. E., *J. Electrochem. Soc.*, **99** 203c (1952).
6. Hamer, W. J., *Advanced Energy Sources and Conversion Techniques*, Vol. I, pp. 41–54, Nov. 1958, Astia No. AD209310, Dept. of Commerce OTS No. PB-151461.
7. Lemman, M. J. H., and Casson, W. E., First International Symposium on Batteries, Paper (J), pp. 1–12 (1958).
8. Bell Telephone Laboratories, "Final Report on Seawater Batteries," NDRC, Office of Technical Services, Dept. of Commerce, Nov. 30, 1945, AD 128639.
9. Jones, K., First International Symposium on Batteries, Paper (ZZ) pp. 1–7 (1958).
0. Gacton, R., Second International Symposium on Batteries, Paper 18, pp. 1–7 (1960).
1. Morse, E. M., *Proc. 9th Annual Battery Res. Dev. Conf.* p. 5 (1955).
2. Id, *Proc. 10th Annual Battery Res. and Dev. Conf.*, p. 23 (1956).
3. Id, *Proc. 11th Battery Res. and Dev. Conf.*, p. 69 (1957).
4. Broglio, E., *Proc. 12th Battery Res. and Dev. Conf.*, p. 88 (1958).
5. Honer, H. N., Final Report Signal Corps Contract DA-36-039-sc-84523 (July 30, 1960), Astia AD 262 008.
6. Patents concerning silver chloride type battery. Meyer, U.S. Pat., 902755, (1908). Axtell, U.S. Pat., 1,226,165, (1917). Lawson, H., U.S. Pat., 2,428,850 (Oct. 14 1947); CA 42,2528. Lawson, H., U.S. Pat., 2,445,306 (July 13 1948); CA 42,7172. Mullens, J. B., U.S. Pat., 2,521,082 (Sept. 5, 1950); CA 44,10555. Harris, L. H., U.S. Pat., 2,543,106, (1951). Taylor, R. L., U.S. Pat., 2,590,584 (Mar. 25 1952); CA 46,5466. Warner, A. J., and Gobat, A. R., U.S. Pat., 2,661,388 (Dec. 1, 1953) CA 48, 3171. Vyselkov, A. A., and Rogova, G. I., U.S.S.R. Pat. 109,345 (Feb. 25, 1958); CA 52,10770. Yuasa Battery Co., Jap. Pat. 5818 (July 3, 1959) CA 54,8376. Haring, H. E., U.S. Pat., 2,988,587 (June 13 1961); CA 55,24330. Mendizza, A., U.S. Pat., 3,004,903 (Oct. 17, 1961); CA 56,3276. Haring, H. E., U.S. Pat., 3,006,821 (Oct. 31 1961); CA 56,3276. Haring, H. E., U.S. Pat., 3,007,993 (Nov. 7 1961). Bur. Ind. Technics, Jap. Pat., 1663 (Apr. 20, 1953) CA 48,3172. Yuasa Battery Co., Jap. Pat., 361 (Jan. 24, 1957) CA 52,3568. Yuasa Battery Co., Jap. Pat., 4616 (July 8, 1957) CA 52,13489. Yuasa Battery Co., Brit. Pat., 123,936 (Mar. 13, 1919); Yuasa Battery Co., Jap. Pat., 120 (Jan. 23, 1959) CA 53,19639.
7. Patents applicable to cuprous chloride type batteries. Wensky, W., Brit. Pat., 49 (1891). Adams, B. N., U.S. Pat., 2,322,210 (June 22, 1943). Fischbeck, A., and Mandel, H. J., U.S. Pat., 2,636,060 (Apr. 21, 1953). Pucher, L. E., and Cunningham, W. A., U.S. Pat., 2,640,090 (May 26, 1953). Pucher, L. E., and Cunningham, W. A., U.S. Pat., 2,640,091 (May 26, 1953). Ellis, G. B., U.S. Pat., 2,655,551 (Oct. 13, 1953) CA 48,60. Chubb, M. F., U.S. Pat., 2,658,935 (Nov. 10, 1953) CA 48,1179. Chubb, M. F., Reissue U.S. Pat., 2,658,935, #23883 (1954). Dines, J. M., U.S. Pat., 2,716,671 (Aug. 30, 1955) CA 49,15570. Salavze, J., U.S. Pat., 2,744,948 (May 8, 1956). Barnard, K. N. and Greenblatt, J. H., Can. Pat., 518,553 (Nov. 15, 1955) CA 52,135. Yuasa Battery Co., Jap. Pat., 3927 (June 19, 1957) CA 52,13489. Yuasa Battery Co., Jap. Pat., 4615 (July 8, 1957) CA

52,13489. Chubb, M. F., U.S. Pat., 2,817,697 (Dec. 24, 1957) CA 52,4360. Yuasa Battery Co., Jap. Pat., 9017 (Oct. 22, 1957) CA 53,2891. Yuasa Battery Co., Jap. Pat., 716 (Feb 8, 1958) CA 53,7830. Yuasa Battery Co., Jap. Pat., 4221 (May 28, 1959) CA 54,2051 Toyo Drycell Co., Jap. Pat., 6571 (July 30, 1959) CA 54,8376. La Faille, J. C., Fr Pat. 1,176,958 (Apr. 17 (1959) CA 54,24031. Pertrix-Union G.M.B.H., Ger. Pat., 1,069,23! (Nov. 17, 1959); CA 55,11147. Ray-O-Vac Co., Ger. Pat., 1,004,690 (Mar. 21, 1957); C/ 55,9119.

18. Patents applicable to both systems. Metallgesellschaft, Ger. Pat., 961,813 (Apr. 1 (1957); CA 53,13849. Black, I. C., U.S. Pat., 2,612,533-7, incl. (1952); CA 47,2615. Fry A. B., George, P. F., and Kirk, R. C., U.S. Pat., 2,712,564, (1955); CA 49,14542. Gobat A. R., U.S. Pat., 2,726,279 (Dec. 6, 1955); CA 50,4681. Shorr, W., U.S. Pat., 2,778, 75 (Jan. 22, 1957); CA 51,5598. Stevens, J. A., U.S. Pat., 2,934,583 (Apr. 26, 1960); C/ 54,12842.

19. Latimer, W. M., Oxidation Potentials, Prentice-Hall, Englewood Cliffs, N.J. (1952).

20. Morehouse, C. K., Advanced Energy Sources and Conversion Techniques, Vol. 1, pp 55–68, (1958), ASTIA AD 209301, Dept. of Commerce, OTS, No. PB 151461.

21. Greenblatt, J. H., J. Electrochem. Soc., 103 539 (1956).

22. Glicksman, R., J. Electrochem. Soc., 106 83 (1959).

23. Casey, E. J., Bergeron, R. E., and Nagy, G. D., Report No. 158.(196-1) Project No. L 52-54-80-08, Defense Research Chemical Laboratories, ASTIA AD 271,168.

24. Robinson, J. L. and King, P. F., J. Electrochem. Soc., 108 36 (1961).

25. Robinson, J. L., Second Quarterly Progress Report, Signal Corps Contract DA-36-039 sc-88912 (Dec. 31, 1961), ASTIA AD 273,775.

26. Specifications: U.S. Army Signal Material Support Agency, MIL-B-10154C- Batteries Primary, Water-Activated (Dunk Type)-BA-253/U, BA-259/AM, BA-292/AM and BA-316/AM. Weather Bureau, Dept. of Commerce, Spec. No. 458.026, Battery Radiosonde, (Water-Activated) (1680 megacycles). Ibid, Spec. No. 458.027, Battery Radiosonde, Water-Activated (72.2 megacycles). Ibid, Spec. No. 458.028, Battery Transponder Radiosonde, Water-Activated. Ibid, Spec. No. 458.029, Battery Radiosonde, Pulsed, Water-Activated, (403 megacycles).

27. Williams, R., Final Report, Phase I (Apr. 10, 1968); Phase II (Nov. 11, 1968); Phase III (July 27, 1970); Signal Corps (USAEC) Contract DA-28-043 AMC-02507(E).

28. Strauss, H. J., Electronics World, 46–48 (Nov. 1968).

29. Faletti, D. W., Electrochem. Soc. Battery Symposium, Extended Abstract No. 43 (Oct 4–8, 1970).

30. Pistoia, G. and Scrosati, B., Electrochem. Soc. Battery Symposium, Extended Abstract No. 45 (Oct. 5-9, 1969).

31. Mathur, P. B., and Venkatakrishnan, N., Electrochem. Soc., Battery Division, Extended Abstracts, No. 58 (Oct. 4–8, 1970).

32. Goodkin, J. and Dalin, G. A., Electrochem. Soc., Battery Division, Extended Abstract, No. 56 (Oct. 9–14, 1966).

33. Kleiderer, C., Modern Plastics, pp. 133, Nov. (1945); cf. Vol. I, Chap. I, pp. 48ff., Ref. 137.

34. Faletti, D. W. and Gackstetter, M. A., J. Electrochem. Soc., 114 299 (1967).

35. Faletti, D. W., Gackstetter, M. A., and Arne, J. A., J. Electrochem. Soc., 116 552 (1969).

36. Faletti, D. W., J. Electrochem. Soc., 117 1524 (1970).

37. Hamlen, R. P., Jerabek, E. C., Ruzzo, J. C., and Swek, E. G., J. Electrochem. Soc., 116 1588 (1969).

38. Faletti, D. W., Herrick, J. W., and Adams, M. F., J. Electrochem. Soc. 116 698 (1969).

39. Faletti, D. W. and Nelson, L. F., Electrochem. Tech. 3 98 (1965).

40. Mueller, C. E. and Bowers, F. M., *J. Electrochem. Soc.* **118** 394 (1971).
41. Sverdrup, H. V., Johnson, M. W., and Fleming, R. H., *The Oceans, Their Physics, Chemistry and General Biology*, Prentice-Hall, Englewood Cliffs, N.J. (1946).
42. American Society for Testing Materials, B 275-63, "Codification of Light Metals and Alloys, Cast and Wrought," June (1965) and (1966).
43. Specification MIL-B-22876A(05), Battery, Water Activated, Mark 67, Mod 1, (4 Mar. 1969).
44. Faletti, D. W. and Gackstetter, M. A., *J. Electrochem. Soc.*, **115** 1210 (1968).
45. Martin, L. F., *Storage Batteries and Rechargeable Cell Technology*, Noyes Data Corp., Park Ridge, N.J. (1974) pp. 346–357.

8.

Nomenclature and Testing Procedures
for Primary Batteries

WALTER J. HAMER

GENERAL CONSIDERATIONS

Testing is necessary in the evaluation of primary batteries and is an essential adjunct to their development. The need for tests arises because the quality of a battery is determined not by its built-in electrical capacity but by that portion of its capacity that provides useful output or service under actual operating conditions. This output is obviously not a fixed quantity. It depends on rate of discharge and cutoff voltage; it is affected by temperature and, during prolonged service, by self-discharge and shelf factors.

Under these circumstances, it is not surprising that there are no simple criteria or nondestructive tests that can be applied to a battery to determine its life expectancy for a variety of service applications. Chemical composition, to be sure, affects service, but in dry cells, for example, it is only in exceptional cases, usually involving discharge to complete or nearly complete exhaustion, that a rough correlation of cell composition and useful life may be obtained, even for a limited range of operating conditions.

Measurements of open-circuit voltage are useful as a means of quality control, serving, for example, in the detection of defective or over-age units, but giving no indications of electrical capacity. Similarly, readings of amperage, though giving a rough indication of internal resistance, bear no relationship to service life. (They can actually be harmful by initiating anode corrosion and accelerating shelf deterioration.)

Internal resistance (cf. Chapter 10), though of value in research or quality control, is also, for several reasons, an unreliable guide to cell performance. Particularly in dry cells, it varies with age, it may change greatly during discharge, and it is seriously affected by service conditions. Nor does it follow that a cell with the intrinsically lower internal resistance will give the better service life, although this, in part, is due to methods of testing, that is, in evaluating batteries by their hours of service on fixed resistance rather than in terms of ampere-hour or watt-hour output. Many years ago, this method of testing gave rise to the fiction that a dry cell actually improved with aging over a period of three to six months, and that deterioration was more rapid in cold storage than at room temperature. What actually happened, of course, was that the normally aged cells had higher internal resistance than fresh or cold-stored (brought to room temperature after storage) units and, on the continuous tests then in use, drifted along for more hours of service, but at lower operating voltages, reflected in reduced ampere-hour and watt-hour capacity.

In short, the only reliable criterion of quality is the behavior of a battery during discharge simulating as closely as possible the conditions of actual service. This situation presents a serious problem because of the wide variety of applications of modern batteries. It is further complicated by the fact that behavior on one test is rarely indicative of overall quality; a cell of high electrical capacity on light drain may be poor on heavy-drain service and vice versa; a cell designed for operation at normal or at elevated temperature may give poor low-temperature service. Nor, generally speaking, have accelerated tests been a satisfactory index of service quality over a significant range of operating conditions.

DEVELOPMENT OF STANDARD TESTS

In the United States, early recognition(1)(2)(3)(4) of the need for reliable and uniform testing culminated, in 1912, in the appointment by The Electrochemical Society of a committee to develop standard methods for dry cells(5). This committee proposed tests for three classes of service: ignition, telephone, and flashlight. The great stress it laid on tests that would simulate conditions encountered in practice has had a profound effect on all subsequent testing procedures—emphasis is invariably placed on end uses in test designs.

Today, testing procedures for civilian applications are standardized, for the most part, by an American National Standards Institute (ANSI) sectional committee on dry cells and batteries working under the sponsorship of the National Electrical Manufacturers Association (NEMA) and for the military by the cognizant military agencies. Internationally, standardization is conducted by the International Electrotechnical Commission (IEC) through a technical committee on primary cells and batteries, formed in 1949. This IEC committee has issued four documents and three amendments(6) pertaining to the nomenclature and testing of dry cells.

The ANSI committee is made up of representatives of the manufacturers, large consumers of dry cells, scientific organizations, consultants, and government departments and, since its organization in 1926, has issued ten editions and nine revisions of its specifications. Originally the ANSI committee was known as the ASA (American Standards Association) and then as the USASI (United States of America Standards Institute) committee and the sponsor for nearly 40 years was the National Bureau of Standards. A history of the activities and output of this committee and earlier work on specifications for dry cells and batteries is given in reference (7). The 10 editions of the standard were issued as American Standards in 1928(8), 1930(9), 1937(10), 1941(11), 1947(12), 1955(13), 1959(14), 1965(15), 1969(16), and 1972(17). Close cooperation between the ANSI (or ASA) committee, the Federal Specifications Board (prior to 1952), and the U.S. General Services Administration (since 1952) has been maintained with the result that Federal Specifications issued in 1931, 1935, 1948, 1954, 1960, 1961, 1970, and 1973(18) have been concordant with the ANSI (previously ASA) specifications, although differing in form.

Periodic revisions of specifications become necessary as a result of changes in the art. New types of and uses for batteries require the design of new testing procedures and the drafting of new specifications, and the improved performance of batteries frequently justifies some increase in requirements(19). With each edition of the ANSI standard, new types of dry cells and batteries have been added, while some types have been removed as being obsolete. For example, in 1937 dry cells for industrial flashlights and hearing aids were added to the specifications. Continuous tests, though simple and rapid, had little or no practical significance as regards service quality, and since they could not be correlated directly with intermittent tests, they were replaced by the latter in 1941. Today, however, some continuous tests, namely, those for alarms, watches, safety flashers, toys, and electric eyes, are required and have been added to the ANSI specifications(14)(15)(17). Continuous tests are obviously required in these applications in order to simulate the service. In the 1947 standard, emphasis was given to the newer flat or miniature batteries and standard types of A/B battery packs were added. In 1955, mercury dry cells(20) and, in 1959, batteries for use with transistorized circuits(21) were added to the specifications. The 1965 specification included, for the first time, specifications on alkaline-MnO_2, zinc-silver oxide, and sealed nickel-cadmium cells(22). All A, B, and C batteries and A/B battery packs were then removed from the specifications—trends to transistorized circuits had made these types obsolete. In 1972, tests were added for toy and photographic batteries(17). Obviously, constant attention must be given to the state of the art in order to keep testing procedures abreast of developments.

Substantially all tests are based on discharge of cells or batteries through fixed resistances and results are recorded in terms of the time required to reach a predetermined cutoff voltage. Such tests give no immediate measure of ampere-hour or watt-hour capacity but these may be calculated if a record is kept of the average current during the discharge. As a first step, the voltage throughout the test should be plotted and the average voltage determined. From this average and the known fixed resistance the average current is computed. This average in turn is multiplied by the total time of actual discharge to obtain the ampere-hour capacity of the cell or battery, from which the watt-hour capacity is obtained by multiplying by the average voltage during the discharge. Tests could also be conducted at constant current rather than constant resistance, but since close control of current over extended periods of time is tedious, dry-cell committees have not recommended their use. In those services where constant currents are encountered, tests based on fixed resistances obviously have limitations in determining service quality for such applications.

The following discussion deals largely with dry cells and batteries, to which most of the work on nomenclature and standardization of tests has been directed.

NOMENCLATURE AND STANDARD SIZES OF DRY CELLS

In designing standard testing procedures it was found that the problem could be simplified if dry cells themselves could be standardized as to size and arrangement within assembled batteries. Not only would fewer tests be needed but also cells and batteries of different makes would be interchangeable between various lines of electrical equipment. Also arguments over the proper test for a particular application become less severe.

In each of the ANSI (or ASA) specifications, a standard nomenclature was adopted to cover various sizes of cells. In Appendix A, standard sizes of Leclanché cylindrical,[1] Leclanché flat, mercury, silver oxide, alkaline-MnO_2 cells, and sealed nickel-cadmium cells,[2] are listed. A cell letter is sometimes preceded by a number indicating the number of cells in series in a battery. When the number follows the cell letter—separated when necessary by a hyphen—it indicates the number of cells connected in parallel. For example, "2D" means two D-size cells in series whereas "D2" means two D-size cells in parallel. For flat cells, a hyphen is necessary when the cells are in parallel, thus "2F12" means two F12-size

---

[1]These standard sizes also apply to Leclanché type cells with magnesium anodes.

[2]These are rechargeable cells but the types given here are used in dry cell applications and, therefore, are included.

cells in series whereas "Fl2-2" means two F12-size cells are in parallel. When a small letter "s" or "d" is used at the end of the code, it indicates either of two structural arrangements as to number and size of cells and electrical connections; "s" indicating a single-row arrangement and "d" a double-row arrangement. For flat cells the ANSI and IEC designations are identical, while for cylindrical Leclanché cells the IEC uses an "R" meaning round in their designations (in many countries Leclanché cells are also made in square or rectangular forms—the IEC specification uses the letter "S" for these types).

For cylindrical Leclanché, flat Leclanché, mercury, and silver oxide cells the dimensions are listed as nominal. The cell designations may apply to other dimensions and shapes of cells when grouped as batteries corresponding to the same standard size in volume or capacity rating. Letters of the alphabet, cell No. 6 excepted, are used to denote cylindrical Leclanché cells while numbers with certain letters (F is chosen to denote flat cells, M for mercury cells, S for silver oxide cells, L for alkaline-MnO_2 cells, and K for Ni-Cd-KOH cells) are used to denote the other types. The numbers chosen for the silver oxide dry cells designated by the letter "S" correspond to those used for mercury cells of about the same volume. The numbers chosen for the alkaline-MnO_2 cells are comparable to the IEC numbers for cylindrical (or round) Leclanché cells except the "L" is used instead of an "R." Since cross-sectional area closely parallels electrical output for flat cells, it rather than volume is given in this case.

Another method of nomenclature is that of the National Electronic Distributor Association (NEDA). The numbers recommended by this association are referred to as the NEDA numbers. The military also has its own system of designating batteries but adheres to the ANSI system for individual cells. Then, too, the manufacturers, in many cases, prefer to use their own method of nomenclature, in some cases because of prior usage. Large consumers (distributors) likewise frequently use a number to their liking. For example, the common D-size Leclanché cell, used widely in flashlights, is designated in at least 39 different ways, as illustrated in Appendix B. Less confusion would exist if a single nomenclature could be adopted. In any case, regardless of the system of nomenclature used, standardization has served to reduce the number of battery and cell sizes, to reduce the volume of testing, and to streamline testing procedures.

GENERAL TESTS

General tests of primary cells and batteries may be classed in three categories:

1. *Initial tests*—tests to determine condition of cells and batteries before they are placed into service.

2. *Service tests*—tests to determine electrical capacity of cells and batteries under various service conditions.
3. *Shelf tests*—tests to determine the rate of deterioration of cells and batteries when on open circuit, that is, when on shelf or in service.

This classification has been acknowledged for several decades (5). Service tests may also include determinations of various operating characteristics of cells and batteries in addition to determination of their electrical output. Service tests also play a part in shelf tests in that they are used in comparing the performance of old cells and batteries with new ones.

INITIAL TESTS

Initial tests include measurements of open-circuit voltage (OCV), flash or short-circuit current (SCC), and internal resistance or impedance, (cf. Chapter 10) including a measurement of initial closed-circuit voltage (CCV). Although these initial tests are frequently recommended as a part of an overall inspection to guard against truly defective cells and batteries, each test has limitations which are set forth in the following paragraphs.

Open-circuit Voltage

The American National Standards Institute specification for dry cells and batteries requires that a voltmeter having an accuracy at least of 0.5 percent of full-scale deflection be used in all voltage measurements. The voltmeter should have a full-scale reading of about 2 V per cell in series and not in excess of 5 V per cell. The scale should have not less than 50 divisions per volt when the voltmeter is used to measure voltage of individual cells and not less than 100 divisions for measurements on batteries of two or more cells. The voltmeter should have a resistance of 1000 ohms/V of full-scale deflection for cells larger than F20 and 10,000 ohms/V or higher for cells of F20 size or smaller.

Values of OCV vary widely among Leclanché cells ranging from about 1.50 to 1.65 V. The higher voltage may indicate that the cell contains synthetic MnO_2. Otherwise OCV gives no information on the electrical output to be expected from a cell under specified services nor the rate of deterioration of cells in storage. Only when the OCV is abnormally low, about 0.05 V or more below the normal OCV for a particular brand of cell, can the quality of the cell be suspect. Accordingly, OCV has principal value only as a check on cell uniformity within any one production lot of a particular brand. These remarks apply equally well to alkaline-MnO_2, silver oxide, and mercury dry cells, in fact to all primary cells and batteries, reserve cells excepted (discussed later).

The electromotive force (EMF) of primary batteries may be measured n the usual way with a potentiometer and galvanometer.[3] For batteries having EMFs above approximately 1.6 V it is necessary, with most nstruments, to measure the difference between the EMF of the battery and an appropriate number of standard cells in series and then calculate the EMF of the battery from the measured difference and the known values of the standard cells. Generally, the difference between OCV and EMF is negligible for most practical purposes unless the internal resistance of the cells is unusually high (e.g., that exhibited by silver chloride dry cells). In this case, the EMF must be measured with a potentiometer or a suitable electrometer (see cells of high resistance presented later).

Initial Closed-Circuit Voltage

Although there is no conformity on measuring the initial closed-circuit voltage, the consensus is that it should be an instantaneous value, if at all possible. However, for comparative and practical purposes the value taken at five seconds after a manual closure of a circuit may be used as there is less controversy on its measurement. This will, however, include the EMF of polarization, but gives a rough indication of the internal resistance of a cell; see Chapter 10. The load used in this test may be varied, depending on the service intended for the cell. For example, for flashlight cells a load of $2\frac{1}{4}$ to 5 ohms is used while for photoflash cells a load of 0.15 to 1 ohm is preferred (cf. Table 8.1.).

Flash or Short-Circuit Current

The flash or short-circuit current (SCC) of a cell is defined as that current indicated by a dead-beat ammeter having a resistance, including that of the leads, of 0.01 ohm, when connected directly across the terminals of the cell. The maximum swing of the needle should be taken as the short-circuit current of the cell. The short-circuit current has value in monitoring the production of a particular brand of cell, in following the changes in internal resistance of cells when the environmental temperature is altered, in spotting complete "duds" among cells, and in giving the maximum current a cell can sustain (Appendix A, the average OCV and SCC of various brands of cylindrical and flat Leclanché, mercury, and alkaline-MnO_2 and of one size of silver oxide (S15) dry cells are given for 21°C in the last two columns. These data were obtained at the National

[3]OCV and EMF are frequently considered as identical. Here, however, EMF refers to that value obtained with a potentiometer when the current is of the order of 10^{-9} amp, whereas OCV refers to that value indicated by a voltmeter having a resistance of 1000 or 10,000 ohms/V as set forth above.

Bureau of Standards). However, the flash current gives no indication o
the ultimate output a cell will give in service or of the relative electrica
capacity of cells of different construction. The flash current of a cell is o
value only when coupled with a familiarity with the brand of cell in
question. If the reading is normal for that brand of cell, it is reasonably
certain that the particular cell is in good condition. However, two cells o
different brand or constructions giving the same flash current may give
widely different outputs under the same service. Also, the cells may
deteriorate on shelf at distinctly different rates.

SERVICE TESTS

Agreement on service tests is not easily attained. The chief difficulty in
devising tests for primary batteries is that batteries may be discharged in a
wide variety of ways and arguments may be offered in the defense of
each. The rate of discharge, the degree of intermittency, and the voltage
at which a battery is considered useless may vary widely. Agreement as to
test methods is most difficult when international specifications are considered. A certain degree of conformity may be found between countries
but marked differences still exist in test schedules from one country to
another. For example, low-voltage (radio) batteries are tested on a
five-day schedule in Europe and Asia but on a seven-day schedule in the
United States. Radio high-tension batteries, in a densely populated
country where transmitting stations are near, will work to a lower average
cutoff voltage than in a country where transmitters are far away. Also,
there are extremes of temperature and humidity between countries which
make international agreement on testing procedures difficult.

Above all, however, as was stated at the start of this chapter, the most
significant single factor leading to the need for *service tests* is that, to
date, no nondestructive battery test has been devised that will give
reliable data on electrical capacity of primary cells when they are fresh,
aged, or partially discharged. It has been shown that OCV, initial CCV,
SCC, and internal resistance or impedance, although they each give
valuable *supplemental information*, do not give information on the *total*
or *residual* capacity of batteries. Likewise, other methods, such as
magnetic susceptibility(23), gassing rates(24), pulse discharge(25), to
name a few, have been disappointing along this line. Accordingly, service
tests are still used to evaluate batteries on the basis of electrical capacity.
These tests are made on a sampling basis by civilian and defense agencies,
testing laboratories, and the manufacturers. Along with all tests, visual
inspections are made to ascertain if the cells and batteries are mechanically sound in all regards and comply with specified dimensions, weights,
and so forth.

Standard Tests

These tests are described in the American National Standards Institute specifications for dry cells and batteries(16)(17). They were arrived at after many conferences between the manufacturers; consumers, including government agencies; and designers of electrical equipment. These standard tests are service tests designed to simulate end uses and are accepted as American Standard. The standard temperature for tests is $70 \pm 2°F$ (21°C); humidity is not specified but is taken as that which results from the control of the temperature at 21°C and usually falls within the range of 45 to 75 percent relative humidity. The International Electrotechnical Commission(6) selected almost the same conditions as standard, namely, $20 \pm 2°C$ for temperature and 45 to 75 percent for relative humidity. The IEC also selected a standard temperature of $27 \pm 2°C$ for use in tropical and subtropical countries where a temperature of 20°C would be impractical. The military(26) has also chosen the ANSI conditions as standard but with a wider tolerance of $\pm 5°F$ for the temperature. In all the standard tests the initial OCV and CCV are taken as well as measurements of CCV at specified, or necessary, intervals throughout the test. Unlike the procedures used for secondary batteries where ratings are given in ampere-hours, ratings for most dry cells and batteries (as discussed at the start of this chapter) are given in time of service to a specified cutoff voltage. Large air-depolarized cells and copper-oxide types for railroad and signal operating are exceptions to this for, in most cases, they are given ampere-hour ratings; even so, they may be tested on some of the standard tests presented here.

In Table 8.1, standard tests, 44 in number, presently recommended for use, are described.[4] Of these, the watch-cell, the high-rate discharge, the electronic-photoflash-cell, the safety-flasher-cell, toy-battery, electronic-equipment, electric-eye, and photo-shutter-warning-light tests are the newest while the one listed second dates from 1912. As stated earlier in this chapter, new tests are added from time to time as new end uses arise and some tests, on becoming obsolete, are removed. Constant attention is required to keep testing procedures abreast of prevailing requirements.

Representative discharge curves are given in the various chapters of Volumes I and II dealing with the individual battery systems. However, most of the testing procedures described have been developed for evaluation of the Leclanché dry cells which for many years were the only

[4]Battery specifications and testing are described in more detail in a publication entitled *American National Standard Specifications for Dry Cells and Batteries*, ANSI C18.1-1972, published by the American National Standards Institute, Inc., 1430 Broadway, New York(17).

commercially available type. In this system, manufacturing techniques can be varied considerably and these variations are often utilized to change the general level and shape of the discharge curve. Thus, a variety of cutoff voltages has been selected, depending on the application for which a cell was designed. Figure 8.1 shows a comparison of the discharge curves of three D-size unit cells on the 4-ohm-light-industrial-flashlight test at 21°C. The industrial cell has a generally higher level of working voltage than the two general-purpose cells included in the comparison. Furthermore the two general-purpose cells behave quite differently. Cell B has an initially higher working voltage while cell C with an initially lower working voltage provides considerably more service to the 0.90 V cutoff.

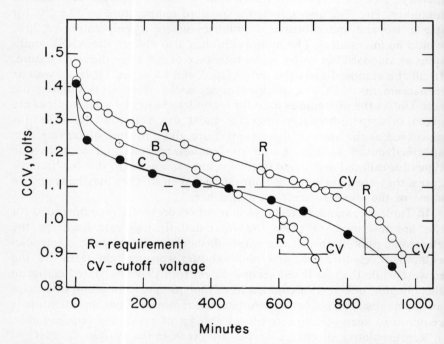

Figure 8.1 Discharge curves for "D" size Leclanché dry cells discharged on the light industrial flashlight cell test at 21°C. A, an industrial cell which gave 710 minutes to 1.10 V and 960 minutes to 0.90 V. The requirements are 600 and 950 minutes respectively to these cutoffs. The data are an average of six cells. B, a general purpose cell which gave 700 minutes to 0.90 V as compared with a requirement of 600 minutes. The data are an average of six cells. C, another general purpose cell which gave 900 minutes to 0.90 V greatly exceeding the 600 minute requirement. The data are an average of three cells. Note Although the industrial cell tests have established cutoffs at both 1.10 and 0.90 V, the general purpose cell on this test has only one cutoff of 0.90 V.

TABLE 8.1. STANDARD TESTS FOR DRY CELLS AND BATTERIES

ANSI No.	Test(8)	Cells per Test	Ohms Res.	Time "On"	Time "Off"	No. Disch. per Day	Total Time "On" per Day	Oper Days per Week	Total Time "On" per Week	Cutoff (V)	Service Reported in Seconds	Minutes	Hours	Days
1	Heavy intermittent	1(4)	2.66	60 min	{ 360 min / 960 min }	2	2 hrs	7	14 hr	0.85			X	
2	Light intermittent(1)	3	20	4 min	56 min	10	40 min	7(1)	260 min	2.80				X
3	Telephone(1)	3	50	4 min	56 min	10	40 min	7(1)	260 min	3.25				X
4	Alarm	1	300	Cont.						0.90				X
5	Alarm	1	500	Cont.						0.90				X
6	Lighting-battery	4	9	30 min	30 min	8	240 min	7	1680 min	2.60			X	
7	Lighting-battery	4	16	30 min	30 min	8	240 min	7	1680 min	3.00			X	
8	Lighting-battery	4	32	30 min	30 min	8	240 min	7	1680 min	3.60			X	
9	Safety flasher	1	15	250 ms	750 ms	86,400	21,600 sec	7	42 hr	0.90				X
10	Safety flasher	1	15	100 ms	900 ms	86,400	8,640 sec	7	16.8 hr	0.90				X
11	General-pur. intermittent	1	2¼	5 min	1435 min	1	5 min	7	35 min	0.65		X		
12	General-pur. intermittent	1	4	5 min	1435 min	1	5 min	7	35 min	0.75		X		
13	General-pur. intermittent	1	5	5 min	1435 min	1	5 min	7	35 min	0.75		X		
14	Light-ind. flashlight	1	2¼	4 min	56 min	8	32 min	7	224 min	0.65		X		
15	Light-ind. flashlight(2)(5)	1	4	4 min	56 min	8	32 min	7	224 min	1.10 / 0.90		X		
16	Heavy-ind. flashlight	1	4	4 min	11 min	32	128 min	7	896 min	1.10 / 0.90		X		
17	Photoflash(3)(6)	1	0.15	1 sec	59 sec	60	60 sec	5	300 sec	0.50 / 0.25	X			

331

Table 8.1. (Contd.)

ANSI No.	Test(8)	Battery or Cells per Test	Testing Schedule							Cutoff (V)	Service Reported in			
			Ohms Res.	Time "On"	Time "Off"	No. Disch. per Day	Total Time On per Day	Oper. Days per Wk	Total Time On per Wk		Flashes	Minutes	Hours	Days
18	Electronic photoflash(7)	1	1	15 sec	45 sec	60	15 min	5	75 min	0.75	×			
19	Hearing-aid-trans. battery	1	625	16 hr	8 hr	1	16 hr	7	112 hr	0.90			×	
20	Miniature-hear.-aid battery	1	1000	16 hr	8 hr	1	16 hr	7	112 hr	0.90			×	
21	Miniature-hear.-aid battery	1	3000	16 hr	8 hr	1	16 hr	7	112 hr	0.90			×	
22	Low-voltage battery	1	5	4 hr	20 hr	1	4 hr	7	28 hr	1.0			×	
23	Low-voltage battery(6)	1	25	4 hr	20 hr	1	4 hr	7	28 hr	1.0			×	
24	High-voltage battery	1	$166\frac{2}{3}$	4 hr	20 hr	1	4 hr	7	28 hr	1.0			×	
25	High-voltage battery	1	1500	4 hr	20 hr	1	4 hr	7	28 hr	1.0			×	
27	Trans. bat.(6)	1	$83\frac{1}{3}$	4 hr	20 hr	1	4 hr	7	28 hr	0.9			×	
28	Trans. bat.(6)	1	$166\frac{2}{3}$	4 hr	20 hr	1	4 hr	7	28 hr	0.9			×	
29	Watch cell	1	0.1 meg	Cont.						1.0				×
30	High-rate-discharge	1	1	Cont.						0.75		×		
40	Toy battery	1	2.25	Cont.						0.80		×		
41	Toy battery	1	4	Cont.						0.80		×		
42	Toy battery	1	7.5	Cont.						0.80		×		
43	Toy battery	1	5	60 min	{360, 960}	2	120 min	7	14 hr	0.80		×		
44	Toy battery	1	10	60 min	960	2	120 min	7	14 hr	0.80		×		
45	Toy battery	1	15	60 min	960	2	120 min	7	14 hr	0.80		×		
46	Motor test-toys and cameras	1	400*	4 min	11 min	32	128 min	7	896 min	1.0		×		

(Values "1.1" and "1.1" appear in the Cutoff (V) region adjacent to rows 22 and 23.)

47	cameras	1	250*	4 min	11 min	32	128 min	7	896 min	1.0	×
48	Motor test-toys and cameras	1	150*	15 min	45	32	128 min	7	896 min	1.1 1.0	×
50	Electronic equip.	1	25	4 hr	20 hr	1	4 hr	7	28 hr	1.0	×
51	Electronic equip.	1	83.33	4 hr	20 hr	1	4 hr	7	28 hr	0.9	×
52	Electronic equip.	1	125	4 hr	20 hr	1	4 hr	7	28 hr	0.9	×
53	Electronic equip.	1	40	4 hr	20 hr	1	4 hr	7	28 hr	0.9	×
62	Electric eye cell	1	13500	Cont.						1.20	
63	Photo shutter-warning light cell	1	250	Cont.						0.80	×

aThis table follows a presentation of similar data given previously(7).

(1) Telephone tests are scheduled at 10 discharges per day for 6 days each week and 5 discharges on the seventh day.

(2) When the LIF test is applied to "C" and "AA" sizes of cells the cutoff voltage is 0.75 V per cell.

(3) When the photoflash test is applied to "C" and "AA" sizes of cells the cutoff voltage is 0.25 V per cell.

(4) Four cells are usually run in series, but not always, in this test.

(5) The LIF (light industrial flashlight) test is used for a variety of cell types and sizes, thus for industrial D size cells on a 4 ohm test cutoffs of 1.10 and 0.90 V are used, whereas for a D size cell of the general purpose type only one cutoff of 0.90 V is applied.

(6) In ANSI Standard C118.1-1969 but omitted in ANSI Standard C18.1-1972.

(7) Number 61 in ANSI Standard C18.1-1972 but with the number of discharges per day not specified.

(8) Tests 1–5 are usually done on No. 6 cells; tests 6–8 on 4F batteries with the resistance and cutoffs given on the basis of a 4 cell (6 V) unit. Tests 9 and 10 apily to multicell batteries such as the 4F type but the resistances and cutoff values are given on a per cell basis. The remainder of the tests, 11–63, apply to a variety of sizes of unit cells.

*Denotes milliamperes. ×Denotes that a cutoff of 0.9 V also specified.

In Appendix C the average performance obtained for typical dry cells and batteries, on some of these ANSI standard tests, are given for the period 1955–1962. The 1972 requirements are also listed. In many cases the earlier performance exceeds the later requirement by a considerable amount.

Delayed-service Tests

The ability of primary cells to retain electrical capacity while in storage that is, their "keeping quality" may be determined by most of the standard tests described above by comparing the capacity obtained when the cell is fresh with that obtained after the cell has been stored for a specified time (compare results of initial and six-month delayed tests in Appendix C). Tests of long duration, for example, the telephone-battery test, 50 ohms (test No. 3) which runs from about 470 to 625 days for No. 6 cells, are obviously unsuitable for delayed-service tests. Longer storage periods are also specified for some dry cells and batteries in the military specifications (26).

Supplementary Service Tests

In addition to the standard tests outlined above, additional information is generally given by the manufacturers on various schedules, at various discharge rates, to various cutoff voltages. Discharges are done either at constant current or with a constant-load resistance. These data give an overall picture of the characteristics of cells and batteries which serve the user in a choice of portable electrical supply and aid equipment manufacturers in their designs. They also make possible the calculation of cell capacity in terms of ampere hours and watt hours. The output obtainable from Leclanché dry cells is almost a linear function of the weight or volume of the cell for low current drains but is nearly independent of weight and volume for higher drains (27).

SHELF TESTS

Shelf tests consist of storing cells and batteries on open circuit under specified environmental conditions and then ascertaining the condition of the cells and batteries at periodic intervals. A temperature of $70 \pm 2°F$ and a relative humidity of 60 ± 15 percent are preferred in shelf tests since these are the conditions chosen for service tests. OCV of Leclanché dry cells is read at intervals not exceeding one month and the service is reported as the number of months on test (on observation) before the OCV falls below 1.45 V per cell. In early ASA standards this test was referred to as the C-battery test.

Although they are time consuming, delayed-service tests constitute the est shelf tests in that they give a measure of the residual capacity in cells or batteries after a specified time in storage. They must, however, be conducted on a sampling basis since the tests are destructive—the cells are expended during the test.

Although, as stated above, OCV and SCC cannot be used to determine electrical capacity they can be used, in a relative way, to follow the changes that take place in cells and batteries on shelf. When the changes are abnormal, trouble can be suspected; in other words, OCV and SCC are used in spot-checking rather than in evaluation. In Appendix D, OCV data obtained for 3B, 5B, 15B, and 200F20 Leclanché batteries are given covering a period of time up to 12 years. They show the characteristics of these cells as voltage references or for grid bias but not as energy sources.

It was common practice at one time to measure SCC at 1, 2, 4, and 8 weeks after cell manufacture and then every 8 weeks, thereafter, and then to report shelf life as the time in months for SCC to reach a specified value, for example, 10 amp for No. 6 cells. Measurements of SCC are probably made less frequently today. In general SCC decreases more rapidly than electrical capacity for cells in storage, especially at higher temperatures (28).

The changing characteristics of dry cells and batteries on shelf may also be determined from measurements of internal resistance. Since relative changes are involved in this case any one of the methods of determining or estimating internal resistance described in Chapter 10 may be employed throughout the time period involved. If one method is selected at the start another method should not be used later for the same cell or battery. As a rule the internal resistance of a cell increases as the cell ages; however, no correlation has been found between this increase and residual capacity.

ELECTROMECHANICAL PROCEDURES IN TESTING

EARLY TESTS

Many of the early tests on primary cells and batteries were conducted in service or in the field, for example, by attaching the test apparatus to the circuits of regular commercial telephones or to actual induction coils used in ignition (5). It was soon found, however, that results of the tests with telephones necessarily varied with the amount of use given the phones and therefore could not be duplicated, and that induction coils changed their characteristics throughout a day of testing. These tests, accordingly, were abandoned and replaced by controlled tests such as those described in Table 8.1.

The tests outlined in Table 8.1 cover a wide variety of conditions; some deal with intermittent current drains with long rest periods intervening others with intermittent drains with short or no rest periods; while still others are for continuous discharges. To conduct these tests simultaneously on a prescribed schedule and for a large variety and quantity of samples, as would be involved in testing by consumers of large quantities of cells, by manufacturers in quality control, or by the government in qualification testing, some type of automation is necessary. Actually automatic testing to a small degree was introduced early in this century(4)(5)(29)(30) or shortly after the volume of production of dry cells became appreciable (see, ref.(7) for changes in the volume of production).

The first automated equipment was designed for cells tested on an intermittent program. The Pritz circuit(4), for example, for testing cell for telephone service included a battery-operated clock whose hand made a complete cycle once each hour. The clock contained four equally spaced contacts which the hand on revolving contacted for a period of two minutes each hour. The current through the contacts magnetized the cores of telegraph relays causing extended armature arms to fall thereby bringing inverted U-shaped fingers into mercury cups, through which the test circuit including the cell under test with an external load was closed. The voltage across the cell under load was read manually with an accurate voltmeter at the mercury cups. This circuit subjected three cells in series to a discharge through 20 ohms for a period of two minutes each hour during 24 hours per day seven days per week until the working voltage reached a cutoff voltage of 2.8 V. The time required to reach the cutoff voltage was determined from a plot of CCV against time of discharge.

In general the early automated circuits were as described above, that is they included a clock, relays, and mercury cups. For example Gillingham(29) stated that "intermittent discharges were operated by clock, for the most part through mercury contacts" while the Dry Cell Committee of The Electrochemical Society reported "when a great many tests are to be made, it is convenient to arrange a clock-operated automatic closing device." Likewise Melson(30) described a circuit containing a weight-driven clock, solenoid relays, mercury troughs, and the test circuit with cell and load. He also described a test rack on which were a number of circuits for cells under test. He measured his voltages with a bifilar galvanometer which had no zero creep and "fast" response.

The advent of automatic equipment meant the passing, to a large extent, of the manually operated switch, crude timing devices, pocket voltmeters, rheostats, lamp resistors, and crude circuitry, with their obvious limitations. The manual determination of cutoff voltages was etained, however. Also by 1919 a standard temperature for testing, namely, 20°C was specified(31); the effect of temperature on the erformance of dry cells had been well recognized(3)(4). The standard emperature for testing was changed to 21°C in 1937(10).

MPROVED APPARATUS AND PROCEDURES

As the volume and variety of testing increased changes were made in utomatic devices and the means of reading voltages. In 1920 Vinal and Ritchie(32) described an apparatus controlled by clock mechanism and onstructed in large part of materials used in the telephone field. The pparatus consisted of two selective relays having three rotating shafts hat made complete revolutions in one hour, one day, and one week, espectively. The coils of the first relay were energized once every minute by a pendulum clock. The coils of the second relay were energized once ach hour by the first relay, thereby eliminating rapidly moving parts and ncreasing the accuracy of the time intervals. Suitably designed ommutators (Pt vs. Pt-Ir) placed on the rotating shafts served to control ny specified periodic test. When the circuit through the contacts on the ommutators was complete, the current closed a master relay. A master elay was provided for each test and was connected to an electric counter which recorded the number of times the circuit was closed. Pilot lights ndicated when the cells were on discharge. Multiple switches (on a test oard) were provided to close 25 (later 31) independent test circuits imultaneously. The contacts were mercury cups with double break. The upper part of the switches was mounted on pivot hinges and was operated by the solenoid. The switch was opened by a counterweight operating on gravity. Cells under test were mounted on a board (or panel) with sliding contacts at the bottom permitting adjustments for any length of cell. A screw with blunt end supported the cell at the bottom terminal and a similar screw with sharp point at the upper end made contact with the upper terminal and clamped the cell in place. This arrangement avoided the use of soldered leads to the cells, a procedure not considered advisable nor recommended. Calibrated resistance coils were mounted over the top screw and potential leads from the cells passed along the back of the board to telephone jacks. Voltage readings were taken rapidly (20 circuits could be read within one minute) on a high-resistance voltmeter by the telephone-plug connection while the cell was discharging. At a later time(33), the apparatus was modified to handle 640 circuits and to include a second clock, so that if one clock should stop the tests would be continued by the other. Both assemblies were housed in a room controlled at 21°C; relative humidity ranged from 45 to 75 percent. Throughout the tests checks were made periodically to determine the

correctness of the circuit resistance and the proper calibration of th
meters. The high-resistance voltmeter had the characteristics outline
previously under *Open-Circuit Voltage*, voltages could be read to 0.01 V
The assembly is shown in Fig. 8.2.

In 1946(34) the National Bureau of Standards installed newer progra
mechanisms (shown in Fig. 8.3) for conducting the annual qualificatior
tests carried out in connection with federal procurement. The battery
operated clock is shown on the right, together with two progra
machines, one of which is open. The master clock was provided with a
Invar pendulum and was wound electrically once every minute; it dro
the two subsidiary clocks at the left. The master clock was regulate
within 10 seconds of true time per month, and was periodically calibrate
in terms of the national frequency standard. Each subsidiary cloc
provided for six programs which could be varied, as desired, b
perforating the paper tapes shown in the open program machine on th
left. The paper tapes made a complete cycle in 24 hours. Electric

Figure 8.2   Early apparatus for testing dry cells. A few of the cells, switches, and coils a
shown at the left. Voltages are read at the telephone jacks at the right.

Figure 8.3 Program machines and master clock used in the testing of dry cells. Electrical contacts to relays controlling test programs are made through perforations in paper tapes shown in program machine which is open.

Contacts made through the perforations actuated relays with gold-plated contacts, controlling the individual test programs. Readings of voltage were made as before using the telephone switchboard. The test rack for housing cylindrical cells was modified, however; the screw arrangement previously used was replaced by a gold-plated pressure contact. When transistor batteries appeared (about 1959), a return (21) was made to the screw arrangement for mounting batteries, except that threaded bolts were used, (cf. Fig. 8.4). This arrangement for mounting was necessary since transistor batteries vary greatly in size.

Apparatus for Tests with High Repetition Rates (Photoflash Cells)

In the ASA specification of 1954(13) a test for photoflash cells was introduced. This was a severe test, operating for exceedingly short periods of time compared with the other tests, and requiring special

Figure 8.4 Panel for testing transistor batteries. Adjustable contacts are used to hol▮ batteries on panel.

apparatus. This test required the discharging of a cell through a resistanc▮ of 0.15 ohm for one second each minute for one hour at 24-hour intervals with the recording of the initial OCV and the CCV on the 1st, 30th, an▮ 60th minute daily. Obviously, automatic testing was essential, especiall▮ for simultaneous testing of a number of cells. The circuit, designed b▮ DeWane(35), for this purpose is shown in Fig. 8.5; a similar arrangemen▮ could be used for electronic photoflash cells and safety-flasher batteries▮ A 24 V storage battery operates the six 24 V, 10 amp aircraft type relay▮ (R-1 to R-6) that close the 12 0.15-ohm discharge circuits. The 24 V circui▮ is closed by the normally open relay R-7. An electrolytic condenser C i▮ shunted across the 24 V circuit to prevent sparking on the R-7 rela▮ contacts; this relay is powered by a $22\frac{1}{2}$ V battery. The $22\frac{1}{2}$ V circuit i▮ closed once every minute by relay R-8 which operates on a 6 ▮ one-second impulse received from a master clock which operates ▮ program machine, P.M. No. 1, of the type described immediately above▮

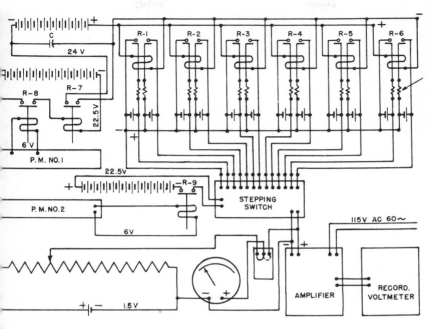

igure 8.5   Circuit for testing a number of cells simultaneously on high repetition rates.

he net result is that the three circuits involved in the discharge operation
24 V, $22\frac{1}{2}$ V, and the discharge circuits) are closed once a minute for the
uration of one second.

Voltage leads from the individual cells are connected to a rotating
tepping switch powered by a second $22\frac{1}{2}$ V battery. This circuit is opened
nd closed by relay R-9 controlled by a second program machine, P.M.
Io. 2, in synchronation with P.M. No. 1. This circuit closes
imultaneously with the discharge circuit but does not open until 20
econds have elapsed owing to the delaying action of a dash pot on P.M.
Io. 2. The stepping switch advances on the break of the circuit, moving
ie voltmeter leads to the next cell to measure its OCV and CCV.

The stepping switch is connected to an amplifier which in turn is
onnected to a recording voltmeter. In parallel with the amplifier-
ecording-voltmeter circuit is a double loop circuit consisting of a
alvanometer key, portable voltmeter, slide resistance, and a No. 6 dry
ell. This arrangement is used to calibrate the recording voltmeter against
ie portable instrument which had previously been calibrated by means of
 potentiometer. The slidewire is adjusted so that the portable voltmeter

registers a voltage obtainable from the No. 6 cell that is close to the expected one-second load voltage of the cells on test. At the moment that the discharge circuit closes, the galvanometer key is depressed transfering the meter from the No. 6 dry cell to the cell whose discharge is being recorded. The voltmeter pointer moves through a very short arc and an accurate reading can be taken since there is no overswing.

The procedure followed in running a day's discharge follows. A discharge relay contacts are first blocked off by fiber inserts except that of cell 1. The stepping switch is in No. 1 position so that the OCV registered on the voltmeters is that of cell 1. This cell is then disconnected and the zero point of the recording voltmeter obtained by a zero-set adjustment of the amplifier. The cell is then reconnected and the variable gain adjusted so that the true OCV of cell 1 is recorded. Switches on the program machine are then closed to place it in operation. As the first minute comes up, discharge relay R-1 closes and the CCV of cell 1 is recorded. At the expiration of 20 seconds, the stepping switch moves the voltmeter circuit to cell 2. The OCV of cell 2 is then recorded and in the meantime the insert on the discharge contacts of R-2 is removed. As the second minute arrives, the initial discharge of cell 2 is recorded and cell 1 receives its second discharge. This procedure is repeated until all the cells on the discharge circuits are working. The initial discharge of each cell is measured with the portable voltmeter and the calibration of the recording meter checked and readjusted if necessary. When the 12th cell is read, the stepping switch moves the voltmeter circuit to cell 1. At this point the stepping switch is moved manually to cell 2 to record its 12th discharge. As a result of this move the 13th discharge on cell 1 is not recorded but puts the system in step so that the 12th discharges of cells 2 to 12 are recorded as well as the 24th, 36th, 48th, and 60th of all 12 cells. The voltage at the 30th discharge can be obtained by interpolation between the 24th and 36th values. After the 60th discharge of cell 1, its discharge relay contacts are blocked off and this is repeated for the succeeding cells so that after 70 minutes and the 71st operation of the discharge circuits all 12 cells have had 60 discharge periods.

A typical record of the test is shown in Fig. 8.6. The recorder was operating at a rate of 12 inches per hour except at A where it was stepped up to 12 inches per minute to show the complete graph of a single discharge. At the slower rate only the minimum discharge voltage is obtained.

Although this test is rather elaborate, it functions well and can be operated by one person working part time during the 70 minutes required per run. For photoflash cells the test of 70 minutes is run only once each day for five days per week.

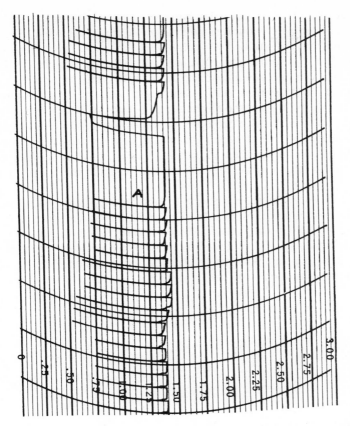

Figure 8.6 Graphical record of tests of photoflash cell. Chart rate, above A—12 inches per hour, below A—12 inches per minute.

TRENDS

The trend in testing equipment is toward the use of semiautomatic voltage recorders, digital voltmeters, and printout equipment. Higher precision in voltage measurements is attained with digital voltmeters and the use of semiautomatic voltage recorders make possible the accurate determination of cutoff voltages of batteries on discharge. In determining cutoff voltages for discharges of long duration the apparatus is frequently designed so that the recorders are placed in a discharge circuit shortly before the cutoff voltage is reached; in other cases where the discharges are short the voltage recorder is kept in the circuit throughout the discharge. The latter procedure is also followed if the *IR* drop, recovery,

and other characteristics of a discharge are desired throughout a complet
discharge. Electronic counters and magnetic tapes are used in recordin
time, especially in studies of activation time, iR drops, and voltag
recuperation. High-precision current stabilizers, zener diodes, and printe
circuits are seen in many measuring circuits. Printout equipment provide
permanent records and with computers aids in the analysis an
summarizing of data and the elimination of spurious data. Computers ma
also be used directly with program machines and readout devices. Thes
trends toward the use of newer measuring devices are seen in automati
quality control as well as in quality evaluation and research. Wit
increased use, these newer devices should broaden the scope of batter
testing.

Three views from their battery testing facility have been furnishe
through the cooperation of Mr. T. A. Reilly of Consumer Product
Division of the Union Carbide Corp. Figure 8.7 shows a view of th
testing laboratory. Figure 8.8 shows an automated testing control an
readout unit. Figure 8.9 shows the testing apparatus used for a variety c
tests on miniature sizes of cells.

Figure 8.7   The battery testing facility at the Consumer Products Division of Union Carbid
Corp. (Courtesy of Union Carbide Corp.)

Figure 8.8 An automated testing control and readout unit for battery tests. (Courtesy of Union Carbide Corp.)

SPECIAL TESTS

In addition to the standard tests described above, several special tests are of interest and need to be noted. With the exception of the gassing test (24) they are the ones recommended in Military Specifications, Batteries, Dry, MIL-B-18D, 29 October 1963(26).

ELECTROLYTE-LEAKAGE TEST

The leakage of dry cells or batteries during operation is of major concern to users. In a majority of cases leakage occurs when the cell or battery is overdischarged, or has been inadvertently kept on closed circuit within an instrument for extended periods of time, or has been overheated. These facts have been taken into account in devising a test procedure for electrolyte leakage. The test is applied only to individual cells, discharged continuously for 24 hours at $70 \pm 2°F$ and 50 ± 15 percent relative humidity through a resistance, the value of which depends on the type and size of cell as follows:

	Leclanché cell		Mercury cell
Size	Resistance, ohms	Size	Resistance, ohms
AA	10	M40	10
C	7.5	M60	8
D	5	M70	5
F	5	M72	5
J	3	D	5

The resistance to be used with other sizes of cells can be estimated from the volumes of the cells and of those for the cells listed above. These discharges bring the cells to a state of complete or nearly complete discharge. The cell is then stored under the same conditions for 15 days during which time the cell is inspected daily for electrolyte leakage on external surfaces with the jacket in place. Electrolyte leakage is considered to have occurred when moisture appears on a piece of absorbent paper when rubbed on the surface of the jacket. When done on a sampling basis, 50 percent of the cells are stored in an inverted position.

GAS-LEAKAGE TEST

This test, as a rule, is applied only to individual cells. A good procedure is that devised by Otto and Eicke (24). A gasometer of the type shown in Fig. 8.10 is used. It is housed in a constant-temperature oven, Fig. 8.11. The cell is placed in cup 5, held to the gasometer by springs 10, and reservoir 6 is nearly filled with mercury. Mercury is also added to manometer 4 until it touches the Pt-tipped tungsten wire 3. Wire 3 and the lower sealed-in wire of the manometer are joined in series with a neon bulb, 0.1 megohm resistor, and 110 ac. While initial temperature equilibrium is being attained the mercury level in the manometer may be disturbed, but is readily readjusted by either adding or withdrawing air at 8 or mercury at 7. As soon as balance is attained, as indicated by the glowing of the neon bulb, stopcock 2 is closed, thereby trapping a quantity of air in reservoir 1 at the constant temperature of the oven and the barometric pressure prevailing at

Figure 8.9   The testing apparatus used for a variety of tests on miniature sizes of cells. (Courtesy of Union Carbide Corp.)

the time. The gasometer is then allowed to stand until the gas evolved by the cell forces a separation of the mercury from the Pt tip by about $\frac{1}{4}$ inch. Mercury is then withdrawn at 7 until electrical contact of the Pt tip and the mercury is remade, again indicated by the neon bulb. From the weight of mercury withdrawn, the volume of evolved gas is calculated at the pressure and the temperature of the gas in the reference reservoir, and then at STP. Otto and Eicke also operated their gasometer in the absence of air, the gasometer being first evacuated through stopcock 2 and inlet 9 and then filled with nitrogen.

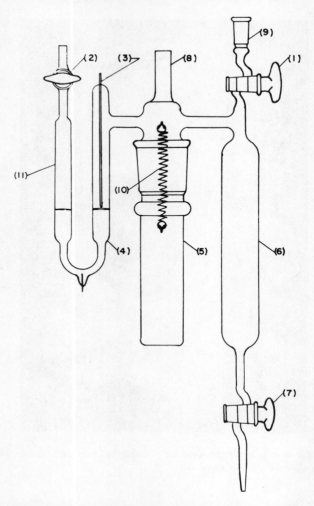

Figure 8.10 Diagrammatic sketch of gasometer used to study the gassing of dry cells.

HIGH-TEMPERATURE TEST (TROPICAL-STORAGE TEST)

In this test, cells or batteries are stored for 90 days at 113 (+2, −8)°F and 50 ± 15 percent relative humidity. The cells or batteries are then cooled to—and discharged on—standard tests at 70 ± 2°F and 50 ± 15 percent relative humidity. Requirements differ for various sizes of cells and

Figure 8.11 Constant-temperature oven for use in testing the rate of gassing of dry cells at constant temperature. Mercury is collected at intervals in bottom compartment without changing the temperature of upper compartment housing the gasometers.

batteries but, on the average, are approximately 50 to 75 percent of the 70°F requirements.

ARCTIC-CAPACITY TEST (36)

In this test, cells or batteries are stored for 48 hours at $-40 \pm 2°F$ and ambient humidity, and then discharged on standard tests under the same conditions. Requirement is approximately 10 percent of the 70°F service for conventional cells or batteries; for special low-temperature batteries made for low-temperature operation the requirement is about 20 to 30 percent. This test is not listed in the 1963 Military Specifications (26).

INSULATION-RESISTANCE TESTS

(a) The battery is stored for 48 hours at $70 \pm 5°F$ and 50 ± 15 percent relative humidity, after which and while at these conditions, the insulation resistance is measured by applying a dc potential of 500 ± 20 V between any two terminals not electrically connected and between all ungrounded terminals and the jacket of the battery. For those batteries having nonmetallic jackets, a 1-inch copper plate is placed with the broad surface against the jacket for electrical contact. The insulation resistance should not be less than 5 megohms.

(b) Another test is the same as above except that the battery is tested at ambient temperature and humidity without previous storage. If the insulation resistance is lower than 5 megohms, when tested at a relative humidity of 80 percent, the battery is retested according to (a) but not at a relative humidity below 50 percent.

VIBRATION TESTS

(a) The battery is rigidly clamped to the platform of a vibration machine in a manner approximating that in which the battery is clamped in use. It is then subjected to simple harmonic motion having a total maximum excursion of 0.06 inch (amplitude 0.03 inch), with the frequency varied at the rate of 1 Hz per minute between the limits of 10 and 55 Hz. The entire range of frequencies and return is transversed in not less than 90 minutes nor more than 100 minutes. The battery is vibrated in three mutually perpendicular directions, one of which is perpendicular to the terminal face. OCV is measured for 30 seconds during the last quarter of each vibration period. Good cells will show no mechanical failure, voltage fluctuations, or other adverse effects during tests.

(b) In an alternate test the simple harmonic motion is applied continuously with the frequency varied from 700 to 3000 Hz in steps of 100 cpm, with the amplitude of vibration and frequency range as follows:

Frequency range (cycles per minute)	Amplitude (inches)	Maximum excursion (inches)
700 to 1500, inclusive	0.030 ± 0.002	0.060
1600 to 2000, inclusive	0.020 ± 0.002	0.040
2100 to 2500, inclusive	0.013 ± 0.001	0.026
2600 to 3000, inclusive	0.009 ± 0.001	0.018

The battery is vibrated in three mutually perpendicular directions one of which is perpendicular to the terminal face and parallel to one side of the battery for 20 minutes at each 100-cycle step for a total of 8 hours in each direction. Good cells will withstand this test without adverse effects.

SHOCK (MECHANICAL) TESTS (FOR MULTICELL BATTERIES ONLY)

(a) The battery is subjected to three mutually perpendicular shocks of equal magnitude applied in a direction normal to any one of the battery faces. The battery, during each shock, is accelerated so that during the first 3 msec the minimum average acceleration is 75 g with peak acceleration being between 125 and 175 g. Good cells will show no adverse effects.

(b) The battery is subjected to three mutually perpendicular shocks of equal magnitude, one of which is perpendicular to the terminal face and parallel to one side of the battery. The battery, during each shock, is accelerated so that during the first 15 msec the minimum average acceleration is 200 g with peak acceleration being between 250 and 350 g and acceleration is applied so suddenly that at least 50 percent of the peak is attained in the first 5 msec. Good cells will show no adverse effects.

(c) The battery is subjected to two shocks of equal magnitude with the primary being a specified one and the secondary being at right angles to the primary. During the impact phase the velocity is changed from 5.5 to 7 ft/sec within 0.2 to 0.4 msec. During the drag phase the battery is accelerated so that during the first 15 msec the minimum average acceleration is 100 g with peak acceleration being between 125 and 150 g. Good cells will show no adverse effects.

(d) This test is the same as (c) except that the velocity for the impact phase is 16 to 20 ft/sec within 0.2 to 0.4 msec and in the drag phase the minimum acceleration is 350 g within 15 msec with peak acceleration being between 450 and 550 g. Good cells will show no adverse effects.

MAGNETIC TESTS

These tests are used only under those conditions where magnetic disturbance would be a problem. Each of the six faces, including terminals, of a rectangular-parallelopiped-shaped battery are placed $\frac{1}{4} \pm \frac{1}{16}$ inch from, and centered relative to, a face of a total-field magnetometer and are oriented in such a way that the background magnetic field vector is perpendicular to and directed into each of the faces of the battery in turn. This gives a total of 36 measurements, six for each of the six faces. The variation of the magnetic-flux density is measured and recorded by the magnetometer. In a background field of not less than 550 mG or more than 650 mG, the change in flux density of the background field from the highest peak to the lowest trough caused by the battery on open or closed circuit is required to be less than 0.10 mG and after idealization less than 0.15 mG.

In the idealization three groups of batteries of about equal number are first placed in position so that the background field of one is mutually perpendicular to that of the other two. Each battery is placed in a uniform magnetic field of flux density of 5 G and a cycled pulsed magnetic field is superimposed on and parallel to the uniform magnetic field. A cycle consists of a square (+) pulse of 1 sec minimum duration and 1 sec minimum off, and a square (−) pulse of equal magnitude and 1 sec minimum duration and finally 1 sec minimum off. The amplitude of the positive pulse for the first cycle is between 50 and 70 G and is reduced by a maximum of 2 G between successive cycles until the amplitude is zero.

For batteries of shape other than that of a rectangular parallelopiped the battery is positioned by assuming the battery to be circumscribed by a rectangular parallelopiped.

TESTS FOR ADDITIONAL TYPES OF DRY PRIMARY CELLS

In the foregoing sections dry cells of the Leclanché, mercury, alkaline-MnO_2, zinc-silver oxide, and sealed nickel-cadmium types were considered. To this list may be added (1) air, (2) mercury dioxysulfate, (3) indium, (4) silver chloride, (5) vanadium pentoxide, and (6) solid-electrolyte cells, all of which may be considered dry types since they are made with non-spillable electrolyte. Zamboni piles or in general batteries made with trace electrolyte, and other solid-ion types high-potential batteries may also be included. Nomenclature and testing procedures for these types have not received the attention given the cells discussed in previous sections.

The first three types may be tested on the appropriate standard tests given above, that is, if the cells are used for flashlights, hearing aids, or watches, they are tested on those tests that simulate these services. Air cells are also given ratings in ampere hours for maximum load resistances for continuous and intermittent discharges to cutoff voltages of 0.80 V and are so tested (see next section for procedures of obtaining ampere-hour capacities). Maximum currents that may be used in momentary discharges (generally not to exceed 15 seconds) are also specified for some types of air cells, especially those for heavy-duty service. These currents and time are determined in testing by following the CCV until marked polarization occurs. The mercury dioxysulfate and indium cells are suitable for electronic-clock and watch service and may be tested on the appropriate test. Three sizes of mercury dioxysulfate cells may be considered as possible standards. These, sometimes, like the silver cells, called "button cells," have maximum diameters of 0.450, 0.591, and 1.000 inch, with the corresponding heights of 0.128, 0.147, and 0.110 inch, and may be tested at currents as high as 0.3 ma.

The other systems, including the high-potential batteries are all of high internal resistance and accordingly, high-impedance instruments, such as electrometers, electrostatic voltmeters, and so forth, must be used in the measurement of their voltage and in tests of their voltage stability. In many cases the accuracy and range of the voltage measurements may be increased, in the usual way, by using a bucking voltage (see later). Their internal resistance may be obtained conveniently by the flash-load method, Chapter 10, wherein the magnitude of the current is about one half of that in the flash-current method. In this method a flash load of proper value is placed across the terminals of the cell or battery so that the CCV is one half the OCV (see equation 2, Chap. 10). The silver chloride dry cell will sustain currents of the order of 0.1 ma while the others will only sustain currents of much lower magnitude (0.1 to 1 μa). Only the silver chloride dry cell is given a rating in ampere hours; the others are used almost exclusively as sources of potential. Determinations of the total ampere-hour capacity of silver chloride dry cells are very time-consuming; they are made, therefore, on a sampling basis.

TESTS FOR WET PRIMARY CELLS

Fewer types of wet primary cells are in use today than was the case many years ago. The copper oxide cell and the air cell, in various forms practically cover the field today.[5] As a result less effort has been devoted to their nomenclature or expended in devising standard tests for them than for the dry types. Furthermore, wet and dry primary cells are used in quite different ways. Wet cells are used in many applications where high currents are required, where service for extended periods of time is needed, and where a knowledge of the ampere-hour capacity is necessary as, for example, in railroad-signal service. Accordingly, wet primary cells with few exceptions are given ampere-hour ratings similar to the practice followed for secondary batteries. The cells or batteries are discharged either at constant current or at constant load resistance to a specified cutoff voltage and ampere-hour capacity calculated therefrom. Testing procedures for wet primary cells and batteries, therefore, are easy and

---

[5]Daniell (gravity), Bunsen, bichromate (Grenet, Poggendorf, and Fuller), and chlorine cells may be found to a limited extent in some applications. Zinc- and cadmium-silver oxide cells are considered under reserve cells since they find wide use in one-shot applications when not constructed for use as secondary cells.

Two exceptions may be cited. One is the standard cell of the cadmium sulfate type, a form of wet primary cell, which is discussed in Chapter 12 Volume I of THE PRIMARY BATTERY. The standard cell, however, differs from other primary cells in not being a source of electric current or power; it is used as a reference voltage or a standard of electromotive force. Another exception is the fuel or continuous-feed cell. Fuel cells are tested, in general, in actual service since they are operated on a continuous or intermittent schedule; accordingly, test procedures are not devised for them, at least not in the same sense as has been followed for other types of primary cells and batteries. Instead they are tested in a *demonstrational sense.*

straightforward. Discharges are made for various currents or at different rates so that the value of the normal and maximum rates may be ascertained and then specified. For copper oxide and air cells, cutoff voltages of 0.50 V and 0.80 V, respectively, are usually specified. The standard temperature for testing is the same as for dry cells, namely, 70°F, but performance data at other temperatures are usually provided by the manufacturers.

Tests for wet primary cells, like those for the dry types, are destructive ones—the cells are expended in the tests. As a consequence, testing is done on a sampling basis and the assumption is made that all cells of a particular production will behave similarly to the cells tested. The same assumption is frequently applied to cells which are of the same general type or construction. Such a procedure is obviously necessary for those cells which last two, three, or more years in continuous operation, such as in current supply for potentiometers or relay controls. Values of OCV, CCV, and internal resistance have the same limited usefulness in assessing electrical capacity for wet cells as they do for dry cells.

TESTS FOR RESERVE CELLS

In this section some typical testing procedures used to determine the characteristics of reserve cells are briefly considered. In many cases the methods discussed for dry and wet cells can be applied to the evaluation of the electrical capacity of these special cells. Only those procedures that are novel or are specific to this class of cells are described here.

Most reserve cells are used in missiles, torpedoes, and like service. Since they are activated at the time of use, *activation time* to peak voltage and *activation stand period* are two important factors in their use. Since activation times may be of the order of seconds or less, voltage measurements must be made with high-speed voltage recorders or similar instruments with voltage amplification. The accuracy of the voltage measurements may often be increased by bucking out[6] a major portion of the signal with a dc-bucking voltage. Batteries offer a very stable source for this bucking voltage. Stopwatches, tape recorders, or chronographs may be used in measuring time. Temperature should be measured throughout the tests since the temperature of many reserve cells, especially the water-activated types, changes rapidly immediately after activation. In stating the ampere-hour or watt-hour capacity of reserve cells it is

---

[6]"Bucking out" means a cell of almost equal and known voltage (but always slightly less) is placed in opposition (or series opposing) to the cell under test, so that it is the difference in voltage rather than the total voltage that is amplified. Details are shown in Fig. 8.12.

necessary to state the temperature precisely.

Tests on activation-stand periods for reserve cells other than the thermal types offer nothing really novel. Delayed capacity to a specified cutoff voltage is simply compared with initial capacity to the same cutoff voltage and at the same temperature. Of course, for each test a newly-activated cell must be used. Since reserve cells decay in voltage rapidly after activation, OCV may be used in evaluating activation stand periods; for water-activated types the OCV must be corrected to a common temperature, otherwise, the cells may appear to have longer or shorter activation stand periods than is actually the case owing to the positive or negative EMF-temperature coefficient of the cells. Also, a load may be placed momentarily across the terminals of a reserve cell and the CCV measured at periodic intervals; the trend in CCV with time gives a good evaluation of activation-stand period. The value of OCV and CCV measurements for reserve cells, thus, is markedly different from that for dry and wet primary cells. Measurements of activation-stand periods for thermal cells are infrequently made, if at all, owing to the short discharges given by thermal cells.

Many reserve cells (zinc-silver oxide, perchloric acid, etc.) are intended for use at high altitudes and a study of their behavior under such conditions becomes necessary. They must pass *internal-pressure, altitude,* and *attitude tests* and withstand *acceleration, vibration,* and *mechanical* and *thermal shock* (37). Individual cells or batteries are taken for each of these 7 tests. With the exception of the thermal-shock test, all tests are conducted during and following a specified activation-stand period.

An unactivated cell is used in the thermal-shock test. The cell is first heated to 165°F and immediately plunged into an air bath at −40°F and allowed to stabilize at this temperature. Satisfactory cells will show no breaking or cracking of the container.

In the internal-pressure test, the cell is subjected, at 80°F, first to an internal air pressure of a specified psi and then discharged at a specified rate. Good cells will show no electrolyte leakage during the test.

In the altitude test the cell is discharged at 80°F at a specified rate in a chamber evacuated to simulate an altitude of 30,000 to 70,000 feet and the test repeated at −54°F. Good cells will not leak, or exhibit flooding, or show voltage fluctuations.

In the attitude test, the cell is first inverted for a specified time after which it is discharged in the normal position but tilted an angle of ±45° from the normal for a specified fraction of the discharge time. No leakage of electrolyte or fluctuations in voltage will be shown by good cells.

For the acceleration, vibration, and mechanical shock tests the cells are

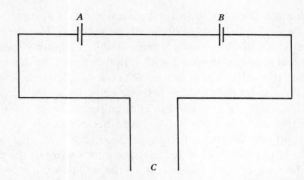

Figure 8.12 Explanation of "bucking out" method of measuring cell voltage. "Bucking out" means connecting a second cell or battery of known voltage (but preferably somewhat less than that of the cell under test), in opposition with (or series opposing) the cell under test, and accurately measuring the difference in voltage between them. The voltage of the cell under test is the sum of the measured difference and the known voltage of the second cell. A, cell under test; B, cell of known voltage; C, leads to potentiometer or voltmeter.

tested along each of the three major axes. In the first, the cell is discharged at 80°F while subjected to an acceleration of 50 g for at least four seconds for each axis. In the second, the cell is subjected at 80°F to a sinusoidal vibration of 0.03-inch amplitude for 30 minutes for each axis over a frequency range of 10 to 55 Hz and returned to 10 Hz in one to three minutes, with the battery being discharged during the last complete cycle of frequency change. In the last test, the cell is subjected to an impact shock of 40 g for a duration of 11 ± 1 msec with the maximum g reached within about $5\frac{1}{2}$ msec and then repeated for a second time. Good cells will pass these tests satisfactorily.

Finally, reserve cells under certain conditions tend to exhibit *electrical noise*. Electrical noise having a frequency greater than 10 Hz can readily be measured using a sensitive ac vacuum-tube or similar voltmeter. These instruments have ranges as low as 1 mV and are available as peak-responding, average-responding, or as rms instruments. The signal can also be observed with an oscilloscope. The voltmeter or oscilloscope is placed directly across the terminals of the cell or battery under test. For noise below 10 Hz an oscillograph or oscilloscope can be used. In either case the dc portion of the signal must be bucked out. If an oscilloscope is used the trace can be photographed to obtain a permanent record. In all cases the discharge resistor should be noninductive and adequate precautions must be taken to minimize both electrostatic and electromagnetic pickup.

Other physical(38) and environmental(39)(40) tests could be applied to ᵊrimary cells and batteries of all classes depending on the circumstances; ᵊowever, the tests given here, although they have no finality, are typical ᵊor the present age and should provide a good overall evaluation of the ᵊnechanical ruggedness and electrical performance of primary cells and ᵊatteries.

REFERENCES

1. Burgess, C. F., and Hambuechen, C., Trans. Am. Electrochem. Soc., **16**, 97 (1909).
2. Loveridge, F. H., Trans. Am. Electrochem. Soc. **16**, 109 (1909).
3. Ordway, D. L., Trans. Am. Electrochem. Soc., **17**, 341 (1910).
4. Pritz, W. B., Trans. Am. Electrochem. Soc., **19**, 31 (1911).
5. Burgess, C. F., Brown, J. W., Loveridge, F. H., and Sharp, C. H., Trans. Am. Electrochem. Soc., **C.21**, 275 (1912).
6. *Primary Cells and Batteries.* 86-1 Recommendations for primary cells and batteries (1957); *General* (1962), Amendment No. 1 (1965); 86-2 *Specification sheets* (1963), Amendment No. 1 (1965); 86-3 *Terminals* (1965), Amendment No. 1 (1967), Bureau Central de la Commission Electrotechnique Internationala, Geneve, Suisse.
7. Hamer, W. J., *Electrochem. Tech.*, **5**, 490 (1967).
8. *Circ. Bur. Standards* 139, 2nd ed. (1927): U.S. Government Master Specification No. 58a; ASA Standard C18-1928.
9. *Circ. Bur. Standards* 390 (1930); ASA Standard C18-1930.
10. *Circ. Bur. Standards* 414 (1937); ASA Standard C18-1937.
11. *Circ. Bur. Standards* 435 (1942); ASA Standard C18-1941.
12. *Circ. Bur. Standards* 466 (1947); ASA Standard C18-1947.
13. *Circ. Bur. Standards* 559 (1955); ASA Standard C18.1-1954, UDC621.352.7.
14. *Natl. Bur. Standards Handbook* 71 (1959); ASA Standard C18.1-1959, UDC621.352.7.
15. ASA Standard C18.1-1965, UDC621.352.7.
16. ANSI Standard C18.1-1969.
17. ANSI Standard C18.1-1972.
18. Federal Standard Stock Catalog, Specification Symbol W-B-101 (March 31, 1931); W-B-101a (May 7, 1935); W-B-101b (February 19, 1948); W-B-101c (November 26, 1954); Interum Fed. Spec. W-B-00101e (November 16, 1960); W-B-101f (April 25, 1961); Interum Rev. Fed. Spec. W-B-101f, W-B-00101G (GSA-FSS) (February 9, 1970); W-B-101H (June 29, 1973).
19. Hamer, W. J., *IRE Trans. Component Parts*, **CP-4**, 86, Fig. 1 (1957).
20. Schrodt, J. P. and Hamer, W. J., *Mag. Standards*, **26**, 132 (1955).
21. Hamer, W. J., *Mag. Standards*, **31**, 81 (1960).
22. Hamer, W. J., *Mag. Standards* **36**, 306 (1965).
23. Selwood, P. W., Eischens, R. P., Ellis, M., and Wethington, K. *J. Am. Chem. Soc.*, **71**, 3039 (1949).
24. Otto, E. M. and Eicke, W. G., Jr., *J. Electrochem. Soc.*, **104**, 199 (1957).
25. Ellis, G. B., U.S. Patent 2,853,676 (September 23, 1958).
26. Military Specifications, Batteries, Dry, MIL-B-18D, 29 Oct., 1963.
27. Hamer, W. J., *Proc. Seminar Adv. Energy Sources and Conv. Techniques*, No. AD209301, Armed Serv. Tech. Inf. Agency, Wash., D.C. (No. P. B. 151461, Off. Tech. Serv., U.S. Dept. Commerce), 1 (Nov. 1958).
28. Hamer, W. J., Schrodt, J. P., and Vinal, G. W., *Trans. Electrochem. Soc.*, **50**, 449 (1946).
29. Gillingham, C. A., *Trans. Am. Electrochem. Soc.*, **34**, 297 (1918).

30. Melson, S. W., *Trans. Faraday Soc.*, **8**, 1 (1912).
31. *Circ. Bur. Standards*, No. 79 (1919).
32. Vinal, G. W. and Ritchie, L. M., *Tech. Papers Bur. Standards* No. 171 (1920).
33. Vinal, G. W., *Comm. Stand. Monthly*, 7, No. 2, 35 (1930).
34. Vinal, G. W., *Primary Batteries*, John Wiley, New York (1950), pp. 115–118.
35. DeWane, H. J., *Mag. Standards*, **26**, 135 (1955).
36. Military Specifications, Batteries, Dry, MIL-B-18B, 1 July 1953.
37. Military Specifications, Batteries, High Capacity, Special Single Discharge, Aircraft Use, MIL-B-7156B (ASG) 23 June 1954.
38. Hamer, W. J., *A.I.E.E. Trans., Part II. Applications and Industry*, **79**, 277 (1960).
39. Military Specifications, Environmental Testing, Aeronautical and Associated Equipment, General Specifications for, MIL-B-5272A, 16 September 1952.
40. Earwicker, G. A., *Proc. Inst. Elec. Eng. (London)*, **103**, pt. A, supplement no. 1, 180 (1956).

APPENDIX A

STANDARD SIZES AND CHARACTERISTICS OF TYPICAL CELLS

TABLE 8.2. LECLANCHÉ CYLINDRICAL CELLS

Cell Designation	IEC Designation (6)[a]	Nominal Dimensions Diameter (mm)	Can Height (mm)	Approximate Volume (cm³)	Approximate Weight (g)	Number of Cells[b]	Average Open-Circuit Voltage (V)	Flash Current (amp)
O	—	11.4	3.3	0.3	0.9	—	—	—
N	R-1	11.2	26.9	2.6	5.4	60	1.63	2.1
AAA	R-03	9.9	42.9	3.3	8.2	60	1.59	3.8
R	R-4	13.5	33.3	4.8	10.4	—	—	—
AA	R-6	13.5	47.8	6.7	15	36[c]-72	1.57-1.58[c]	4.7-6.6[c]
A	R-8	16.0	47.8	9.5	21	—	—	—
B	R-12	19.1	54.1	15.4	35	—	—	—
C	R-14	23.9	46.0	20.6	45	36[c]-72	1.58[c]-1.60	4.6-10.7[c]
D	R-20	31.8	57.2	45.2	100	48[c]-120	1.59[c]-1.62	7.8-14.7[c]
E	R-22	31.8	73.2	57.8	132	—	—	—
F	R-25	31.8	87.4	69.2	159	168	1.61	10.3
G	R-26	31.8	101.6	80.5	181	—	—	—
J	R-27	31.8	149.4	118	272	—	—	—
No. 6	R-40	63.5	152.4	483	998	18-114	1.61-1.63	32.4-43.1

[a]IEC dimensions approximate those given here.
[b]Number of cells on which the data of columns 8 and 9 are based.
[c]Photoflash cell.

TABLE 8.3. LECLANCHÉ FLAT CELLS

Cell Designation	IEC Designation (6)[a]	Nominal Dimensions			Approximate Cross-Sectional Area (cm²)	Number of Cells[b]	Average Open-Circuit Voltage (V)	Flash Current (amp)
		Length (mm)	Width (mm)	Thickness (mm)				
F12	—	—	16.0[d]	7.1	2.0	48	1.64	0.18
F15	F15	14.2	14.2	3.0	2.0	720	1.58	0.24
F17	—	—	18.3[d]	7.4	2.6	—	—	—
F20	F20	23.9	13.5	2.8	3.2	1800	1.59	0.47
F22	F22	23.9	13.5	7.1	3.2	48	1.77	0.48
F24	—	—	23.1[d]	6.6	4.2	84	1.72	0.93
F25	F25	22.6	22.6	5.8	5.1	616	1.61	0.75
F30	F30	31.8	21.3	3.3	6.8	1800	1.59	1.11
F40	F40	31.8	21.3	5.3	6.8	6450	1.61	1.11
F45[c]	—	38.1	21.3	2.8	8.1	—	—	—
F50	F50	31.8	31.8	3.6	10.1	—	—	—
F60	F60	31.8	31.8	3.8	10.1	840	1.64	1.35
F70	F70	43.2	43.2	5.6	18.6	240	1.61	2.13
F80	F80	42.9	42.9	6.4	18.5	210	1.69	2.37
F90	F90	42.9	42.9	7.9	18.5	1080	1.60	3.04
F96	—	54.1	44.5	5.3	24.1	1270	1.63	3.70
F100	F100	60.5	45.2	10.4	27.4	660	1.64	3.43

[a]IEC dimensions approximate those given here.
[b]Number of cells on which the data of columns 8 and 9 are based.
[c]Not included in ANSI Standards C18.1-1969 but in ANSI Standard C18.1-1965.
[d]Diameter.

Cell Designation	Approximate Leclanché Equivalent	IEC Designation (6)	Maximum Dimensions		Approximate Volume (cm³)	Approximate Weight (g)	Number of Cells[b]	Average Open-Circuit Voltage (V)	Flash Current (amp)
			Diameter (mm)	Overall Height (mm)					
M4	—	—	5.6	3.3	0.1	0.3	—	—	—
M5	—	MR-41	7.9	3.6	0.2	0.6	48	1.44	0.12
M6	—	MR-48	7.9	5.3	0.2	1.1	—	1.44	0.17
M8	—	MR-08	11.6	3.3	0.3	1.4	12	1.44	0.27
M10	—	—	11.6	3.4	0.3	1.1	84	1.44	—
M11	—	—	11.6	4.2	0.7	1.7	—	—	—
M12	—	—	12.6	7.3	1.0	2.0	24	1.49	0.24
M15	—	MR-07	11.6	5.3	0.5	2.3	96	1.47	0.32
M17	—	—	15.2	3.8	0.7	2.3	—	—	—
M20	—	MR-09	15.6	6.1	1.1	4.3	84	1.47	0.63
M22	—	—	15.0	7.8	1.3	4.8	—	—	—
M23	—	—	25.5	2.8	1.5	4.3	—	—	—
M25	—	MR-01	11.6	14.5	1.5	5.1	24	1.45	0.47
M26	—	—	17.4	7.8	1.8	7.4	—	—	—
M30	—	—	15.9	11.2	2.1	7.9	24	1.46	0.72
M34	—	—	23.1	6.0	2.5	11.3	—	—	—
M35	—	MR-1	11.9	28.7	3.3	11.3	84	1.44	1.46
M40	N	MR-7	15.9	16.8	3.3	12.2	84	1.44	1.12
M55	AA	MR-6	13.9	49.5	7.5	29.8	36	1.42	3.53
M60	—	MR-17	25.0	16.8	8.2	26.4	84	1.46	0.85
M70	—	MR-8	16.3	49.5	10.3	39.7	84	1.43	4.14
M100	D	—	32.5	60.7	50.5	165.8	3	1.35	8.85
S3	—	—	5.6	3.3	0.1	0.3	—	—	—
S4	—	SR-41	7.9	3.6	0.2	0.6	—	—	—
S5	—	SR-48	7.9	5.3	0.26	1.1	—	—	—
S10	—	—	11.6	4.2	0.44	1.7	—	—	—
S15	—	—	11.6	5.3	0.56	2.3	12	1.58	0.56
S16	—	—	11.6	5.6	0.59	2.3	—	—	—

[a]These types are not covered by IEC.
[b]Number of cells on which the data of columns 9 and 10 are based.

TABLE 8.5. ALKALINE MANGANESE DIOXIDE CELLS

Cell Designation[a]	Approximate Leclanché Equivalent	Maximum Dimensions		Approximate Volume (cm³)	Approximate Weight (g)	Number of Cells[b]	Average Open-Circuit Voltage (V)	Flash Current (amp)
		Diameter (mm)	Overall Height (mm)					
L10	—	11.6	14.5	1.5	5	—	—	—
L15	—	23.1	6.0	2.5	11	—	—	—
L20	N	11.9	28.7	3.3	10	12	1.49	2.6
L25	—	15.9	16.8	3.3	11	—	—	—
L30	AAA	10.3	44.5	3.8	11	12	1.49	3.1
L40	—	14.3	50.0	8.0	23	72	1.53	8.8
L70	C	26.2	50.0	26.9	65	36	1.54	16.8
L80	$\frac{1}{2}$D	34.1	30.7	28.2	82	36	1.56	8.5
L90	D	34.1	61.1	55.9	130	24	1.56	25.4
L95	F	34.1	94.5	86.3	194	—	—	—
L100	G	34.1	110.7	101.4	227	12	1.57	38.4

[a]These types are not covered by IEC.
[b]Number of cells on which the data of columns 8 and 9 are based.

TABLE 8.6. SEALED NICKEL-CADMIUM CYLINDRICAL CELLS

Cell Designation[a]	Approximate Leclanché Equivalent	Maximum Dimensions		Approximate Volume (cm³)	Approximate Weight (g)	Capacity[b]	
		Diameter (mm)	Overall Height (mm)			Five-Hour Rate (ma-hr)	One-Hour Rate (ma-hr)
K20	—	25.7	6.4	3.3	9.6	85	—
K23	—	25.7	6.9	3.6	9.9	135	—
K25	—	15.9	16.3	3.3	8.5	—	—
K28	—	25.7	9.0	4.6	12.2	200	—
K30	—	14.3	27.8	4.4	15.9	200	—
K32	—	34.7	5.4	5.1	14.2	300	210
K40	AA	14.3	50.0	8.0	22.7	400	280
K45	—	34.7	10.0	9.5	23.5	435	315
K46	A	15.9	40.3	10.0	36.9	550	400
K47	—	26.2	26.3	14.3	22.1	800	560
K60	—	22.9	42.8	17.5	51.6	1000	700
K65	—	50.9	10.3	21.0	56.7	875	630
K70	C	26.2	50.0	26.9	77.4	1300	980
K72	—	26.2					
K75	—	50.9	15.2	31.0	99.2	2100	1470
K80	½D	34.1	37.3	34.2	93.6	1750	1500
K85	—	26.2	84.9	45.7	113.4	2800	1960
K90	D	34.1	61.1	55.9	153.1	3500	2450
K95	F	34.1	91.4	83.7	237.9	5000	3500
K100	G	34.1	107.1	98.0	285.2	7000	5040

[a]These types are not covered by IEC.
[b]Based on the D-Size manufacturers' test results.

APPENDIX B

TABLE 8.7. VARIOUS NOMENCLATURES FOR ANSI D-SIZE CYLINDRICAL LECLANCHÉ CELL

	Low-Voltage Battery	Industrial and Lighting
General		
American National Standards		
Institute, C18	D	D
Federal Specifications, W-B-101f	D	D
Military, MIL-B-18C	—	BA-30
NEDA	13	813
International Electrochemical		
Commission, P86	R-20	R-20
U.S. Manufacturers		
Bright Star	10M	10M or 0197
Burgess	230	2
Union Carbide	A100	905 or D99
Mallory	M-13R	M-13F
Marathon	123	121 or 122
Ray-O-Vac	13	2LP or 5LP
U.S. Consumers (Distributors)		
Gamble	17-439	17-435 or 17-428
Mastercraft	SF51B	SF51B or SF49
Montgomery Ward	23	2340
Philco	P920	P907
RCA	VS336	VS036
Sears	6445	4650 or 4662
Western Auto (Wizard)	7D9013 or 7D8013	7D8013 or 7D7013
Zenith	Z4NL	Z2NL

APPENDIX C

TABLE 8.8. AVERAGE PERFORMANCE OF SOME TYPICAL DRY CELLS AND BATTERIES ON THE STANDARD TESTS OF THE AMERICAN NATIONAL STANDARDS INSTITUTE[a,b]

	1972 Requirement	(1956–62) Performance
Heavy intermittent test, $2\frac{2}{3}$ ohms (Test No. 1)	Hours	Hours
No. 6 general-purpose cells	70	79
After 6 months	65	73
No. 6 industrial cells	100	106
After 6 months	90	100

TABLE 8.8 (*Contd.*)

	1972 Requirement	(1956–62) Performance
Light intermittent test, 20 ohms (Test No. 2)	Days	Days
No. 6 general-purpose cells	200	253
No. 6 industrial cells	310	357
No. 6 telephone cells, special grade	370	394
No. 6 telephone cells, regular grade	300	311
Alarm battery test, 300 ohms (Test No. 4)	Days	Days
No. 6 alarm cells	300	383
Lighting battery test, 16 ohms (Test No. 7)	Hours	Hours
4Fd lantern battery[c]	20	21
After 6 months	15	19
Lighting battery test, 32 ohms (Test No. 8)	Hours	Hours
4Fd lantern battery	45	48
After 6 months	40	44
General-purpose intermittent test, $2\frac{1}{4}$ ohms (Test No. 11)	Minutes	Minutes
General-purpose D-size cells	400	550
After 6 months	375	492
General-purpose intermittent test, 4 ohms (Test No. 12)	Minutes	Minutes
General-purpose C-size cells	325	394
After 6 months	275	345
General-purpose AA-size cells	80	120
After 6 months	65	103
General-purpose intermittent test, 5 ohms (Test No. 13)	Minutes	Minutes
General-purpose AAA-size cells	50	87
After 6 months	40	78
Light-industrial flashlight cell test, 4 ohms (Test No. 15)	Minutes	Minutes
General-purpose D-size cell	600	718
Industrial D-size cells	550,[d]850	686,955
Heavy-industrial flashlight cell test, 4 ohms (Test No. 16)	Minutes	Minutes
Industrial D-size cells	400,[d]800	470,844
After 6 months	300,750	389,691
Photoflash cell test, 0.15 ohm (Test No. 17)[g]	Seconds	Seconds
D[·h] size photoflash cells	800	1568
After 6 months	650	1013
C-size photoflash cells	700	1222
After 6 months	550	830
AA-size photoflash cells	150	341
After 6 months	120	289
Hearing-aid transistor battery test, 625 ohms (Test No. 19)	Hours	Hours
Containing M5 cells in series	17	22
Containing M10 cells in series	30	37

TABLE 8.8 (*Contd.*)

	1972 Requirement	(1956–62) Performance
Hearing-aid transistor battery test, 625 ohms (Test No. 19)	Hours	Hours
Containing M12 cells in series	60	67
Containing M15 cells in series	75	84
Containing N cells in series	150	188
Containing M20 cells in series	160	166
Containing M25 cells in series	160	173
Low-voltage-battery test, 5 ohms (Test No. 22)	Hours	Hours
F4d batteries	90,[e]115	105,126
After 6 months	85,110	101,121
Low-voltage-battery test, 25 ohms (Test No. 23)[g,h]	Hours	Hours
3D batteries	50, 70	79, 81
After 6 months	40, 60	72, 84
3F, 4Fd batteries	120, 140	140, 163
After 6 months	110, 130	136, 159
3G batteries	140, 170	167, 203
After 6 months	130, 160	157, 193
4F2s[f] batteries	240, 325	311, 382
After 6 months	230, 310	294, 349
High-voltage-battery test, $166\frac{2}{3}$ ohms (Test No. 24)	Hours	Hours
Containing F25 cells	30	37
After 6 months	25	36
Containing N or F40 cells	30	48
After 6 months	25	45
High-voltage-battery test, 1,500 ohms (Test No. 25)	Hours	Hours
Containing F20 cells	100	163
After 6 months	90	161
Containing F30 cells	300	356
After 6 months	250	353
Containing N or F40 cells	525	621
After 6 months	475	581

[a]The "requirements" above were taken from ref. (16) and (17).
[b]Performance data on specific sizes and types of cells are constantly being revised by battery manufacturers and the interested reader is referred to the Technical and Engineering Handbooks which are issued by various battery manufacturers. Ref. (7) provides a summary of the changes in service levels for the period 1932–1964, some as far back as 1916, for important sizes of cells on several important tests.
[c]The "d" means double-row arrangement.
[d]First number is for 1.10 V cutoff while second is for 0.90 V cutoff.
[e]First number is for 1.1 V cutoff while second is for 1.0 V cutoff.
[f]The "s" means single-row arrangement.
[g]In ANSI standard C18.1-1969 but omitted in ANSI standard C18.1-1972.
[h]Requirements of ANSI standard C18.1-1969.

APPENDIX D

TABLE 8.9. OPEN-CIRCUIT VOLTAGE (OCV) OF 3B, 5B, 15B AND 200 F20 LECLANCHÉ BATTERIES DURING SHELF TESTS AT 21°C

Time in Months	3B Batteries[b]			5B Batteries[b]			15B Batteries[b]			200 F20 Batteries[b]		
	Number of Batteries[a]	Average OCV per Battery (V)	Average OCV per Cell (V)	Number of Batteries[a]	Average OCV per Battery (V)	Average OCV per Cell (V)	Number of Batteries[a]	Average OCV per Battery (V)	Average OCV per Cell (V)	Number of Batteries[a]	Average OCV per Battery (V)	Average OCV per Cell (V)
Initial	28	4.66	1.55	56	7.77	1.55	42	23.3	1.55	6	329	1.65
3	28	4.66	1.55	56	7.75	1.55	42	23.3	1.55	6	319	1.60
6	28	4.64	1.55	56	7.72	1.54	42	23.2	1.55	6	316	1.58
12	28	4.62	1.54	56	7.68	1.54	42	23.1	1.54	6	312	1.56
24	24	4.59	1.53	54	7.58	1.52	37	22.7	1.51	6	310	1.55
36	11	4.58	1.53	39	7.58	1.52	18	22.9	1.53	5	311	1.56
48	8	4.59	1.53	25	7.54	1.51	14	22.7	1.51	5	309	1.55
60	8	4.54	1.51	13	7.58	1.52	7	23.0	1.53	5	307	1.54
72	6	4.52	1.51	8	7.61	1.52	3	22.9	1.53	5	298	1.49
84	5	4.49	1.50	7	7.57	1.51	2	22.4	1.49	4	298	1.49
96	2	4.44	1.48	5	7.49	1.50	—	—	—	—	—	—
108	—	—	—	2	7.62	1.52	—	—	—	—	—	—
120	—	—	—	2	7.53	1.51	—	—	—	—	—	—
132	—	—	—	2	7.44	1.49	—	—	—	—	—	—
144	—	—	—	1	7.40	1.48	—	—	—	—	—	—

[a]Number of batteries on test decrease with time because the OCV of some of them falls below 1.45 V per cell, at which point they are removed from test.

[b]See Appendix A and text for explanation of cell sizes and types.

9.

Reversibility of Battery Systems

R. J. BRODD and R. M. WILSON

With the ready availability of battery charging devices and the associated advertising, the technologist may be tempted to assume that all batteries can be charged. This assumption is not correct but should be revised to read "primary batteries with reversible electrochemical systems generally may be recharged a limited amount." Although the rechargeability of battery systems has not been discussed in the literature to any great extent, this topic is covered in depth in this chapter. The important factors governing the rechargeability of a battery system are described, some for the first time, to clearly assist the reader in this important phase of battery technology. Sealed secondary systems, a major development in rechargeable batteries, are described in the examples cited.

HISTORICAL

Secondary cells (storage batteries, accumulators) are battery systems which, after discharge, can conveniently and repeatedly be recharged, to deliver power again at reasonable efficiency. They have many features in common with primary cells; the resemblances are fundamental, the differences are superficial, and their evaluation should lead to a better appreciation of both species.

As early as 1801, Gautherot(1) noted the back electromotive force developed when saline solutions were electrolyzed. In 1805 Ritter(2) showed that a strong discharge could be obtained after a charging current had been applied to a cell consisting of a bibulous separator, moistened with sodium chloride solution, between copper plates. Similar observations followed in due course, mostly of secondary currents of short duration, and a committee was formed to investigate the possibilities of the "secondary or charging pile."

Faraday(3), in 1834, found that the electrolysis of lead acetate solutions yielded lead at the cathode, lead dioxide at the anode. He commented on the good electrical conductivity of PbO_2, and later credited Muncke(4) and de la Rive(5) with observations on its strongly cathodic nature. However, he seems not to have noted the possibility of using this system as a battery, either primary or secondary. Grove(6), the pioneer in the work on gas electrodes, was fully aware of the possibility of using the hydrogen:chlorine as well as the hydrogen:oxygen cell as a reversible system.

As long as secondary cells could be recharged only by primary batteries, there was comparatively little incentive for their commercial development. Yet, in 1859, Planté(7) announced the system which was to change the outlook for the storage battery. His cell was simply two spiral-wound lead sheet electrodes in sulfuric acid electrolyte, alternately charged, discharged, and recharged until layers of lead and lead dioxide had been built up on the anode and cathode, respectively. Though it took two Bunsen primary cells in series to recharge, and capacity was relatively low,[1] the high discharge rate capability of the Planté cell opened up new areas of usefulness for the accumulator. The advent of Faure's cell construction(9), in which the plates of lead were covered with pastes of red lead or litharge with sulfuric acid and "formed" by a charging current, greatly simplified the assembly and substantially increased the capability. The use of lead grids to support the porous lead anodes and PbO_2 cathode reactant masses followed in due course. This work was the forerunner of the modern lead storage battery. The advent of cheap central power in the 1880s assured the future of the secondary battery.

The early 1880s seems to have been a period of great interest in storage batteries, and many different systems were proposed. Tommasi(10), in his *Traité des Piles Électriques*, published in 1889, mentions the following *inter alia*, as cells given serious consideration:

$Zn:PbO_2$ with acidified $ZnSO_4$ electrolyte (2.37 to 2.5 V)
$Cu:PbO_2$ with acidified $CuSO_4$ electrolyte (1.26 V)
$Fe:PbO_2$ with $FeSO_4$ electrolyte
$Zn:MnO_2$ with dilute H_2SO_4 electrolyte
$Zn:Br_2$ with dilute acid electrolyte (1.95 to 2.2 V)

Among the alkaline systems, attention seems to have been originally directed largely to the Lalande–Chaperon Zn:caustic alkali:CuO cell despite its limitations. However, an important advance was marked about the turn of the 20th century by the development of two batteries; the

---

[1]As late as 1880, Niaudet(8) wrote "On voit clairement que la pile secondaire ne peut donner que des effets de courte duree...."

SEALED, MAINTENANCE-FREE BATTERIES

ron:nickel oxide cell generally credited to Edison(11), and the
:admium:nickel oxide cell proposed by Jungner(12) (who also described
ι cadmium:silver oxide system). Applications as rechargeable units have
»een found, particularly in recent years, for such systems as
Ξn:KOH:Ag₂O and Zn:KOH:MnO₂. The Drumm (Zn:KOH:NiOOH)
»attery(13) is reportedly still in use on railroad cars.

Edison(14) conceived the first sealed nickel-cadmium cell in 1912
ιtilizing a platinum catalyst to recombine the gasses generated on
»vercharge. In actual practice this technique failed as hydrogen and
»xygen are rarely generated in stoichiometric quantities during over-
harge. Present sealed cell technology rests on the principles described by
Ψeumann(15) and Dassler(16) and others(17) for oxygen recombination
.nd proper balancing of electrode capacity. The use of a third electrode to
ecombine oxygen on overcharge or to signal end of charge has been
uggested but has not found wide commercial acceptance.

The development of sealed rechargeable systems constitutes a mile-
tone in secondary battery technology. The cells are leakproof, not
ensitive to orientation, and extremely rugged. Areas of usefulness
»reviously reserved for primary batteries are now open to rechargeable
ystems. The convenience of primary throwaway batteries may be offset
»y the lower overall cost for a rechargeable system. Often times the
ealed rechargeable system permits new battery usages, especially at
ιigh-rate drain applications. The rechargeable systems really extend the
phere of application rather than replace primary batteries. A discussion
»f the commercially significant sealed rechargeable systems follows.

SEALED, MAINTENANCE-FREE BATTERIES

NICKEL-CADMIUM BATTERY

The nickel-cadmium battery, available in button, rectangular, and round
:ell constructions, is the most widely used of all small, sealed or
*maintenance-free" rechargeable batteries. Although the principles of
onstruction are entirely different, the basic principles of operation of the
ιickel-cadmium system are identical for all cell constructions. The cell is
'ery rugged and can tolerate overcharge and deep discharge. It can stand
or extended periods of time in the discharged state without serious
Ιamage. It has excellent cycle life and can be charged with inexpensive
onstant-current chargers.

The chemical characteristics of the nickel-cadmium system have been
·xtensively discussed elsewhere(18)(19); therefore, only the basic
»rinciples of operation will be discussed here. The overall cadmium anode

reaction may be presented as

$$Cd + 2OH^- \underset{\text{charge}}{\overset{\text{discharge}}{\rightleftharpoons}} Cd(OH)_2 + 2e^- \qquad (1$$

One of the outstanding characteristics of the nickel-cadmium system i its ability to withstand extended overcharge. The reactivity of cadmiur prevents excessive oxygen pressure buildup in the cell since the metal in the charged anode has a high surface area and reacts readily with an removes the oxygen produced from the fully charged positive electrode.

$$Cd + \tfrac{1}{2}O_2 + H_2O \rightarrow Cd(OH)_2 \qquad (2$$

The $Cd(OH)_2$ is then reduced to cadmium electrolytically and the cycl can repeat. On overcharge, the cell is essentially on an oxygen cycle. It i possible that the oxygen is readily reduced on the exposed nickel surface and a local corrosion couple between cadmium and oxygen is responsibl for the rapid recombination.

Recombination is achieved by adjusting the electrolyte level in the cel so that just enough electrolyte is added to provide good conductivity bu not enough to completely fill all the void spaces in the separator. It is als necessary to adjust the cell balance so that the positive becomes full charged before the negative does or else hydrogen evolution occurs.

Hydrogen generation in sealed cells is very dangerous. Hydroger evolution may be a problem when charging at low temperatures Apparently at low temperatures a modification of the crystal form of the cadmium hydroxide occurs which inhibits nucleation of cadmium metal thus favoring hydrogen evolution[20]. There is no recombinatio mechanism for hydrogen so that the gas accumulates in the cell Eventually, the hydrogen pressure buildup will cause the cell to vent Whenever the cell vents, water is lost and the cell balance changes. Both water loss and changing cell balance can eventually lead to cell failure.

The cathode in the nickel-cadmium cell is nickel hydroxide in conductive matrix of graphite or nickel. The electrode reaction is no clearly understood; however, a tentative mechanism indicating the complexity of the positive electrode reaction[21] is

$$\beta\text{-NiOOH} \xrightarrow[\text{overcharge}]{\text{extended}} \gamma\text{-NiOOH}$$

discharge ↑↓ charge   discharge ↑↓ charge   (3

$$\beta\text{-Ni(OH)}_2 \underset{\text{strong caustic}}{\overset{\text{stand}}{\longleftarrow}} \alpha\text{-Ni(OH)}_2 \cdot 3H_2O$$

The voltage of the nickel oxide electrode is such that oxygen evolutio begins before the electrode becomes fully charged. The charging reaction is very temperature-sensitive. For example, at 25°C, the charging efficiency is about 85 percent, while at 40°C the efficiency is about 40 percent. It is normally recommended that at room temperature a fully discharged cell be given about 140 percent charge to achieve full capacity.

Another consequence of the high cathode potential and low oxygen overvoltage is the increased self-discharge and poor high temperature stability of the cathode. The charged nickel oxide is thermodynamically unstable and reacts directly with water to produce oxygen or hydrogen peroxide. These are consumed at the anode to provide a chemical short in the cell. Although the charged nickel hydroxide is thermodynamically unstable, its spontaneous-decomposition forming oxygen and a lower valent hydroxide is very slow. When charged nickel hydroxide is mixed with a conductor (e.g., graphite or nickel), the rate of self-discharge increases markedly. Evidently the conductor acts as the anode of a local corrosion cell to accelerate the self-discharge process(22). Impurities such as iron will increase self-discharge rates by lowering the oxygen overvoltage. Lithium and cobalt are sometimes added to decrease the rate of self-discharge.

Another cause for self-discharge is the presence of materials such as nitrates which can set up a redox couple in the cell. Nitrate is a common impurity in the nickel positive electrode as a result of the manufacturing process. It is critical for long shelf life to exclude nitrates as well as other redox systems from the cell.

The electrolyte in the nickel-cadmium cells is usually about 7 N KOH. The actual concentration will depend on the manufacturer and the intended use for the cell. The overall electrode reactions do not involve hydroxide ions, so the system is said to have a nonvariant electrolyte.

Proper cell balance is critical for long cycle life of nickel-cadmium cells. Electrode uniformity is necessary to achieve the proper cell balance and cell capacity. This is especially true in multicell batteries which are to be deep discharged. Separator oxidation or corrosion of the positive current collector can cause a change in cell balance during the life of a cell, especially on prolonged overcharge. Overcharge may also cause a recrystallization of the surface layers of the active materials, thus causing a temporary decrease in cell capacity.

The so-called "memory effect" appears to be related to the recrystallization phenomenon. If a nickel-cadmium cell is cycled repeatedly on a shallow discharge, for example, 25 percent of the capacity, and then given a deep discharge, the cell will deliver, at a useful voltage, only that 25 percent which it "remembered" delivering. The total capacity can be restored by a number of deep discharge cycles. The cause for this loss of useful capacity is thought to be associated with self-discharge of the unused nickel oxide and recrystallization of the active materials.

Cycle life limitations are primarily shorting as a result of separator failure due to oxidation, cadmium dendrite growth, or failure of the seal.

Nickel-cadmium cells have three common types of construction: prismatic, jelly roll, and button. The prismatic construction (shown in Fig. 9.1) is generally reserved for higher capacity cells. The electrodes are generally flat, and the capacity depends on the thickness and surface area of the plate (amount of active nickel). The electrodes are stacked with alternating polarity (e.g., negative, positive, negative, etc.) until the desired capacity is reached. The negative and positive electrodes are respectively connected in parallel, usually by welding a nickel strip to current collector tabs on each individual electrode. The electrodes are placed in a stainless steel or nickel-plated steel can under compression, electrolyte added, and the cell closed.

The jelly roll construction (shown in Fig. 9.2) is usually reserved for small cylindrical cells. Its name comes from the familiar cake which is rolled up with jelly between the layers. The two electrodes (positive and negative) are placed in juxtaposition with a separator between them. The electrodes are fastened to a spindle and wound around as one would wind up a string. After winding, the electrodes are inserted into the cylindrical can, usually with some compression provided by the can so that the jelly roll is not permitted to relax its tension completely. The can is normally closed with a radial compression seal utilizing a compressible plastic as the insulator and sealing material. The can forms one terminal and the cell

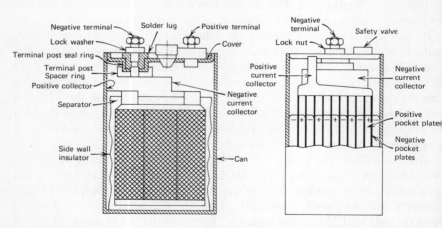

Figure 9.1 Cutaway view of typical standard rate pocket plate rectangular cell.

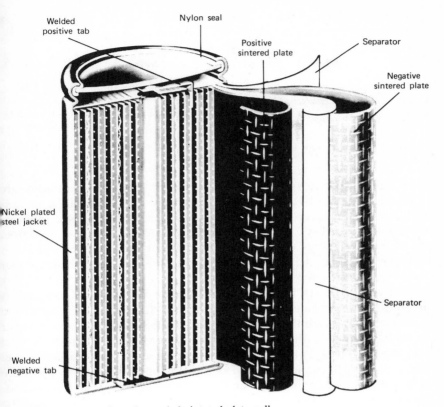

Figure 9.2 Construction of a sealed sintered-plate cell.

cover the other. The cells are provided with a safety vent to prevent explosions from excessive gas generation caused by improper cell operation. The vent may or may not be resealable.

The separator materials are normally a felted-plastic material. The material must be stable in the KOH electrolyte and inert to strong oxidants and reductants. It must be strong enough to prevent physical contact between the electrodes and be electronically nonconductive so that internal short circuits are avoided. A felted-nylon fabric is most common. The exact thickness, porosity, and electrolyte retentivity depend on the particular cell design and compression. Polypropylene, a more inert material, is preferred for long-life, high-reliability cells.

Button cells, whose shape is suggested by their name, are available in sizes up to 1 amp-hr. Their construction as shown in Fig. 9.3 is very similar to that of the prismatic cells. The electrodes are stacked

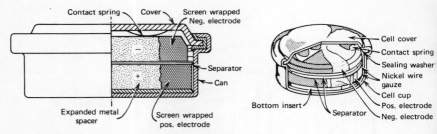

Figure 9.3 Cutaway view of standard rate button cell.

alternately (negative, positive, negative, etc.) and like electrodes are
connected together by a common current lead. The electrode spacing is
maintained by a layer or two of separator material. The electrolyte is
added and the cell is closed with a compression seal using an insulating
plastic ring to separate the positive terminal from the negative terminal. A
leaf-type spring is usually included inside the cell to provide proper
operating pressure for the electrodes.

Sintered electrodes were first developed about 1940 in Germany. A
sintered type electrode is basically a high surface area and conductive
metal matrix in which the active materials can be dispersed in a thin layer
in intimate contact with the metal matrix. The base metal matrix is formed
from carbonyl nickel powders. The nickel powder is spread over a nickel
screen or perforated nickel-plated steel strip and heated to a temperature
sufficient to form a bond between the particles of nickel powder; that is
the powders are sintered. Porosity is usually attained by adjusting the
bulk density of the powders or by adding water and/or a bulking agent to
form a slurry of the nickel powders. The process leads to electrodes of
about 75 to 80 percent porosity.

The active materials are normally dispersed in the matrix by vacuum
impregnation techniques. Aqueous nickel nitrate or cadmium nitrate
solutions are used. The more concentrated solutions require higher
temperatures to prevent the solutions from solidifying. The electrodes are
vacuum filled with the appropriate salt solution and, after removal from
the solution, the electrodes are allowed to stand until the excess liquid is
drained from the electrode. The electrodes are then immersed in caustic
(NaOH; it is less expensive) to convert the nitrate to battery-active
hydroxide materials. The electrodes are rinsed in deionized water to
remove caustic and nitrates, and then dried. The impregnation process is
repeated until the desired capacity per unit volume ratio is attained.

Pressed powder electrodes are constructed by pressing the mixtures of
active material and inert conductor into pellets. The mixtures for negative

electrodes may be composed of cadmium, cadmium oxide, or cadmium hydroxide, and nickel powder. The mixtures for positive electrodes are composed of nickel hydroxide, graphite and cadmium oxide, or cadmium hydroxide for reversal protection. The mixtures also may contain a bulking agent to provide porosity and enhance electrolyte permeation into the electrode.

The mixtures are pressed into pellets whose size is dependent on the desired capacity for the cell. The molding pressure may vary, but generally approaches 30,000 psi. After molding, the pellets are generally wrapped in nickel screen which acts as the conductor. It is also possible to prepare pressed powder electrodes with the conductor in the center of the mass.

Cadmium anodes are also manufactured by an electrodeposition technique (24). In this method, a loosely structured porous mass of cadmium is deposited on a nickel screen from a caustic bath at relatively high current densities. The performance of this electrode is similar to the pressed powder electrode.

Electrochemical deposition of $Ni(OH)_2$ or $Cd(OH)_2$ may provide an improved method of impregnating sintered-nickel plaques to form either positive or negative electrodes (25). Cells made from such electrodes should be very reliable and also should exhibit excellent high-rate performance.

LEAD-ACID BATTERY

The development of relatively small sealed or maintenance-free, lead-acid batteries occurred in the mid-1960s (26), even though the system itself had reached a high state of development for automotive and industrial applications. In these applications, the problems of gassing on charge, self-discharge, grid growth, and corrosion have been successfully dealt with to give batteries with satisfactory shelf and cycle life. It is possible that the maintenance-free, lead-acid battery may dominate the small rechargeable battery field just as it presently dominates the field of larger batteries.

The chemical characteristics of the lead-acid system are well known. The overall lead anode reaction is given by

$$Pb + SO_4^= \rightarrow PbSO_4 + 2e^- \qquad (4)$$

It may be postulated that the mechanism involves a soluble intermediate. The anode is a high surface area spongy lead on a supporting matrix of lead. The matrix is usually pure lead or a lead-calcium alloy. The anode should be fully charged to prevent build up of inert nodules of $PbSO_4$. If $PbSO_4$ particles are allowed to build up, the negative is said to be "sandy" and gives poor performance. $PbSO_4$ build up also causes buckling of the plates due to molar volume differences between lead and $PbSO_4$. Organic expanders are sometimes added to negatives to retard $PbSO_4$ formation and growth.

The overall cathode reaction is given by

$$PbO_2 + 4H^+ + SO_4^= \rightarrow PbSO_4 + 2H_2O - 2e^- \qquad (5$$

The exact mechanism of the cathode electrode reactions are not known Typical charge-discharge curves are shown in Fig. 9.4.

The control of the charging procedure, especially overcharge, is critical to the successful operation of the maintenance-free batteries. In the batteries, reserve water is depleted during overcharge by the gassing reactions. It matters little whether the overcharge occurs all at one time or by a number of small overcharges, because when the reserve water is lost there is not sufficient electrolyte for proper battery operation. On the other hand, in order to prevent sulfation and to promote increased battery life, recharging to full capacity is a necessity. However, excessive overcharge should be avoided as it increases positive grid corrosion and tends to loosen the active material in the grids.

Water loss is the prime cause for failure of maintenance-free, lead-acid batteries. The charging procedures to control the charge and amount of overcharge are critical in that the amount of overcharge is usually correlated with water loss. The control of the water loss controls the useful cycle life of the battery. Most sealed or maintenance-free batteries require the use of a constant voltage charge with a cutoff from a trickle charge at a predetermined voltage and/or coulombic input.

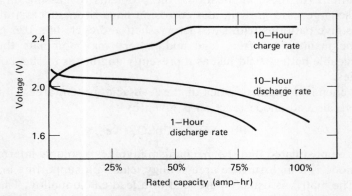

Figure 9.4 Typical charge-discharge characteristics of the maintenance-free, lead-acid system. A VLTC-charge is shown(23).

The general features of the construction of small lead-acid batteries are shown in Fig. 9.5. The anode and cathode lead-acid battery electrodes are made by very similar processes. There are two types of methods: the Planté and the Faure (or pasted grid). The Planté plates are manufactured from pure lead sheets. The active layer of Pb and PbO_2 is built up by cycling the electrodes in sulfuric acid. This method is usually reserved for low rate batteries. By far, the more important method is the pasted-grid technique. In this method, a pasty mixture of litharge (PbO), red lead, lead, H_2SO_4, and H_2O is forced into the open network of a lead grid. The litharge is normally made by air oxidation of lead. The lead is contained in the litharge due to incomplete oxidation of lead. The water and acid both

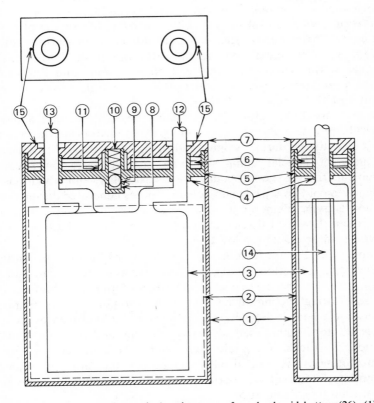

Figure 9.5 Cutaway view of a typical maintenance-free, lead-acid battery(26). (1) Container, (2) unisep, (3) negative plates, (4) rubber O-ring, (5) inner cover, (6) neutralizer, (7) top cover, (8) check valve seat, (9) check valve ball, (10) spring, (11) baffle, (12) negative post, (13) positive post, (14) positive plate, (15) vent.

control the density, porosity, and the consistency of the paste. Various additives are also incorporated in the paste, depending on whether they are intended for positive or negative electrodes. Positive plates generally contain litharge and red lead mixed with Sb_2O_3 or Co_2O_3. Negative plates usually contain litharge with about 1 percent blanc fixé ($BaSO_4$) and about 5 percent organic expanders. The pasted plates are set aside for drying During the drying process, complex chemical reactions occur which convert the potential active materials into lead sulfate, basic lead sulfate etc. The exact composition of the "cured" plates is not certain.

Curing may be accomplished by allowing the plates to stand in the open air for several days, or it can be accomplished more rapidly in a humidified oven. This curing process is a critical step in the battery production because it is at this point that the bond between the grid and active materials is established. Also, the crystal structure of the paste which results will control the capacity and cycle life of the plates to a great extent. After curing, the plates are immersed in sulfuric acid and charged. This "forming" charge converts the paste into the active materials of the battery; spongy lead on the anode and PbO_2 on the cathode. Formation is usually performed in an electrolyte lower in density than that of the battery electrolyte.

The grids are the skeleton which contains the active material and makes electrical contact with it. The important grid parameters are the volume of contained active material, weight, and thickness. The grid is normally a lead alloy of special composition to give it corrosion resistance mechanical strength, good conductivity, castability, and which has high resistance to cold flow when subjected to forces generated during cycling Common alloying elements have included antimony, arsenic, tin, silver and calcium. Low self-discharge and low-maintenance batteries usually use pure lead or calcium alloy grids.

Antimony is the most common alloying constituent. The antimony enhances the formation of a good bond between the paste and grid greatly improves the casting characteristics of the lead, and acts as a hardener. During cycling of the battery the antimony slowly migrates to the negative where it deposits, thus lowering the hydrogen overvoltage This leads to a rapid self-discharge of the negative via hydrogen evolution. This loss of capacity can reach 1 percent per day. For this reason antimony is not usually added to the grid of sealed or maintenance-free, lead-acid batteries. Cobalt salts are sometimes added to retard antimony migration.

The separator systems today are either microporous plastics such as polyvinylchloride or "Fiberglas" mats. The separator should have: (1) small pore size to prevent lead dendrites from growing through the separator and shorting the cell, yet pores large enough to permit ready ion transport and diffusion; (2) good mechanical strength to resist the crushing forces of plate expansion during cycling; (3) good chemical stability to resist oxidation and reduction from contact with PbO_2 and lead, respectively, as well as resistance to 35 percent H_2SO_4; (4) a low electrolyte resistance path between the plates, and (5) low electronic conductivity.

The capacity of the cell will depend on the amount of active material in the plates. The life expectancy will depend on the type of service, depth of discharge, overcharge, and so forth. As the cell is discharged, both the Pb and PbO_2 are converted to $PbSO_4$. The density of the reaction products is less than the reactants, thus on discharge there is a volume increase. The volume of the negative increases 168 percent while the volume of the positive increases 92 percent on discharge. The volume change places severe mechanical stress on the grid unless proper care has been taken to assure sufficient porosity, and so on in the plates.

The electrolyte is sulfuric acid with a density of approximately 1.3. The acid may contain a thixotropic gel of highly dispersed oxides of Al_2O_3 and SiO_2 to prevent flow of the electrolyte from between the plates. Careful attention must be given to particle size and separator properties so that the thixotropic agents do not penetrate into the porous battery plates.

Silver Oxide-Zinc Battery

The silver-zinc battery has the highest energy density of any commercial rechargeable system. The battery in its present form was developed in the early 1930s by H. André (27) who introduced a semipermeable membrane as the separator system. The silver-zinc is used in aerospace, electronic, and military applications where energy density requirements outweigh cost considerations (28). It does not possess good cycle life on deep discharge, so that the energy density of the battery in actual use may be considerably less than might be expected at first glance. A 100-Whr/lb, silver-zinc battery operating at 10 percent depth of discharge would be equivalent to a 10-Whr/lb, nickel-cadmium battery operating at 100 percent depth of discharge.

The silver electrode reactions are unique in that two valence states of the active silver oxides behave as separate entities during electrode operation. Other oxides, for example, manganese and nickel, have variable valence states during their charge and discharge but do not exhibit the separate and distinct discharge behavior that is observed at the silver electrode. The silver electrode reaction may be written as

$$2AgO + H_2O + 2e^- \underset{\text{charge}}{\overset{\text{discharge}}{\rightleftharpoons}} Ag_2O + 2OH^- \tag{6}$$

and

$$Ag_2O + H_2O + 2e^- \underset{\text{charge}}{\overset{\text{discharge}}{\rightleftharpoons}} 2Ag + 2OH^- \qquad (7$$

or simply,

$$AgO \underset{\text{charge}}{\overset{\text{discharge}}{\rightleftharpoons}} Ag_2O \underset{\text{charge}}{\overset{\text{discharge}}{\rightleftharpoons}} Ag \qquad (8$$

The distinct nature of the electrode reactions is reflected in the two voltage plateaus of about 1.8 to 1.6 V observed during cell operation. These voltages correspond to the divalent and monovalent silver reactions, respectively. Typical charge-discharge curves are shown in Fig. 9.6.

If a constant voltage discharge or flat discharge curve is required, it will be necessary to operate on either the AgO or Ag_2O plateaus.

The silver electrode undergoes a change in its structure on cycling. After only a few cycles, the particle size of the silver oxides has increased over that originally present. This increased particle size of the active materials leads to a decrease in available capacity and in high-rate discharge capability. The change in structure may be the result of the inherent solubility of silver oxides in KOH electrolyte.

The silver oxides are reducible by hydrogen even when wetted with cell electrolyte. Thus, hydrogen produced at the zinc electrode as a result of corrosion processes as well as hydrogen formed during low rate charging is effectually removed from the battery case by reaction with the silver oxides. This reactivity permits the construction of sealed cells without fear of pressure buildup.

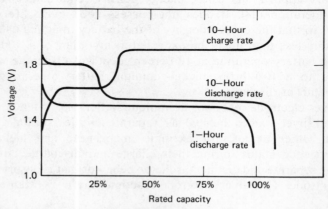

Figure 9.6 Typical charge-discharge characteristics for the silver-zinc system.

The zinc negative electrode has the highest voltage of all rechargeable anodes in aqueous electrolyte. It has a high exchange current and thus has a high reaction rate. Zinc electrodes used in commerce are amalgamated to reduce corrosion.

The zinc electrode reaction is fairly complex but may be written as

$$Zn \rightarrow Zn^{++} + 2e^-$$

$$Zn^{++} + 2OH^- \rightarrow Zn(OH)_2 \qquad (9)$$

$$Zn(OH)_2 + 2OH^- \rightarrow Zn(OH)_4^-$$

The form of the product $Zn(OH)_2$ will depend on the conditions under which the product formed, for example, current density, concentration of KOH, and so forth. The equilibrium with the solid product is attained so slowly that it is possible to find several species, for example, ZnO, γ-$Zn(OH)_2$, α-$Zn(OH)_2$, and so on, present simultaneously on the electrode surface.

In addition to the zinc corrosion problem mentioned above, there are two main disadvantages found in the zinc anode on repeated cycling: dendrite formation and shape change (29). Both of these phenomena are the result of the solubility of zinc hydroxide in the battery environment.

Zinc dendrite growths are capable of penetrating the cellulosic separators by deposition of zinc within the separator membrane. When a dendrite has grown through the separator, a short circuit develops within the cell. It is postulated that dendritic growths occur on points on the electrode which penetrate into the zincate diffusion layer at the surface. The rate of growth of the dendrite depends on the overvoltage, limiting current density, the thickness of diffusion layer, and the onset of spherical diffusion to the point.

Attempts to prevent zinc dendrite formation have led to electrolyte additives, complicated charging techniques, and extensive separator development. Electrolyte additives are partially successful in changing the nature of the zinc deposits. The more effective additives include surface active agents, lead and tin salts, and $Ca(OH)_2$.

Charging techniques include intermittent or pulsed charging as well as asymmetric ac-charging methods. The goal of these techniques is to allow sufficient time between charging pulses to provide for zincate ions to diffuse to or reaccumulate at the zinc surface. The current density for charging should be as high as possible without permitting diffusion control to be established.

The development of newer organic membranes to replace the cellulosic base membranes has led to separators with increased resistance to oxidation by soluble silver oxides as well as increased resistance to dendrite penetration. There are presently available grafted polyethylene membranes which are extremely resistant to dendrite penetration and which effectually prevent cell failure due to cell shorting via chemical degradation, silver or zinc dendrite penetration(28). Failures due to zinc dendrites can be eliminated by multiple layers of the membrane. The resulting longer cycle life is obtained at a sacrifice in current density and energy density (higher IR from the membrane).

The problem of shape change at the zinc electrode during cycling is not well understood. On repeated cycling, the cycled plate thickens somewhat below the center and thins out on the perimeter. One mechanism postulates that gravity is responsible since zincate solutions are denser than the normal cell electrolyte. However, it has been shown that this shape change is independent of cell orientation, ruling out gravity forces as the culprit. The postulation of nonuniform current density across the surface of the plate appears to have some merit. The current density is usually greater at the edges of the plate, leading to the formation of a local action corrosion cell as a result of the potential differences between the edges and center portion of the plate. Since there is good electronic contact through the electrode matrix, zinc slowly accumulates in the center portion of the plate. This zinc relocation leads to a decrease in effective electrode capacity and an increased thickness in the center of the plate.

The silver-zinc cell has a lower operating limit of about 0°C. The low temperature limit is due to the poor charging characteristics of the zinc electrode below 0°C. Excessive amounts of hydrogen are formed unless the charging is carried out at very low currents. Hydrogen generated at the zinc electrode on charge and on stand may be, as previously mentioned, recombined slowly at the silver electrode. Palladium is sometimes added to the silver electrode to catalyze the reaction. However, palladium has a small solubility and, should it diffuse to the zinc, would result in increased zinc corrosion. Oxygen generated on overcharge from the positive is recombined readily by metallic zinc. The rates of both H_2 and O_2 recombination are relatively low due to the semipermeable membrane separator in the cell.

The activated shelf life of the battery is quite good. It has about 80 percent capacity after one year(29). Failure is due primarily to separator oxidation and degradation. If stored dry (unactivated), the shelf life is several years.

Precise charge control of silver-zinc batteries is needed for good cycle life. The overcharge should be limited to about 102 to 105 percent of the capacity removed. A voltage-limited taper charge is normally employed. It is necessary to have close, stable voltage control in the charging circuit 'or good cycle life. On stand or extended trickle charge, the high voltage step disappears from the discharge curve.

Zinc anode plates are manufactured by two general types of processes: pressed powder, and electrodeposition. The two types have equal performance at low rates, but the electrodeposited electrodes are preferred for high-rate batteries.

Pressed powder-type anodes are constructed by pressing at high pressures mixtures of powders of the electrode materials into a conductor mesh. The mixture consists primarily of high purity zinc oxide and of 1 to 10 percent mercuric oxide. A small amount of inert binder such as polyvinyl alcohol, "Teflon," or a fibrous material may also be included in the mixture. Care must be exercised to avoid contamination by iron, nickel, and so on, which lead to excessive zinc corrosion due to their low hydrogen overvoltage. The mixture is usually pressed onto a silver screen or expanded mesh. A silver-plated copper mesh may be used in place of the pure silver mesh. A zinc mesh is also used. After pressing, the electrodes are electroformed in dilute caustic to produce an amalgamated zinc electrode. The electrodes produced by this process have approximately 60 percent porosity. The porosity can be increased for high-rate performance by controlling the particle size of the active materials, the amount of binders, the molding pressure, and the like.

The electroformed electrodes are prepared by electrodeposition of zinc either from a caustic bath of a slurry of zinc oxide or from a conventional cyanide bath. Care must be exercised to remove impurities from the bath. The zinc plates must be thoroughly washed and dried before incorporating them into a battery. A silver mesh or screen is normally used for current collection, although a copper or silver-plated copper mesh is also used. The porosity of the plates approaches 75 percent depending on current density, bath composition, temperature, etc., used for deposition.

Silver electrodes are manufactured by one of two general techniques: sintering, and pressed powder. Whichever technique is used, the electrodes are electroformed prior to use.

The sintering technique has two main variations: slurry pasting, and sintering. The two methods are very similar, with the difference lying in the initial method of preparing the silver plaques. For slurry pasting, powders of silver, Ag_2O, and Ag_2O_2[2] are mixed with water and binder, for example, CMC or polyvinyl alcohol, to form a paste mixture. The particle size of the powders and amounts vary. The paste slurry is then forcibly spread over a silver or nickel grid. The pasted sheets are then pressed to the desired thickness and dried (at 70°C). The plates could be used at this

[2]The formula for divalent silver oxide has been given in the literature as AgO and Ag_2O_2.

point but are very fragile to handle. Normally, the plates are dried and sintered at about 400°C to convert the silver oxides to silver by thermal degradation, during which time the organic binders are burned off. The plates are electroformed in dilute caustic (5 percent KOH) at relatively low current densities to achieve good conversion to Ag_2O_2. The plates are thoroughly washed and dried before incorporating into batteries.

A variation in the sintering process involves the direct sintering of silver powder. The particle size of the silver powder is carefully controlled, as this is the primary control on the resulting density of the finished electrode. The grid material is usually expanded metal, for example silver, silver-plated copper, or nickel. The grid is placed in a mold and the silver powder sprinkled in. A pore-forming powder (e.g., NH_4HCO_3) or resin may be included to increase the porosity of the final electrode. The electrodes are sintered at between 400° and 800°C. The sintered electrodes are then electroformed in dilute KOH, washed, and dried.

The pressed powder electrodes are manufactured by a "dry" process. Silver or silver oxide powders are blended with a binder, as for the slurry-pasting technique. The powders are placed in a mold along with a grid material, usually a screen of silver or silver-plated copper or nickel. The electrodes are formed by pressing at high pressures, for example, 30,000–100,000 psi. These electrodes are then ready for incorporation into cell manufacture without forming. This process has the advantage that it is simpler, avoids formation, and permits close control of the electrode composition. However, the electrodes must be handled carefully as they are very fragile until placed in the battery and cycled.

The separator is the key to successful operation of the silver-zinc battery. Without a suitable separator, the slightly soluble silver species diffuses to the zinc electrode where it is removed by reaction with metallic zinc. The separator also prevents dendritic zinc growths from shorting out the cell on charge.

The first material to partially satisfy the requirements was a cellulose film or a fibrous sausage casing. These materials were susceptible to oxidation and penetration by the soluble silver and did not completely prevent dendritic growths from shorting the cell. It was common practice to wrap each electrode with several layers of the films.

In recent years, thin polyethylene films which have been grafted and cross-linked for thermal and chemical stability and hydrophilicity have been developed. These films are very inert to silver oxidation and essentially eliminate shorting by zinc dendrites.

The assembly process for the silver-zinc battery is similar to that described for the nickel-cadmium cells shown in Figs. 9.1 and 9.2. Only a few cells use the jelly roll construction. Most cells use the prismatic construction with flat rectangular plates and a rectangular cell case. The battery case is a molded plastic material such as polystyrene, but various metals and plastics are utilized.

The finished positive and negative electrodes are wrapped in separator material. The electrodes are stacked such that the cells fit snugly into the cell case. Upon addition of electrolyte, the separator material absorbs electrolyte and swells. The swelling action produces a pressure on the electrodes for better cell performance. Without proper pressure on the electrodes, the cell voltage is erratic and the plates would disintegrate.

The electrolyte is usually 30 to 45 percent KOH. A vent is used to prevent ingress of CO_2 and O_2 which adversely affect the electrolyte cell balance and the operation of the zinc electrode. The vent is set to open at a predetermined pressure so that the internal pressure does not rupture the cell case. Most batteries are given a formation cycle and capacity check routine in the factory.

The batteries are normally designed to be positive-limiting with a large excess negative capacity. The end of life is limited by a gradual loss of capacity with cycling and inability to accept charge at high rate. These limitations are usually associated with the silver positive electrode performance. Short cycle life is also associated with shorting by zinc dendrites or mechanical pressure exerted from the shape change on the zinc electrode.

MANGANESE DIOXIDE-ZINC BATTERY

The MnO_2-Zn system(61) potentially represents the lowest cost, sealed, secondary battery system, with the possible exception of the lead-acid system. In available systems, the cycle life of the MnO_2-Zn is limited and represents the largest drawback to more general use of this system. Although the electrolyte is alkaline, the general principles of cell chemistry are related to the neutral or slightly acid Leclanché-type electrolyte.

The zinc anode reaction in alkaline solution has been discussed in the section on silver-zinc batteries. The reactions and problem areas are applicable to both battery systems.

The MnO_2 cathode reaction is very complex. Normally, the electrode is cycled in a controlled fashion to prevent deep discharge and an irreversible compound formation. The cell reaction is thereby controlled to prevent reduction to the $+2$ state, so that the cathode reaction is

$$MnO_2 + H_2O + e^- \underset{\text{charge}}{\overset{\text{discharge}}{\rightleftharpoons}} MnOOH + OH^- \qquad (10)$$

On charge, care must be exercised to prevent overcharge. If the electrode potential increase is too high, a soluble species is formed.

$$MnO_2 + 4OH^- \rightarrow MnO_4^- + 2H_2O + 3e^- \qquad (11$$

Typical charge–discharge curves are shown in Fig. 9.7.

The construction of primary and secondary alkaline MnO_2-zinc bat teries (shown in Fig. 9.8) is essentially the same, with only minor change: in electrode formulation and cell design. For example, thicker separator: are used in the rechargeable version to reduce dendrite formation.

A conductive matrix formed by mixing zinc powder, sodium carboxy methylcellulose, and electrolyte is used in most cells. The zinc i amalgamated to reduce corrosion. The zinc anode has a high surface are: and is capable of good high-rate performance.

The MnO_2 cathode is prepared by molding the cathode mix under high pressure into a steel can which serves as the cell container and positive terminal. The cathode mix generally consists of MnO_2 and graphite plus : suitable binder.

Cell assembly involves placing a separator inside the molded cathode and filling the center cavity of the cell with the anode gel, thus forming the so-called "inside-out" construction. A brass current collector is used for the anode. A nylon or rubber gasket is used to insulate the anode cap from the cell. The cell is closed by forming a radial compression seal. The gasket may also be used as a pressure relief vent in case of a buildup of internal pressure.

For good cycle life, secondary MnO_2-zinc batteries should not be discharged below 1.2 to 1.0 V per cell to prevent irreversible compound formation. Likewise, the cells should not be charged above 1.75 V per cell

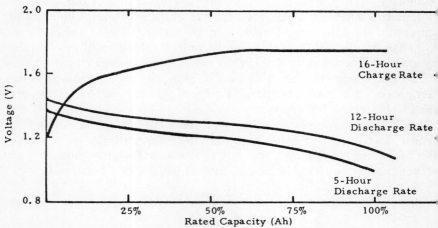

Figure 9.7 Typical charge-discharge characteristics of the manganese dioxide-zinc system A VLTC-charge is shown.

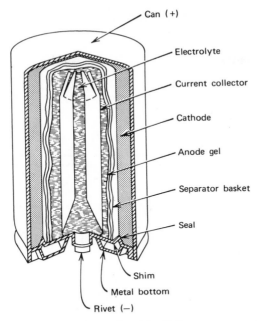

- Can (+)
- Electrolyte
- Current collector
- Cathode
- Anode gel
- Separator basket
- Seal
- Shim
- Metal bottom
- Rivet (−)

Figure 9.8   Cutaway view of rechargeable alkaline-manganese dioxide cell.

with a charge return of at least 100 but less than 120 percent to prevent formation of soluble manganese compounds. Alkaline MnO_2-zinc batteries have excellent charge retention. It is the only secondary system normally assembled in the charged condition.

SILVER-CADMIUM BATTERY

The silver-cadmium battery is the newest commercial rechargeable battery system. It became available in the early 1960s, although it has been recognized as a potential system since about 1900. The system represents a compromise between the high energy density silver-zinc system with its low cycle life, and the low energy density nickel-cadmium system with its long cycle life. The silver-cadmium is the most expensive of the battery systems.

The electrode reactions of the cadmium negative and silver oxide positive have been discussed previously in the section on nickel-cadmium and silver-zinc systems. The overall cell reactions are

$$2AgO + Cd + H_2O = Ag_2O + Cd(OH)_2 \tag{12}$$

$$Ag_2O + Cd + H_2O = 2Ag + Cd(OH)_2 \tag{13}$$

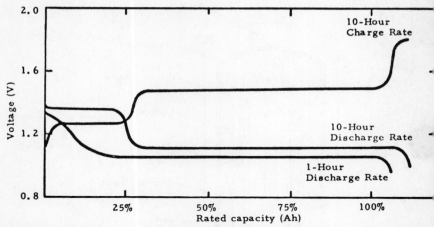

Figure 9.9   Typical constant-current charge-discharge characteristics of the silver-cadmium system.

The cell voltages corresponding to the two reactions are 1.38 and 1.16 V Typical charge-discharge curves are shown in Fig. 9.9.

The silver-cadmium system will operate at higher temperatures (70°–80°C) than the nickel-cadmium (40°C) and has good-activated shelf life. The capacity loss at the end of one-year stand is 10 to 20 percent. Because of the solubility of the silver in the positive active material, a semipermeable separator must be used. Although separator materials effectively block silver migration, they also block the oxygen recombination on overcharge. Consequently, overcharge at rates greater than about C/100 represents a potential problem due to pressure buildup. Therefore, close control of the charge current on overcharge is needed. If taper-charge techniques are used, close control of the voltage limit is needed. This also places a restriction on multicell batteries as the cells must exhibit very uniform capacities and close cell matching. Otherwise, the low capacity cell will be driven into reverse on deep discharge. Since recombination is poor, the gassing reactions on cell reversal can cause excessive pressure buildup. The negative is usually constructed with a high charge reserve.

On stand, separator oxidation by the soluble silver specie leads to carbonate formation. This appears to be a limiting factor on shelf life of the system, also carbonate decreases overall performance of the battery.

The cadmium negative is manufactured in a manner similar to cadmium anodes for the nickel-cadmium cells. Anodes for use in nonmagnetic applications have been fabricated by pressed powder or pasted techniues using silver grids. A mixture of finely divided cadmium metal, admium oxide, and cadmium hydroxide is blended with silver oxide powder. A water solution of a binder (e.g., polyvinyl alcohol) is mixed in o form a paste. This paste is then applied to the silver grid, dried, and pressed.

The silver positive is manufactured as described for the silver-zinc attery.

The separator system used in the silver-cadmium cell is a combination of that used in the silver-zinc cells and nickel-cadmium cells. A emipermeable-type membrane is required to inhibit silver migration to he negative and to resist chemical attack by the soluble silver oxides. The rafted polyethylene membranes have also been used. A felted nylon or bsorbent cellulosic material is used next to the cadmium electrode.

Cells are constructed by procedures similar to those described for ickel-cadmium and silver-zinc batteries. The electrodes are stacked with eparators in place. The lead wires are usually spot-welded together and he assembly placed in the case and the top sealed. Electrolyte (usually 0–40% KOH) is added. The volume of electrolyte in sealed cells is estricted so that recombination of oxygen can occur. As a result, the estricted-electrolyte cells can tolerate a certain degree of overcharge.

RECHARGEABLE LECLANCHÉ SYSTEM

Although the conventional Leclanché flashlight battery is not normally lassified as a secondary battery, it can be recharged successfully (30). 'he electrode reactions are reversible, and a significant gain in output can e obtained if careful control of the charge–discharge conditions is xercised. Otherwise, undesirable side reactions will occur which negate he charging process. Deep discharge, for example, in excess of 20 to 35 ercent of rated capacity, gives rise to irreversible reactions and must be voided.

The overall cell reaction may be written as

$$Zn + 2MnO_2 + 2H_2O = Zn^{++} + 2MnOOH + 2OH^- \qquad (14)$$

After discharge, and especially on open circuit stand, several irreversible hemical reactions can occur which essentially remove material from urther participation in electrochemical reactions. Examples of these eactions are the formation of zinc diammine chloride and basic zinc hloride:

$$Zn^{++} + 2NH_4Cl + 2OH^- = Zn(NH_3)_2Cl_2 + 2H_2O \qquad (15)$$

$$Zn^{++} + OH^- + Cl^- = Zn(OH)Cl \qquad (16)$$

and the hetaerolyte reaction in the cathode

$$Zn^{++} + 2MnOOH = ZnO \cdot Mn_2O_3 + 2H^+ \qquad (17$$

The products of these three reactions are very insoluble compounds. The precipitation of these compounds in the separator, in the bobbin and the pores of the MnO_2 particles not only makes them unavailable for recharge but also effectively increases internal resistance by plugging up the porous matrix. Therefore, using the strict definition, the Leclanché cell probably should not be classed as a secondary cell, as it cannot be restored to its original state by reversing the current.

The physical construction of the common Leclanché cell is not designed for rechargeability. The cathode geometry is not suited for charging, and there is a very nonuniform reaction zone concentrated near the central carbon collector. The separator system in the Leclanché cell is intended only for primary use and is not stable to drastic changes in pH and salt concentration which occur on charge and discharge. The separator does not inhibit zinc dendritic growths from shorting the cell during charge.

In spite of these drawbacks, the Leclanché cell can be repeatedly recharged with careful control. In order to minimize the irreversible nature of the electrode reactions, several techniques can be used to successfully recharge or extend the life of the Leclanché cell. The recharging process should occur immediately after the discharge is completed. Any delay in recharging increases the likelihood of the occurrence of unwanted chemical reactions.

The discharge should be limited to 1.0 V per cell, and preferably to 1.1 V per cell. Discharge below these voltages seriously decreases the chances for efficient recharge. Apparently the deeper discharges (in excess of 25 percent of rated capacity) of MnO_2 enhances the irreversible formation of insoluble reaction products mentioned above and must be avoided. Short high-rate discharges remove less energy from the cell making it easier to recharge, partly because the internal resistance contributes a substantial portion of the voltage loss in the cell, thereby indicating that the cell has reached the voltage cutoff prematurely.

The recharging process is more efficient on new cells than for cell which have been stored for an appreciable time. The exact cause of the decreased rechargeability is not clear, but it may be related to the availability of water in the cell. The necessity of adequate water for the cell reaction is noted in equation (14). As a cell is cycled it appears to slowly dry out. The cathode mix becomes hard and dry even though there is no water loss detected by weight measurements. It appears that the water in the cathode mix is slowly absorbed either by hydration of the various lower valence manganese oxides, or as the result of greater porosity created by a greater dispersion of the manganese oxides and acetylene black. In any event, conditions which lead to water loss (e.g., shelf stand) appear to shorten the cycle life.

Cells that have been recharged have shorter shelf life than do undischarged cells. The cells should be used soon after recharging for optimum performance. Also, cells that have synthetic ores appear to have better recharging characteristics than cells made from natural ores.

To summarize, very shallow depth of discharges with immediate recharge result in the most efficient mode for recharging of the Leclanché cell. Long discharges allow time for the secondary chemical reactions along with water loss to occur which limit rechargeability. Unless Leclanché cells are used regularly and for fairly constant periods, and unless they can be recharged soon after discharge under carefully controlled conditions, their use as secondary batteries is not profitable.

Improper charging may result in shorter rather than longer battery life. Recharging Leclanché cells greatly increases their tendency for leakage of corrosive fluids. Should the normal vents clog with the fluids during charge, the cell may rupture or explode. The high voltage of the cells immediately after charge could damage sensitive electronic instruments and will shorten bulb life. Chlorine gas generation on overcharge may also present a hazard.

Recharging Leclanché cells can be hazardous. Unless proper precautions are taken and care in the use of the cells is exercised, the gain may not be worth the bother.

CRITERIA FOR RECHARGEABILITY

GENERAL CONSIDERATIONS

Most battery electrodes are reversible, although some only partially. There is no clearcut distinction between primary and secondary systems, and their classification must be based primarily on practical and economic factors rather than theoretical considerations. Some systems are definitely irreversible. A cell with magnesium anodes, for example, cannot be recharged because the metal cannot be deposited from aqueous solutions. Thus the magnesium cell is clearly classified as a primary system.

The reversible nature of a cell is determined by the exchange currents for the cell-electrode reactions. The exchange current is the dynamic equilibrium rate of a reaction expressed as a current flow. For example, a zinc electrode immersed in a zinc solution will, at equilibrium, have a rate of reaction for zinc metal dissolving to form zinc ions, and zinc ions depositing as zinc metal, of about 10^{15} atoms/sec/cm^2, or 10^{-8} eq/sec/cm^2(31). The exchange current density, i_o, is approximately 10^{-3} amp/cm^2(31). The exchange current density, a fundamental kinetic parameter, is a direct measurement of the reversibility of a system.

Previously, Yeager(32) discussed the fundamentals of electrode kinetics as they apply to batteries. The expression relating exchange current to polarization, η, is

$$I = I_o \left[e \frac{-\alpha\eta}{b} - e \frac{(1-\alpha)\eta}{b} \right] \tag{18}$$

In this equation, $b = RT/F$ and α is a constant characteristic of the process. Also, $I_o = i_o s$ where i_o is the exchange current density and s is the surface area.

Examination of equation (18) shows that the polarization, or irreversibility at a given current is greater the smaller the I_o. Thus, less reversible reactions have smaller exchange currents and more reversible reactions have larger exchange currents. Since $I_o = i_o s$, battery reactions can be made more reversible by increasing the electrode surface area.

Most commercial battery electrodes exhibit a fairly large exchange current. One measure of the exchange current is the flash or short circuit current. In fully charged secondary batteries the flash currents are limited by the internal resistance of the cell rather than electrode kinetics. In fact shorting a large silver oxide-zinc battery can generate sufficient heat energy inside the battery to cause an explosion.

Efforts to exploit the Lalande (Zn:CuO) alkaline battery(33) as an accumulator generally have given unsatisfactory results. There is a tendency under charging conditions to form soluble copper complexes which deposit irreversibly on the anode and cause wasteful zinc corrosion. Even more serious is the difficulty of replating zinc as uniform adherent deposits, rather than as sponge or short-circuiting dendrites.

One might cite other theoretically reversible-systems which have proved unsatisfactory as accumulators. For example, the recharging of an exhausted Daniell cell(Zn:ZnSO$_4$::CuSO$_4$:Cu) could hardly be accomplished without mixing anolyte and catholyte, thereby leading to the deposition of copper on the zinc electrode. The difficulty of restoring the Poggendorff diaphragm cell(34), Zn:H$_2$SO$_4$::K$_2$Cr$_2$O$_7$ + H$_2$SO$_4$(C), to its original condition by recharging is another example that appears to rule out all two-fluid cells or units with soluble reactants as secondary systems.

An examination of the features of sealed rechargeable cells leads to a set of criteria for the quality of an accumulator. The categories that nclude the most important parameters are listed below.

1. Ability to recharge and deliver power;
2. Cycle life;
3. Mechanical and chemical stability;
4. Self-discharge;
5. Overchargeability;
6. Charge time;
7. Cell reversal;
8. Shape of discharge curve;
9. Temperature range of operation;
10. Cost.

These topics are interrelated and thus cannot be discussed with complete independence. However, every attempt has been made to cover the important aspects of each category and its relationship to other phenomena.

ABILITY TO RECHARGE AND DELIVER POWER

In order to qualify as a secondary battery, the electrochemical cell must be able to serve as a reversible storage place for electrical energy. This criterion is basic to secondary batteries and serves as their definition. A more formal definition of a storage battery given by the AIEE(35) is: "A storage cell is an electrolytic cell for the generation of electric energy in which the cell after being discharged may be restored to a charged condition by an electric current flowing in the direction opposite to the flow of current when the cell discharges."

The key phrase is "restored to a charged condition." Many cells can be discharged but resist recharging. To be a true secondary cell, the recharging process must return the electrodes and the active materials to their original charged state with good efficiency. *This implies that the electrode reactions are reversible, and the charge reaction is the reverse of the discharge reaction.* In other words, the discharge products are restored to the charged condition. The coulombic efficiency for many reactions approaches 100 percent.

A measure of the ability of the battery to function in a reversible manner can be judged in a practical sense by the current and voltage characteristics of the cell during repetitive charge and discharge. The generalized behavior of batteries on charge and discharge is shown in Fig. 9.10. The departure from the theoretical or open circuit voltage (OCV) is the result of irreversible behavior of the electrode processes. The overall

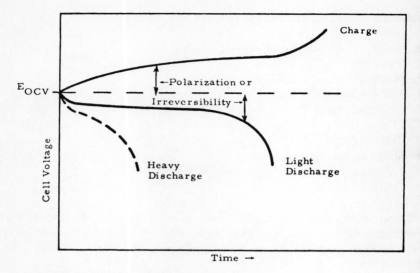

Figure 9.10   Typical charge-discharge curves.

storage efficiency of the battery is defined by

$$\theta = \frac{\int E_d i_d t}{\int E_c i_c t} \qquad E_d < E_{\text{ocv}} < E_c \tag{19}$$

where E_d, E_c, i_d, i_c are voltages and currents on discharge and charge, respectively, and t is the time.

The overall storage efficiency of a cell includes the effects of electrode polarization, internal resistance as well as the current efficiency. In Fig. 9.10 the increased polarization or energy loss with increased current drain is noted. Also, it may be noted that on charge the voltage and total coulombic passage may be significantly larger than that on discharge. It is typical in most rechargeable cells to have the charge input exceed the discharge output by 20 to 40 percent. This is due in part to the desire to ensure that the cells are fully charged. As a result, considerable gas may be evolved near the end of charge, which may result in an increase in polarization and an increase in the irreversible behavior of a cell. In cells with free electrolyte, gassing also serves to stir the electrolyte and equalize any concentration gradients that may have arisen during the charging process.

The thermodynamic efficiency, ϵ_T, of battery systems or the efficiency of converting the total energy available (ΔH) into the net available electrical energy (ΔG) for a reaction is

$$\epsilon_T = \frac{\Delta G}{\Delta H} \times 100 = \left(1 - \frac{T\,\Delta S}{\Delta H}\right) \times 100 \tag{20}$$

where

$$\Delta G = \Delta H - T\Delta S \tag{21}$$

The overall thermodynamic efficiency is about 90 percent since the entropy contribution, $T\,\Delta S$, is usually less than 10 percent for battery systems.

The electrochemical efficiency, ϵ_E, of the system to deliver power is given by

$$\epsilon_E = \frac{\int^E d^{idt}}{\Delta G} \times 100 \tag{22}$$

The electrochemical efficiency is in a large measure a function of the current drain for a cell. At low currents ϵ_E approaches 100 percent, while at very high rates of discharge ϵ_E can fall to 50 percent or less. The effect of current drain is shown in Fig. 9.10 where the effect of increased polarization at high current drain is noted.

The current or ampere-hour efficiency, ϵ_I, is also of interest and is defined by

$$\epsilon_I = \frac{\int i_d\, dt}{\int i_c\, dt} \times 100 \tag{23}$$

The reactions that occur in charging a battery, in fact almost any electrochemical reaction, involve a direct competition between several possible electrode processes. Most anode materials are more negative than hydrogen in the EMF series so that hydrogen evolution competes with the metal deposition reaction during charge. Likewise, the cathode is usually more positive than oxygen in the electromotive series and there is a competition between oxygen evolution and the cathode-charging reaction. Successful charging of a battery, therefore, involves the suppression of the undesired reaction (gas evolution) in favor of the desired reaction (battery charging). It is in this regard that the term "charge acceptance" or the ability of an electrode to accept charge to restore the starting material, is used.

Since the charging voltages of most batteries are above the decomposition voltage for the electrolyte, either hydrogen or oxygen evolution or both, depending on cell balance, can occur. Redox reactions can also occur but are less common. Since electrode reactions will follow the path of least resistance, competition between reactions is the heart of charge acceptance. The factors which have greatest influence on charge acceptance are gas overvoltage characteristics, temperature, charging rate,

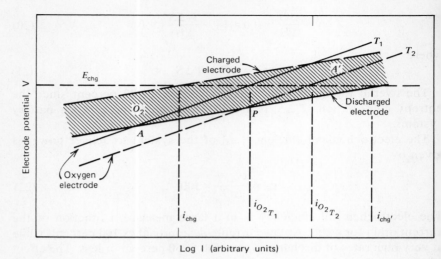

Log I (arbitrary units)

Figure 9.11 Polarization curves for charging and oxygen evolution reactions at a positive electrode charging at a potential E_{chg}, $T_2 > T_1$. At the potential E_{chg}, i_{chg} and $i_{chg'}$ are the charging currents for a discharged and charged electrode respectively; and $i_{O_2T_1}$ and $i_{O_2T_2}$ are the oxygen evolution currents at T_1 and T_2 respectively.

state of charge, and electrolyte composition. The influence of electrolyte composition is more subtle and will be discussed later.

The principles of charge acceptance are best illustrated by considering the typical case(36) shown in Fig. 9.11. The competitive nature of the reactions is characteristic for all systems. An electrode is said to have poor charge acceptance when the charge passed through the discharged battery does not result in increased electrode capacity. Conversely, good charge acceptance indicates a good efficiency in converting the electrical energy to chemical energy stored in the electrode.

In Fig. 9.11, point A represents the crossover point where the current paths for oxygen evolution and charging of a discharged electrode are equal. At currents or potentials below this point, the charging is very inefficient and the majority of the current goes into oxygen evolution $i_{O_2} > i_{chg}$. At larger currents and higher potentials—a larger fraction of the current goes to charging, $i^1_{chg} > i_{O_2}$. The charging is more efficient and the electrode is said to have good charge acceptance.

Increasing the temperature has a greater effect on the gas overvoltage than on the battery electrode process. Raising the temperature to T_2 has the effect of moving point A to point P. In this instance, large charging currents are necessary for the electrode to accept charge. This may be

elf-defeating as large currents and gas evolution can generate considera-
le heat, raising the battery temperature, lowering the overvoltage, and
ausing a greater fraction of current to go to gas evolution.

Let us consider an example wherein the electrode is charged at
onstant potential, E_c. At the lower temperature, the charging reaction is
avored over gas evolution at E_c and the ratio of current for the charging
eaction to the oxygen evolution, i^1_{chg}/i_{O_2}, is large. If the charging potential
s higher, then a greater fraction of the current goes to the charging
eaction. This predicts that it is more efficient to charge the nickel oxide
lectrode at high current, that is, to attain a better charge acceptance.

At the higher temperature, the ratio of i_{chg}/i_{O_2} is smaller, that is, poor
harge acceptance. This illustrates the importance of temperature control
or an efficient charging operation.

As an electrode approaches a fully charged condition, its potential
ncreases. The top of the shaded portion represents the final charged state
s shown in Fig. 9.11. It should be noted that at a given potential the
xygen evolution is favored on a charged electrode to a much greater
xtent than on a partially charged electrode.

The electrolyte composition has a secondary effect. As the concentra-
ion increases or decreases, the gas evolution and battery reaction are
oth affected. Generally the battery reaction is influenced the most by
oncentration changes. For instance, in alkaline systems higher electro-
yte concentrations generally favor high charge acceptance, but may
nake the anode discharge products more soluble.

Additives or impurities may also influence charge acceptance. For
xample, *lignin* added to the lead-acid cell increases the hydrogen
vervoltage and facilitates lead deposition. Nitrates in nickel-cadmium
ells set up a redox "shuttle" which effectively competes with the
harging reaction and lowers charge acceptance.

A serious problem on overcharge is the increased heat generation,
vhich leads to a more rapid deterioration of the cell components and can
horten cycle life of the cell. There are two significant sources of heat: the
hermodynamic entropy production ($T \Delta S$), and the irreversible heat
rom polarization (η). The amount of heat per anode, q, generated is given
y

$$q = T \Delta S - nF(\eta) \tag{24}$$

vhere the values of ΔS and η must be taken for the processes occurring
n the cell. The irreversible behavior results in more heat generation in the
ells than would theoretically occur. The reversible heat production
$T \Delta S$) is low for most battery systems. At higher currents the reactions
end toward increased irreversible behavior. Equation (24) shows that the

greater the irreversibility, the greater is the heat production. At high currents and especially during gas evolution on overcharge, significant amounts of heating can occur.

An example of the heat generation in charging is given below. The charging of a 1.2 amp-hr sealed sub-C-size nickel cadmium battery is carried out with 100 percent overcharge. The temperature coefficient for the charging reaction is about 2.5×10^{-4} V/deg(37). On charge, $\eta \cong 0.05$ V and since $\Delta S = nF(dE/dT)$, the heat during charge is

$$q = 298 \left(\frac{1.2}{26.8}\right)(2.5 \times 10^{-4})(23060) - \left(\frac{1.2}{26.8}\right)23060(0.05) \cong 26 \text{ cal}$$

Since q is positive there is a small cooling effect during the initial charge of the Ni-Cd cell. On overcharge $\Delta S = 0$, since there is no net electrode reaction; oxygen is produced at the positive and consumed at the negative electrode. The charging voltage is about 1.5 V so that on overcharge $\eta = 1.5$. Then,

$$q = 0 - \left(\frac{1.2}{26.8}\right)23060(1.5) \cong -1550 \text{ cal}$$

Here, q is negative, so heat is evolved on overcharge.

On overcharge where oxygen recombination occurs, a significant amount of heat is liberated inside of the cell, which may become quite warm. The internal temperature can be estimated by an analogy to heat transfer from a pipe. A value of 2.2 Btu/hr/ft^2/°F (about 1.1 cal/hr/cm^2/°C) may be assumed for heat loss from a pipe(38) and, by analogy, also a cylindrical cell. The 1.2 amp-hr cell has about 20 cm^2 surface area. If the overcharge continues, the temperature rise in the cell may be shown to be approximately 45°C for an ambient 25°C. The internal cell temperature can reach 70°C (160°F). Since the heat is released uniformly over the time period, the cell will not reach 70°C until well into the charge period. If a cell has a specific heat of about 10 cal/°C, the cell could reach the steady state value of 70°C in 20–30 minutes. It may be noted in passing that the specific heat for several battery systems is about 0.23 cal/g/°C(39). While many approximations are involved, the calculation does serve to point out the significant amount of heat that a cell must withstand. In this case overcharge at the one-hour rate may damage the cell, especially if the cell is enclosed in a battery case or has a plastic jacket. The internal heating of this severity places an additional stress on the mechanical and chemical stability of the system and generally leads to shortened battery life.

CYCLE LIFE

The number of times that a battery can be successfully charged and discharged is referred to as the "cycle life" of the battery, usually rated at

ℓe number of cycles to, for example, 75 percent or 50 percent of the rated
ɹapacity of the battery. Although failure may occur abruptly on charge or
ʤischarge, it is more common to have the battery slowly decrease in
ɹapacity as the battery is cycled.

The cycle life of a battery is determined primarily by its chemical and
ℓechanical stability. Factors such as depth of discharge, dendritic
ɹrowth, shape change, overcharge, and temperature place a stress on
ℓtability and decrease the cycle life. Cells that receive only shallow
ℓischarges, say 10 percent depth of the rated capacity, proportionally
ℓave a much longer cycle life than cells which are completely discharged
ℓach cycle. The effect of depth of discharge on cycle life is noted in Fig.
ℓ.12 for several battery systems. The increased depth of discharge
ℓncreases the possibility of either redeposition of the active material in an
ℓnusable form or in electronic isolation from the electrode structure.
ℓeep discharging also increases the release of impurities trapped in the
ℓarticles of the active materials, for example, NO_3^- in the Ni-Cd cell, Cu^{2+}
ℓn the MnO_2-Zn cell, which can lead to increased self-discharge and
ℓorrosion.

A slow, steady loss in capacity, fading, with cycling can be the result of
ℓeveral different phenomena. The most common cause of fading is the
ℓesult of changes occurring in the morphology of the active materials.
ℓhere is generally a growth in particle size and change in location of the
ℓctive materials in the electrode matrix. The discharge of larger particles
ℓs less efficient(42) than that of smaller particles. These effects will be
ℓonsidered in more detail under mechanical and chemical stability.

Fading may also be the result of a physical loss of contact of active
ℓaterial with the matrix such as found in the MnO_2 and NiOOH pressed

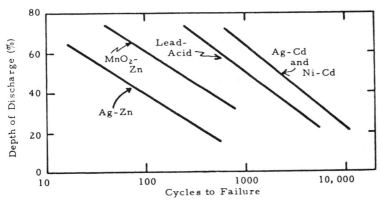

Figure 9.12   Cycle life of various battery systems (23, 40, 41).

powder cathodes. Also, either the volume changes in the active materi
or oxidation of the conductive matrix can disrupt the integrity of th
cathode. A physical break in an electrode with resulting decrease i
capacity is usually fairly abrupt. The loss of electrolyte or poor electro
lyte distribution can lead to loss of capacity either from increased IR o
loss of reactant water.

Overcharge is another important factor affecting cycle life. Generall
overcharge is detrimental, although it may be necessary in order to full
charge the battery. The cycle life of most battery systems decrease
directly as the overcharge increases. While there is no well-establishe
basis for this observation, Carson(43) has proposed that

$$N = N_0 k \int f(i)\, dt \qquad (2!$$

where N is the cycle life, N_0 is the cycle life in the absence of overcharge
k is a constant, $f(i)$ is a function related to current flow or rate o
overcharge.

Since overcharge is damaging, primarily at the positive electrode som
means of control should be incorporated in the charging.

Some electrodes, for example, manganese dioxide, tend to form solubl
higher valent compounds on overcharge, others evolve oxygen, generall
involving the production of highly active oxygen compounds. Thes
compounds oxidize the positive collector and conductive matrix as we
as the separator material. The separator materials (usually plastics) ar
particularly susceptible to the strongly oxidative environment in the are
of the cathode, and their deterioration can lead to short circuits throug
the weakened material. Heat generated during oxygen evolution an
recombination further enhances the oxidizing environment.

Oxidation of the conductive matrix and separator materials leads to a
increase in the state of charge of the negative electrode, which will alte
the cell balance and can contribute to premature failure of sealed cell
The products from the oxidation of the conductive matrix and separator
can also affect the electrolyte in a cell. For example, oxidation of a carbo
matrix in caustic electrolyte leads to the formation of carbonate and a n
loss of water which is attributed to the formation of a hydrated produc
This drastically affects the characteristics of the KOH solution an
decreases the integrity of the electrode structure:

$$C + O_2 + 2KOH \rightarrow K_2CO_3 + H_2O \qquad (2\epsilon$$

$$K_2CO_3 + 2H_2O \rightarrow K_2CO_3 \cdot 2H_2O \qquad (2\gamma$$

Construction also plays an important role in the overall batter
performance. Lead placement, matrix conductivity, electrode length an
separation are important in determining the current distribution. Mos

fficient operation occurs with more uniform current distribution. Uniorm electrolyte distribution prevents small portions of the electrode rom furnishing the majority of the current.

Many cell failures can be traced to an internal short circuit, commonly 1e result of manufacturing defects such as a burr or metal chip on the urface of the electrode or in the separator. As the cell is cycled the 1ternal stresses cause the burr or chip to cut through the separator. The 1echanical stresses which develop on charge and discharge as a result of olume changes of the various forms of active material are substantial 1nd may warp, bend, or break the electrode structures. If the conductive 1atrix disintegrates, it may slough off and sift into the separator, thereby 1creasing the opportunity for shorting.

Internal short circuits can also occur from penetration of the separator y the anode material. This penetration results from either a general rowth of the anodes into the separator or dendrite formation. Special recautions, for example, semipermeable separator materials, must be 1ken when the electroactive materials have a tendency to develop endrites. Additives are sometimes used to control dendrite formation 1nd anode growth.

If high pressure builds up inside a cell so that a cell vent mechanism is ctivated, gas escapes from the cell. While the performance of most cells ; not affected by a single venting action, the venting does indicate 1ternal troubles that can lead to limitations on cell life. Once a cell has ented it usually has a limited cycle life. It is, of course, necessary to have good seal on the cell in order to achieve satisfactory life. This problem is lso discussed in overchargeability.

Excessive pressure buildup can lead to rupture of the cell case as well s the restraints; in a large prismatic cell of $100 \, cm^2$ surface, $100 \, psi$ 1ternal pressure could exert a force of about 10^3 pounds on the cell estraints. Seals may develop leaks or lead to short circuits as a result of 1e stress of excessive internal pressure.

If the cells in a multicell battery are not uniform in capacity, the low apacity unit will reach full charge before the others. If there is no vercharge protection, the cell could explode from gas generation. If a arametric charge sensor is used in the cell, an early charge cutoff will be ignaled. Consequently, the battery will not reach full charge. Whatever 1e result, the cycle life of the battery is considerably shortened when the ells in a battery have variations in capacity.

MECHANICAL AND CHEMICAL STABILITY

Mechanical stability of the components of a cell, essential to long cycle fe and reliability, is frequently affected by dimensional changes in lectrodes during cell operation. The reactants and products of most electrode reactions have different densities and molar volumes, and as the cell is cycled, this density difference leads to a constant expansion contraction pressure on the matrix as either products or reactants are formed. Electrode structures must be able to resist these changes or hold them to a minimum. The positive electrode in the lead-acid battery is good example of this phenomenon. The $PbSO_4$ reaction products have about 90 percent larger volume than the reactant PbO_2. The rate of expansion of the electrode varies with the composition of the active material, rate and amount of charge or discharge, and so on. During discharge the lead grid is distorted by the expansion pressure. The ability of the lead grid to withstand or give with the distortion is essential to maintain contact to the active material and good cycle life of the battery When accompanied by grid corrosion some lead electrodes expand or "grow" enough to circumvent the separator and short out the cell. Volume changes occur to a greater or lesser extent in all of the rechargeable electrode systems. Allowance must be made for this expansion and contraction in the design of the electrode. This problem of mechanical strength of the electrode is less critical in primary cells where no provision need be made for recharging.

As electrode materials are cycled, there is always a tendency to redistribute the active ingredients. While there are exceptions, the general tendency on cycling is toward the formation of larger crystallite sizes; that is, a sort of sintering effect on the active material. This is especially true if the electrode reaction involves a soluble intermediate The larger particles exhibit poorer high-rate performance because of lower surface area which reduces the availability of the active material for reaction. The material in the center of the particles does not react at useful voltages because of loss of contact or increased internal resistance and reduced electrolyte diffusion (42). This effect is manifested as a slow steady decrease in cell capacity on cycling.

The active material may move about and physically change location during cycling. For instance, the cadmium in sintered cadmium negative electrodes has been shown to migrate toward the surface of the matrix (44). The zinc metal in the negative electrodes tends to accumulate in the center portion of the electrode. This phenomenon is referred to as shape change. The cause for the movements is not clear but appears to be associated with variations in current density at the electrode surface coupled with a local internal corrosion couple resulting from concentration variations along the surface. The forces involved can be quite large and result in rupture of the cell case by pressure from the zinc accumulation.

Battery seals are also subject to mechanical instability. Electrolyte creepage around the seal can lead to battery failure due to electrolyte loss r damage to the surrounding equipment. Compression seals using lastics are susceptible to cold flow over a period of time. Seals that are atisfactory when initially fabricated can develop leakage several years ter. Thermal cycling is known to accelerate seal failure. Differences in oefficients of expansion lead to excessive pressure or strain on joints, nd the like, which cause parts to change shape over a period of time.

Corrosion of the electrode current collector may be important in cycle fe and overchargeability, especially if two dissimilar metals are used in onstruction. For instance, nickel-plated steel used as a collector for ositive Ni-Cd electrodes usually contains small pinholes, and over a eriod of time the underlying steel will oxidize, leaving only a conductive hell. Stress corrosion can occur at welded joints.

Impurities in the cell can also lead to corrosion problems by enhancing 1e self-discharge of an electrode. The lead positive electrode grids ·equently contain antimony and arsenic for strength and castability. 'hese metals slowly dissolve from the positive on cycling, and migrate to 1e negative electrode where they deposit. The hydrogen overpotential of 1ese metals is much lower than on lead. As a result, the negative lead lectrode spontaneously reacts with the sulfuric acid electrolyte to roduce hydrogen and lead sulfate. The negative electrodes become ischarged and sulfated. Some older automotive batteries will self-ischarge in a week or so if they are not charged. Similar problems occur 1 the alkaline cells with zinc as the anode. Impurities are liberated slowly uring cycling which diffuse to the zinc, deposit, and enhance the elf-discharge of the zinc anode. The act of cycling itself increases lectrode porosity and facilitates the solution of impurities locked inside f the structure.

ELF-DISCHARGE

elf-discharge, the loss of electrochemical capacity during open circuit torage of a battery, limits the usefulness of the battery while standing lle. Many secondary batteries self-discharge at fairly high rates (2 to 5 ercent per week). Primary batteries, on the other hand, have excellent helf life with very low self-discharge rates (about 5 to 10 percent per ear). For instance, the mercury cell will give essentially full performance fter five years. Leclanché cells will give 90 percent or more of the riginal capacity after one year stand at room temperature. This would orrespond to a self-discharge rate of about 0.2 percent or less per week. 'hus, there is a noticeable difference between primary and secondary ystems in their capability to retain their capacity during prolonged pen-circuit stand in the charged condition.

The cause for self-discharge resides in the reactivity of the active materials, the very thing that makes them useful as battery reactants. Th active materials usually react with water in the electrolyte, with th evolution of either hydrogen or oxygen, or both.

Hydrogen generation during self-discharge is the result of a corrosio process on the anode, and is more likely to occur on older or cycled cell in which the various metallic impurities from the electrolyte have had a opportunity to deposit on the anodes. Since many of the impurities hav low hydrogen over-voltages, local cell action produces hydrogen genera tion and anode dissolution. Since gaseous hydrogen does not react readil with the cathode, its formation results in water loss and pressure buildu in sealed systems, and a change in the cell balance.

Oxygen is evolved on stand when active cathodes react with water, an it is also possible that hydrogen peroxide formation may occur instead o oxygen evolution. Whichever is formed, it can escape from the system o diffuse to the anode where it reacts to discharge the active material of th anode. Oxygen generation, fortunately, usually does not result in harm ful effects. Oxygen reacts readily with all common anode material (recombines) so that the gas does not normally accumulate in seale systems unless, of course, there is severe blocking of the gaseou diffusion path. Because of the ability of oxygen to recombine with th anode, water loss does not occur.

Self-discharge that is associated with a cathode-electrolyte reactio generally does not alter the internal cell balance, since the oxygen carrying species produced is usually electroactive and, upon diffusion t the anode, loses an equivalent charge by oxidation of the charge negative. This can be a one-step process or a cyclic one, depending o whether or not the reduced species is capable of diffusing back to th cathode and being reoxidized. The cyclic process, also called chemica short or redox shuttle, is equivalent to a low rate discharge of the cell.

The rate of self-discharge from a redox shuttle can be estimated fairl accurately if it is assumed that the material is reduced immediately upo arrival at the anode. The following diffusion equation is applicable:

$$i = nFD\left(\frac{dc}{dx}\right) \tag{28}$$

where n is the number of electrons in the electrode reaction, F is th Faraday constant, D is the diffusion coefficient, and dc/dx is th concentration gradient.

Nitrate in the Ni-Cd cell can be used for a sample calculation. In Ni-C cells, the nitrate can be present in quantities up to 10^{-4} molar o 10^{-7} eq/cm^3, especially if there is inadequate washing of sintered elec trodes. If it is assumed that the electrode separation is 0.003 inch o

.0076 cm, it follows from equation (28) that

$$i = (1)(10^5)(10^{-6}) \frac{10^{-7}}{7.6 \times 10^{-3}} = 1.3 \times 10^{-5} \, amp/cm^2$$

t was assumed for the purposes of this calculation that $n = 1$ and $D = 1 \times 10^{-6}$. If this were a 1 amp-hr, nickel-cadmium battery with 00 cm² electrode area, the rate of self-discharge by diffusion could be quivalent to approximately 1.3 ma. The cell would essentially lose its seful capacity in a little over one month.

OVERCHARGEABILITY

n order to ensure that the battery electrodes are fully charged, it is ecessary to pass charge through a cell in excess of the charge removed rom the cell during discharge. This is the result of the competition etween gassing reactions and the charging reaction. The extra charge is alled the overcharge, and thus the term "overchargeability." Overcharge educes the current efficiency for the system, but it is needed to ensure ull capacity. Overcharge also protects or preserves the integrity of the lectrodes. If an electrode is only partially charged, the morphology of the ctive material will slowly change by local cell action and digestion. In his case, both the capacity and cycle life begin to deteriorate. If, for xample, the positive of the lead-acid battery is partially charged, pockets f uncharged $PbSO_4$ remain imbedded in the PbO_2. With time, the $PbSO_4$ rystals grow in size, thus removing active material from participation in he charge–discharge process. To prevent this type of growth, a large vercharge (boosting charge) is sometimes used to prevent permanent amage to the battery plates.

During overcharge the potential and active material availability is such hat gas evolution occurs. The ability of a cell to accept overcharge epends on both the rate and identity of the gas generated on overcharge. he identity of the gas can be controlled by controlling the cell balance, hat is, the capacity of the positive and negative electrodes. Cell balance is epicted in Fig. 9.13 for two conditions of a typical Ni-Cd battery. In the rst case, the sealed cell, only oxygen is formed on overcharge, while in he second case, the open cell, both hydrogen and oxygen are liberated on vercharge.

Gas evolution can occur before a cell reaches full charge because of the ontiguous nature of the voltage for gas evolution and cell charging eaction. In the case of the nickel-cadmium battery, oxygen evolution ccurs on the positive electrode at about 80 percent of full charge. During igh rate overcharge oxygen pressure can build up to several hundred

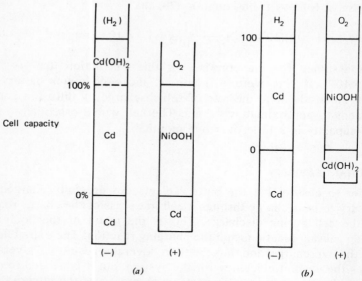

Figure 9.13 Cell balance. (*a*) Sealed cell (charge). (*b*) Vented cell (charge).

pounds and a means for reacting or venting the gas is necessary or the cell will explode.

Most sealed cells are balanced to generate oxygen on overcharge. On stand, the oxygen usually reacts (recombines) with the negative electrode materials. Hydrogen evolved on charge does not readily recombine with positive electrode materials, other than silver oxide, and accumulates in the cell. Thus, hydrogen accumulation when coupled with the varying oxygen pressure can lead to venting even though the oxygen pressure is well below the vent pressure. Hydrogen accumulation may also be the result of corrosion processes on the anode, co-deposition during charging, overcharge of the anode, or cell reversal on discharge. In recent years small fuel cell electrodes have been built into cells to remove any hydrogen that forms. However, this approach has not been adopted for use in commercial cells.

Since gases are generated by electrolysis of water, a loss of gas from a cell corresponds to a water loss from the electrolyte. Continued water loss increases the internal resistance and further enhances gas generation in the cell. The cell balance also changes since the gases do not react with the active materials as would normally be expected. For instance, in the Ni-Cd cell, when oxygen gas which would normally react with cadmium in the negative is vented, more cadmium is formed in the negative and

apacity of the negative is increased. Continued venting will result in a
ully charged cadmium electrode. When this occurs, hydrogen will be
evolved from the negative in addition to the oxygen from the positive.
The cell is said to run over the end of the negative during charge. This
condition is depicted in Fig. 9.13b. The loss of oxygen by venting results
in an increase in the state of charge of the cadmium electrode, thereby
altering the cell balance. It is possible that both hydrogen and oxygen
would be generated in the new cell balance. If the oxygen were prevented
from escaping, the oxygen would react with the cadmium, thus maintaining the original cell balance.

The nickel-cadmium cell is the only truly sealed commercial cell, all
other cells having low pressure vents to permit the escape of gases
generated on overcharge. Commercial Ni-Cd cells constructed with the
cell balance shown in Fig. 9.13a contain vents, but they are merely safety
valves. The recombination of oxygen in the Ni-Cd cell is facilitated by
careful control of the amount of electrolyte added to the cell and by the
separator material used. The separator must have a porous open structure
that contains sufficient electrolyte to provide good conductivity, but not
so completely filled that the gaseous oxygen diffusion to the anode is
blocked. If the separator in a Ni-Cd cell is "flooded" with electrolyte, and
oxygen is forced to diffuse through the solution, the rate of diffusion
would be too slow to permit overcharge at any appreciable rate without
excessive pressure buildup. Oxygen is consumed at the anode via either a
chemical or electrochemical reaction which converts an equivalent
amount of cadmium metal to $Cd(OH)_2$. This may be termed a chemical-
shorting mechanism wherein on overcharge the cathode product is
transferred to the anode and consumed.

Positive electrode	$4OH^- - 4e^- \rightarrow O_2 + 2H_2O$	(29)
Negative electrode	$2Cd + O_2 + 2H_2O \rightarrow 2Cd(OH)_2$	
	$\underline{2Cd(OH)_2 \rightarrow 2Cd + HOH^- - 4e^-}$	
or	$O_2 + 2H_2O \rightarrow 4OH^- - 4e$	(30)

The overall cell reaction is oxygen generation at the positive and
oxygen reduction at the negative electrode. Note that the net reaction at
the negative is the reverse of the reaction at the positive. This is the
mechanism which prevents a pressure buildup in the cell.

The use of restricted electrolyte also reduces the life and capacity of
the electrodes. In the Ni-Cd system the "starved" electrolyte conditions
may reduce the capacity of electrodes approximately 10 percent. The
need to control gas evolution by electrode balancing also reduces cell
capacity.

As previously noted, the cycle life of a cell is directly related to the amount of overcharge given a cell. It was pointed out also that overcharge may lead to water loss and premature failure. Overcharge also leads to a strong oxidizing environment in the area of the positive electrode. The positive electrode collector is slowly oxidized and eventually electrical contact to the active materials is lost. The separator materials, usually polymeric organics, are also oxidized and suffer a loss of strength. The weakened materials then develop weak or open spots which lead to internal shorting of the cell.

Ion migration within a cell is necessary to carry the current and to maintain electroneutrality when current flows in a cell. This applies to the discharge, charge, and overcharge of a cell. Thus, the net effect of the electrode reactions and ion migration can be a shift in the location of water in a cell. Although ion migration is necessary in all battery systems only one example (the Ni-Cd system) is shown below. For continuity migration on charge and discharge as well as overcharge are shown.

It is assumed in the KOH electrolyte commonly used in the Ni-Cd system that the migrating species are $K^+(4H_2O)$ and OH^- and that the relative rate of migration is 1 $K^+(4H_2O)/3OH^-$. Therefore, the net change in the anolyte and catholyte can be shown for the passage of four equivalents of electricity.

Charge Reactions

Positive Electrode	Negative Electrode
$4Ni(OH)_2 + 4OH^- \rightarrow$	$2Cd(OH)_2 + 4e^- \rightarrow$
$4NiOOH + 4H_2O + 4e^-$	$2Cd + 4OH^-$

$4e^- \rightarrow$ in external circuit
$K^+(4H_2O) \rightarrow$ ion migration in solution
$\leftarrow 3OH^-$ ion migration in solution

The net changes in the cell can be seen below.

	Changes at the Positive (in equivalents)			Changes at the Negative (in equivalents)		
	Reaction	Migration	Net	Reaction	Migration	Net
H_2O	+4	−4	0	0	+4	+4
OH^-	−4	+3	−1	+4	−3	+1
K^+	0	−1	−1	0	+1	+1

The electrode reactions and changes during discharge of a Ni-Cd cell will be the reverse of those shown for charging.

It was shown previously that there is no net chemical change in a Ni-Cd cell operating on the oxygen cycle when being overcharged. The same approach used to show the change during charging can be used to show that, even though there is no net chemical change on overcharge, there is a continuing shift in the electrolyte. Note that the reaction for oxygen consumption is the net reaction.

Positive Electrode   Negative Electrode

$4OH^- \rightarrow 2H_2O + O_2 + 4e^-$ $2H_2O + O_2 + 4e^- \rightarrow 4OH^-$

$4e^- \rightarrow$ external circuit

$K^+(4H_2O) \rightarrow$ ion migration in solution

$\leftarrow 3OH^-$ ion migration in solution

	Changes at the Positive (in equivalents)			Changes at the Negative (in equivalents)		
	Reaction	Migration	Net	Reaction	Migration	Net
H_2O	+ 2	− 4	− 2	− 2	+ 4	+ 2
OH^-	− 4	+ 3	− 1	+ 4	− 3	+ 1
K^+	0	− 1	− 1	0	+ 1	+ 1

Though the example used was the nickel-cadmium cell, the same type of material balance can be carried out for all cell systems. When semi-permeable membranes are used as separators, considerable osmotic pressure can arise across the membrane due to cell reaction processes.

Gas generation can be controlled, in addition to cell balance, by the use of a constant voltage or a voltage-limited charger to keep the electrodes from reaching the potential for vigorous gas evolution. Because of the delicate balance between the cell voltage for full charge and gas evolution, precise voltage control is necessary, requiring a complex charging unit and adding substantially to the overall complexity of the battery unit. Most battery systems use this approach to control gassing in the cells.

CHARGE TIME

The proviso that a secondary or rechargeable battery must be restored to its original charged condition does not state how fast or at what rate the battery is recharged. The time for recharge is important as it controls the "turn-around" time or the time between useful discharges of the battery. A battery which can be charged and discharged three to four times per day has a broader scope of application than a battery that can be used only once a day.

The ability of an electrode to accept charge at a rapid rate is determined primarily by the exchange current and morphology of the electrode deposit on recharge. Larger exchange currents indicate more reversible reactions which can handle the stress of high charging currents with minimum polarization (but does not rule out dendritic formations, etc). For fast recharge it is necessary that the materials be continually available for reaction. In the case of solid reactants, availability generally means close physical position or contact with the electrode structure. Sufficient electrolyte must be contained within or near the pore structure of the electrode for the reaction and for conductivity. Since the reaction at high rates utilizes only the electrolyte in the pores, and since the electrolyte in the separator does not have time to diffuse into the electrode, the pore structure of the electrode becomes an important factor. If the electrolyte is directly involved in the reaction, it must be able to diffuse to the reaction site in the alloted time. Soluble reactants generally do not yield themselves to good rechargeability.

A large exchange current and reactant availability are not sufficient for rapid charging; the morphology of the deposit or charged material must also be considered. If the electrode can be recharged rapidly but the deposit is loosely adherent to the electrode structure (e.g., dendritic or mossy), much of the capacity may be unavailable due to lack of contact with the current collector. The electrode may not be reformed in its original structure at very high rates. The rapid recharge may also deposit material on the surface and block or plug the internal pore structure thus preventing the interior or the electrode from contributing fully to the cell capacity. If the electroactive materials do not reform in their original condition at high rates, a slow recharge may be necessary. A slow recharge favors more uniform current distribution, allows time for redistribution, recrystallization, diffusion, and so on, to occur and results in good cycle life.

Shephard's equation (45) (see the section on the shape of the discharge curve) can be applied to battery charging with reversal of the appropriate signs. In this equation the important factors for charging are the exchange current, reactant availability, internal resistance, and charging current, and it includes a term associated with the initial buildup of polarization. This equation points out two interesting phenomena: first, faster recharges are favored for shallow discharge cycles on partially charged batteries due to the effect of the reactant availability term; second, a method of charging whereby the battery is charged by a series of pulses of short duration may be advantageous. The exponential term results in lower cell polarization at short times. Thus, extremely high currents may be employed without excessive polarization as long as the pulse time is short.

There are three main types of battery chargers: constant current, constant potential, and voltage limited taper charge (VLTC) which limits the initial current flow. The constant current charge, as its name implies, passes a constant current through the battery and generally utilizes time as the cutoff, the amount of overcharge being included in the calculation for charge time. The constant current charge can be simple and inexpensive. The constant potential charge holds a constant voltage across the battery terminals. Extremely high currents can pass during the first moments of a constant potential charge and can lead to thermal runaway. As the battery voltage approaches the set charging voltage, the current tapers off until at full charge only a very small current is flowing. This method has the advantage that no cutoff due to time is needed. The constant potential charger is more complex than the constant current charger. The VLTC combines the good features of both constant current and constant potential chargers. During the initial portion of charge, the current is limited to prevent cell damage by the high currents of a constant potential charge. After a time, the battery voltage approaches the charger voltage and the current tapers off to a small fraction of its original value.

Since the amount of overcharge of a cell is important, it is necessary to match the type charger with the individual battery system and application. For example, during overcharge of nickel-cadmium cells the internal heating is proportional to the overcharge current or gassing rate and, thus, it is necessary to control the overcharge to avoid thermal runaway or the so-called "vicious cycle" (46). High-rate charging can generate a significant temperature rise inside the battery which could damage the cell and shorten the cycle life. Various charge control devices such as coulometers, temperature or pressure sensors, or auxiliary (3rd) electrodes have been studied (47)(48), but all of these add to the complexity and cost of the battery charger system.

It must be recognized that there is no one optimum charging method for any battery systems. Factors such as the time of charge and discharge, the use characteristic of the battery (e.g., depth of discharge), and power level of overall operation must be considered in selecting a charging method. In general, if a problem is anticipated in this area, the user and battery manufacturer should jointly select the best charging method. If a multicell battery is to be rapidly recharged, some consideration must be given to the uniformity of capacity of the cells in the battery. If all the cells are not identical, the cell with the lowest capacity can reach full charge and go into overcharge before the majority of the cells reach full charge. The cell on overcharge can heat up significantly. Should the cell malfunction (e.g., boil out the electrolyte, vent, or explode from internal pressure), the battery could develop an open circuit and become nonoperative. Also, if the bad cell develops an internal short from melting of the separator, electrode thermal expansion or dendrite formation, and the like, the energy contained in all the other cells could be catastrophically released into that cell. The same danger exists on deep discharge of a multicell battery, since it could also lead to cell reversal.

Rapid recharging, while desirable, places severe restrictions on cell quality and reliability. It is necessary to carefully design the complete system; electrodes, separator, vent, and charge cut-off mechanism to produce a satisfactory fast-charge battery.

CELL REVERSAL

Cell reversal occurs when a cell is completely discharged yet current continues to be forced through it in the discharging direction. The terminals change polarity, and if a voltmeter were connected to the cell its leads must be reversed. Thus the term "cell reversal" indicates a reversal in sign of the electrodes in a cell and also indicates a shift in the chemical reactions inside the cell to correspond to the polarity.

Most commonly, cell reversal occurs in a multicell battery when one cell has less capacity than the other cells. On deep discharge, the weak or low-capacity cell is completely discharged before any of the other cells. The others continue to supply current into the external circuit in a normal fashion. Since current is forced through the completely discharged cell, the battery reactions do not contribute, and electrolysis occurs, resulting in a reversal of the cell polarity. Reversal of a cell subtracts from the battery voltage. More importantly, in the reversed cell electrolysis usually generates gas internally that leads to excessive pressure and temperature buildup. Problems of a similar nature are discussed under the section on overchargeability.

The increased availability of the "do-it-yourself" battery chargers, with the associated risk of putting the batteries in the charger backward, introduces the possibility of cell reversal to single cells as well as cells in a series-connected, multicell battery. The effect of cell reversal depends on the cell balance; that is, the discharge capacity of the anode and cathode. Reversal, in general, is harmful to sealed cells in that the tendency is to evolve hydrogen gas at the cathode and oxygen at the anode. Thus, in the absence of a recombination device for hydrogen and oxygen, a pressure buildup will occur. Recombination devices for hydrogen gas, oxygen gas, or both, are known but have not achieved general use in commercial cells.

The lead-acid system is unique in that the discharge product at both electrodes is the same (lead sulfate), and that continued oxidation or reduction is theoretically possible without gas generation. However, reversal, or even deep discharge, is not recommended for the lead-acid systems.

The most commonly used reversal protection is the so-called "antipolar mass (APM)" in the nickel-cadmium system(16)(49) in which cadmium oxide or hydroxide is incorporated into the positive nickel hydroxide electrode. In addition, the cell balance is designed to be positive-limiting. When the positive is completely discharged, the cadmium hydroxide present is reduced to the metal prior to hydrogen evolution. The cadmium is available to react with the oxygen generated at the anode when the negative electrode is completely discharged and goes into gas evolution, i.e., the cell is reversed. Reversal protection is therefore the same type of oxygen cycle as is used for overcharge protection. The major problems are maintaining the desired cell balance throughout the life of the battery and the rate at which the cell can tolerate reversal. It is not practical to incorporate a sufficient quantity of cadmium hydroxide in the positive to make high-rate reversal possible.

Reversal protection can be obtained by placing a diode across the cell terminals(48). The diode would conduct when the cell voltage reaches zero or a small negative value. This technique can effectively bypass the cell reversal problem but adds significantly to the cost of the multicell battery.

It would seem to be better to avoid cell reversal by the design of the apparatus into which batteries are placed rather than attempt to incorporate reversal protection in the cells.

SHAPE OF THE DISCHARGE CURVE

The shape of the discharge curve is directly related to the ability of a battery to transform chemical energy into electrical energy. The work which a system can produce is given by

$$w = \int E(I)I \, dt \tag{31}$$

where the voltage (E) is a function of the current as is the product It.

The performance of single electrodes or, more specifically, the electrode polarization can be predicted with good accuracy if the kinetic parameters are known. However, in batteries the constructional effects make the predictions based on kinetic parameters difficult. Typical discharge curves are shown in Fig. 9.14. Each method of presentation has its advantages yet presents essentially the same information in a different fashion. Most of the discharge curves for rechargeable batteries are relatively flat at low currents. However, at high currents they tend to have a significant slope. Flat discharge curves are generally preferred because the energy output is maximized and the need for voltage regulation is minimized. The sharp voltage inflection at the end of discharge is also preferred. A sloping discharge curve may result in an appreciable fraction of the energy delivered at a voltage below that which is useful. Cells with sloping discharge curves usually have a serious inefficiency in the ability of a system to deliver energy. The stable voltages in discharge or charge are usually associated with reversible electrode reactions.

In many instances, the set of discharge curves shown in Fig. 14 may not illustrate a discharge at the current of greatest interest to a particular application. It may be necessary to estimate the performance or to replot the curves in different form. To eliminate this troublesome exercise and to report system characteristics in a minimum of space, several equations have been developed to predict battery performance in terms of a few constants.

The first attempt to correlate battery data was proposed by Peukert(50). The original equations were developed for lead-acid batteries, but the equation can be applied to any battery system. The equation is given by

$$I^n t = c \tag{32}$$

where I is the current, t is the time, and n and c are constants. The time is measured to the knee or inflection in the curve. If the curve is sloping, the intersection of tangents drawn to the curve can be used. The curve is usually plotted as $\log I$ versus $\log t$. The behavior of several systems is shown in Fig. 9.15. The sensitivity of capacity to the current is indicated by the value of n. When n is much larger than one, mass transfer and activation polarization make an important contribution to cell polarization, thereby decreasing cell performance.

The Peukert equation is empirical and has not been widely used. Often it does not adequately describe the performance over a wide range of discharge rates. However, the equation can be used to characterize performance. For instance, in Fig. 9.15, the characteristics of two Ni-Cd cells with the same nominal capacity are shown. The value of n is 1.3 and 1.07 for the low-rate and high-rate cells, respectively. The electrodes in the low-rate cell are relatively thick and have lower surface area as compared to the electrodes in the high-rate cell. It would be expected that the thicker low surface area electrodes would be more rate-sensitive. The thicker electrodes, especially at higher current drains, should have greater mass transfer and activation polarization. These problems are characterized by a significant deviation from the Peukert equation for the low-rate cell at high current. There is some question whether the equation adequately describes the performance of this cell. Yet, the equation can be useful in characterizing cells and related cell performance.

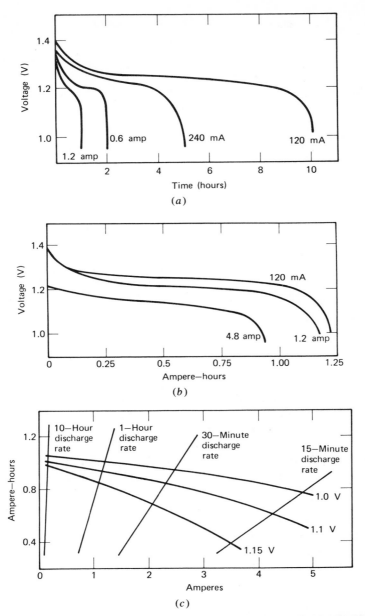

Figure 9.14(*a*) Discharge curves of a 1.2 amp-hr nickel-cadmium cell(23). (*b*) Discharge curves of a 1.2 amp-hr nickel-cadmium cell(23). (*c*) Capacity of a 1.2 amp-hr nickel-cadmium cell as a function of the discharge rate(23).

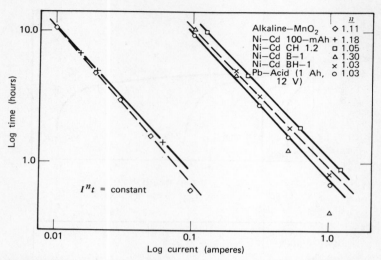

Figure 9.15 Comparison of observed capacity at various discharge rates.

Another means of representing discharge and charge curves has been given by Shephard(45). The equation has the form

$$E = E_s - K \left(\frac{Q}{Q - it} \right) i - Ni + A e^{-\frac{Bit}{Q}} \tag{33}$$

where A and B are constants, E_s is a constant potential, Q is the capacity of the cell, N is the internal resistance, and K is a polarization parameter. The constant K is called the polarization coefficient and is related to the exchange current and has the dimensions ohms/cm². The term $[Q/(Q - it)]i$ is related to the current density on the active material and is inversely proportional to the unused active material. The last term, an exponential, is used to describe the initial portion of the curve. After a short period of discharge time, this term can be neglected. This equation satisfactorily fits the discharge of common battery systems.

Equation (33) can be made applicable to the charge reaction by changing the sign on the last three terms. The values for the constants are not the same for charge and discharge. Although largely empirical in practice, the equation does identify the important factors which determine the nature of the curves; that is, flat or sloping. These factors are the internal resistance N, the polarization parameter (exchange current) K, and the surface area of the unused active material, $[Q/(Q - it)]$. Sloping curves are found whenever the values of these constants are large or at extremely high discharge rates. Conversely, flat discharge curves indicate low internal resistance, large exchange currents, and large surface area electrodes.

Other representations have been given by Selem and Bro(51) and by Selas and Russell(52). By using these equations one is able to predict battery performance over a wide range of combinations with a minimum of experimental data. It also permits reporting experimental data in a very condensed form.

TEMPERATURE RANGE OF OPERATION

Most commercial cells are used at or near room temperature as they are contained in devices in an ambient condition where people use the devices. However, there are instances where batteries are used below freezing or exposed to the direct sunlight in the summertime. Therefore, cell operation at very low or very high temperatures can be important. Some systems are limited by the temperature characteristics imposed by the electrolyte, while others are limited by the polarization of the electrode reactions.

The electrolyte used in the system is the primary limiting factor. Cell operation is restricted to the range between the boiling point and freezing point of the electrolyte. In general, these limits are about 100°C and −40 to −50°C, respectively, for water-based electrolytes. Variations from these limits depend on the type and concentration of the solute and the stability of the electroactive components.

A limitation imposed by the electrolyte is associated with the electrolyte conductivity. The temperature coefficient of resistivity is about 2 percent per degree. The electrolyte resistivity will not affect the high temperature performance but can significantly reduce the low temperature performance. At −10°C, the resistivity of aqueous KOH is about twice that at +30°C. The resistivity increases dramatically at the freezing point. Freezing usually causes the cell to cease functioning. The increased electrolyte resistance also indicates an increase in the viscosity of the electrolyte and a decrease in the diffusion rate of the ions. The slower diffusion rates will increase concentration polarization and may possibly lead to electrode passivation, especially at high rate discharges. Increased internal resistance also raises the cell voltage on charge. If a constant voltage charge is used, allowance must be made for the changing cell voltage with temperature.

The temperature effect on electrode polarization is reflected by the exchange current for the electrode. The exchange current decreases with decreasing temperature; thus, during charge and discharge, the activation polarization will be larger at the lower temperatures. The temperature effect is not necessarily the same for both electrodes in a cell. The poor low temperature performance of the silver oxide-zinc battery is generally associated with silver electrodes and not to increased IR. The increased polarization not only decreases the operating voltage but also increases the charging voltage. If a constant voltage charger is used at low temperatures, the cell may not be fully recharged unless allowance is made for polarization.

The increase in charging voltage is also noted in the theoretical cell EMF which generally increases as the temperature rises. Most reactions involving gas evolution have large temperature coefficients in the direction that would indicate that charging of negative electrodes should be more efficient at lower temperatures. This is not always the case, and the cadmium electrode is an especially good example. At low temperatures the stable crystal structure of the cadmium hydroxide changes, nucleation is hindered, and it becomes easier to form hydrogen than cadmium metal. This effect is temporary, and the cadmium electrode will recover after sufficient time at room temperature. At low temperature, the use of a constant potential charge must allow for increased IR, electrode polarization, and theoretical cell EMF.

The effects at high temperature are in contrast to the low temperature effects. At high temperature, material stability is the prime factor affecting cell operation. Stability is important for both the negative and the positive electrodes. Generally, electrolyte resistivity and diffusivity are not problems, although at extremely high-rate discharge the boiling point of the electrolyte may be a problem. If the rate is high enough, the cell may be considered adiabatic in the first approximation and all the released heat is contained in the cell. At high discharge rates, electrode polarization can release enough heat to boil the electrolyte. On boiling, a nonuniform electrolyte path usually results in increased heating. Soon the cell is destroyed from lack of electrolyte or from melting of plastic parts.

In general, increasing temperature increases chemical reaction rates and decreases electrode polarization. It also leads to increased tendency for corrosion and possibly a rearrangement of the morphology of the electrode structure. At the positive, increased temperature leads to an increased self-discharge rate and poor charge acceptance. The poor charge acceptance, or self-discharge, of the positive is understood by examining Fig. 9.11. The EMF of the electrode is only slightly affected by temperature, but the oxygen electrode overvoltage is lowered considerably. Therefore, at a given potential more current goes to oxygen evolution and very little to charging. On stand, the oxygen is liberated spontaneously by reaction of the cathode active material with water of the electrolyte. Of the common cathode materials, MnO_2 and Ag_2O are the only materials which do not spontaneously liberate oxygen from the electrolyte. The nickel positive electrode is perhaps the most sensitive to increased self-discharge with temperature. At 45°C, the rate of self-discharge is about seven times the rate at room temperature.

Temperature also affects the materials of construction. The thermal stability of plastics may limit the temperature range. At higher temperatures the plastics can soften and flow. This can lead to degradation and shorting through separator materials as well as to failure of the plastic cases or components of compression seals. At low temperatures, the plastic can become brittle and shatter under slight impact.

COST

The cost of each battery system is determined by the cost of the raw materials and the labor to assemble the materials into a commercial entity. Each of these costs may be dependent on the play of the marketplace. If there is a shortage of a particular material, for example, nickel, the cost of the raw material can double in the period of a year or so. Labor costs are variable, depending on the degree of automation in the manufacture of the product. In automated mass production, labor contributes less, percentagewise, to the cost of the battery, than for batteries which involve many hand operations. The lead-acid battery is a case in point. By automating the production lines to reduce labor, a company can compete with a smaller local company with low overhead. The smaller local company, unable to build mass-production assembly plants, must rely on more hand operations with higher labor costs. The cost of shipping is a significant portion of the total battery cost. The low shipping costs of the local company offset the manufacturing advantage of the larger company. However, because of the play of the marketplace, both engage in selling pressure which lowers the cost to the consumer.

The cost comparison for some of the common battery systems is listed in Table 9.1. The approximate price used in the calculation is also given in Table 9.1. It was assumed that the current efficiency was 100 percent in making the calculations. No allowance was made for cell balance.

The cost of the basic raw materials when coupled with labor and marketing does not always reveal the cost of the rechargeable systems. In a rechargeable battery the best way to judge is the cost in the end-use. Therefore, expected average cycle life of the system should be included in cost considerations so that the cost-per-cycle or cost-per-useful-ampere-hour can be calculated. Of course, if the battery will not be used to the limit of its cycle life, then other considerations may be important; for example, leakage, reliability, and service maintenance.

A cost in rechargeable batteries that is often overlooked is the cost of the charger. It is very possible that the charger will exceed the battery

TABLE 9.1. COST COMPARISON-SEALED SYSTEMS

System	Material Cost $/amp-hr[a]	Charger Type	Cycle Life	Approx. Battery Cost($/amp-hr)	Charger Cost[b] ($/amp-hr)[b]	Cost per Cycle($)
Nickel-cadmium	0.033	Constant current	1000	1.50	0.10	0.0015
Lead-acid	0.0027	Voltage limited	200	0.40	0.30	0.0045
Alkaline-MnO$_2$	0.0018	Voltage limited	75	0.40	0.30	0.0093
Silver-zinc	0.12	Voltage limited	25	3.00	0.30	0.13
Silver-cadmium	0.14	Voltage limited	1000	3.00	0.30	0.0033

[a]Assumed cost per pound: Cd, $4.50; Ni(OH)$_2$, $2.00; Pb, $0.135; PbO$_2$, $0.155; Zn, $0.155 MnO$_2$, $0.20, and AgO, $23.80.
[b]Based on 10 amp-hr battery with a charger cost of $1.00 for constant current and $3.00 for voltage-limited, tapered-charge charger.

cost if very sophisticated charging techniques are used. The overall manufacturing costs of batteries and chargers are difficult to judge since much depends on the production volume. Sometimes the charger is built into the appliance. Larger transformers for high-rate charging increase nonlinearly in price and size as the power requirement increases. For example, a small constant current 10-hour rate charger may cost $0.50 to $1.00. A constant current $\frac{1}{2}$-hour rate charger may cost three to five times as much, especially if a pressure or temperature-sensing device is included.

SUMMARY

The various properties of small sealed or maintenance-free batteries are summarized in Table 9.2 and are discussed in the appendix. The properties of more advanced systems in various stages of development are summarized in Table 9.3. These systems may or may not become important battery systems in the future. In Table 9.4, some promising battery systems still in the research stage of their development are listed. Included in this list are several high temperature systems which may not be considered as small sealed units. However, they represent the highest energy systems available for use. Not included in these tables are mechanical, thermal, or photoregenerative systems. Each of the systems is capable of efficient energy storage. Also not included in the tables is the Leclanché battery or common carbon-zinc flashlight cell.

TABLE 9.2a. PROPERTIES OF SMALL SEALED OR MAINTENANCE-FREE BATTERIES (18)(23)(40)(41)(53)(54)

Variable	Ni-Cd	Pb-Acid	Alkaline-MnO$_2$	Silver-Zn	Ag-Cd
Whr/lb(theor.)	109	80	153	230	150
Whr/lb(actual)	12–15	12–15	6–10	20–55	24
Whr/in³	1.5	1.1	0.6	1.1–3.5	2.8
Cycle life (a) Shallow	3000–5000 (sinter) 500 (press. powder)	300	50–100	25–50	3–5000
Cycle life (b) Deep discharge	300–2000 (sinter) 100–250 (press. powder)	100–300	50	25	700
Ability to over-charge	Good	Fair	Poor	Very poor	Poor
Shape of discharge curve	Flat	Flat	Sloping	Two-level	Two-level
Oper. temp. range	−20° to +40°C	−20° to +50°C	−20° to +70°C	0° to +70°C	−40° to +70°C
Efficiency (Whr) percent	60–65	75	60	70–75	70
Nominal rechg. time	1 hr (sinter) 10 hr (press. powder)	10 hr	14 hr	5–10 hr	1 hr
% Self-discharge at 6 months	60% (sinter) 30% (press. powder)	35%	4%	10%	10%
Charger type	Constant current	Voltage limited	Voltage limited	Voltage limited	Voltage limited
Charge time	1 hr	14 hr	14 hr	5–10 hr	1 hr/5–10 hr
High-rate discharge	Good	Medium	Poor	Excellent	Good
W/lb(max.)	500	250	20–40	600	500

TABLE 9.2b. PROPERTIES OF SMALL SEALED OR MAINTENANCE-FREE BATTERIES

System	Advantages	Disadvantages	Applications
Nickel-cadmium	Ruggedness Overchargeability Long discharge Good cycle life Flat curve Simple charger Good pulse capability	Higher initial cost Poor high temperature performance Memory effect	Appliances Portable electronics Motor starting
Lead-acid	Inherently inexpensive Flat discharge curve	Complex charger Poor low temperature performance	Emergency lighting Appliances Portable electronics Motor starting
Alkaline-MnO$_2$	Inexpensive Excellent shelf life	Poor cycle life Complex charger Sloping discharge curve	Portable electronics
Silver-zinc	High energy density Excellent pulse capability	Poor cycle life No overchargeability Complex charger H$_2$ evolved on charge and stand Two-step discharge	Electronics Space Military
Silver-cadmium	Nonmagnetic Good cycle life Good energy density Good shelf life Good high temperature capability	Complex charger Low cell voltage Two-step discharge Very limited overchargeability	Space Electronics

TABLE 9.3. DEVELOPMENTAL SYSTEMS(18)(54)(55)(56)

Systems	Theoretical (Whr/lb)	Estimated (Whr/lb)	Estimated Cycle Life	Electrolyte	Problem Areas	Comments
Nickel-zinc[a]	150	25–35	300	Aqueous KOH	Hydrogen evolution Cycle life of zinc	Lower cost and higher energy density than Ni-Cd
Nickel-iron[a]	120	20	500	Aqueous KOH	Hydrogen gassing on charge Poor low temperature performance	Needs considerable overcharge Lower cost than Ni-Cd
Mercury-cadmium	67	5–10	300–500	Aqueous KOH	Low voltage cell	
PbO$_2$-Cu	88	5–12	300	Aqueous H$_2$SO$_4$	Copper corrosion	
Zinc-oxygen	540	20–50	300	Aqueous KOH	Oxygen electrode recharging Cycle life of zinc Recharging under pressure in sealed system	Can use air as O$_2$ source
Hydrogen-oxygen	1620	20–50	200	Aqueous KOH	Recharging damages gas electrodes Pressure regulation	High cost system

[a]Included as sealed batteries; vented cells have been in commercial production for many years.

TABLE 9.4. RESEARCH SYSTEMS(56)(57)(58)(59)(60)

System	Open Circuit Voltage	Theoretical (Whr/lb)	Operating Temperature	Electrolyte	Comments	Applications
Lithium-chlorine	3.5	1270	650°C	Molten LiCl	Poor material stability High power output	Electric autos Stand-by power
Lithium-sulfur	2.25	700	300°C & room temperature	β-Al$_2$O$_3$ and nonaqueous	Energy density depends on discharge products	Electric autos
Lithium-silver chloride	2.8	230	Room temp.	Propylene carbonate-LiAlCl$_4$	SO$_2$ to protect lithium	Electronics Space
Lithium-copper chloride	3.1	500	Room temp.	LiAlCl$_4$	Poor lithium rechargeability	Electronics
Sodium-sulfur	2.1	680	350°C	β-Al$_2$O$_3$ separator	Energy density depends on discharge product High power output	Electric autos
Lithium-cadmium chloride	2.2	290	Room temp.	Cyclic lactone LiClO$_4$ LiAlCl$_4$		Electronics Space

426

REFERENCES

1. Gautherot, N., *Phil. Mag.*, **24**, 183 (1806).
2. Ritter, J. W., *Phil. Mag.*, **23**, 51 (1805).
3. Faraday, M., *Exp. Researches in Electricity*, Ser. VII, No. 750 (1834).
4. Muncke, G. W., *Bibliothèque Universelle*, **1**, 160 (1836).
5. de la Rive, A. A., *Bibliothèque Universelle*, **1**, 162 (1836).
6. Grove, W. R., *Phil. Mag.* III **14**, 388 (1839); Ann. Chim. Phys., **58**, 202 (1843).
7. Planté, G., *Recherches sur L'Électricite*, 2nd ed., La Lumiere Electr., Paris (1883); *Compt. Rend.*, **49**, 402, (1859); **50**, 640 (1860).
8. Niaudet, A., *Pile Electrique*, Paris, 2nd ed., p. 240 (1880).
9. Faure, C. A., *Electrician*, **6**, 122 (1881), Compt. Rend., **73**, 890 (1871).
10. Tommasi, D., *Traité des Piles Électriques*, G. Carré, Paris (1889).
11. Edison, T. A., U.S. Pat. 692,507 (1902).
12. Jungner, W., Swed. Pat. 15,567 (1901); Ger. Pat. 163,170 (1901).
13. Drumm, J. J., U.S. Pats. 1,955,115 (1934); 2,227,636 (1942), and Brit. Pat. 365,125 (1930).
14. Edison, T. A., U.S. Pat. 1,016,874 (1912).
15. Neumann, G. and Gottesmann, U., U.S. Pat. 2,571,927 (1951).
16. Dassler, A., U.S. Pat. 2,934,581 (1960).
17. Rublee, G., U.S. Pat. 2,269,040 (1939).
18. Falk, S. U. and Salkind, A. J., *Alkaline Storage Batteries*, John Wiley, New York (1969).
19. Milner, P. C. and Thomas, U. B., in *Advances in Electrochemistry and Electrochemical Engineering*, Vol. 5, Ed. C. W. Tobias, Interscience, New York, p. 1 (1967).
20. Okinaka, Y. and Whitehue, C. M., *J. Electrochem. Soc.*, **118**, 583 (1971).
21. Bode, H., Dehmelt, K., and Witte, J., *Electrochim. Acta*, **11**, 1079 (1966).
22. Kronenberg, M. L., *J. Electroanalyt. Chem.*, **13**, 120 (1967); Brodd, R. J., unpublished data.
23. *Eveready® Battery Applications and Engineering Data*, Union Carbide Corporation, Consumer Products Division, New York (1968).
24. Globen, M. and Mueller, G. A., U.S. Pat. 3,320,139 (1967).
25. Kandler, Brit. Pat. 917291 (1963); Mueller, G. A., U.S. Pat. 3,203,879 (1965); McHenry, E. J., *Electrochem. Tech.*, **5**, 275 (1967); Beauchamp, R., U.S. Pat. 3,573,101 (1971).
26. Elberto, K., in *Power Sources 2*, Ed., D. H. Collins, Pergamon, London, p. 69 (1968); Thomas, J. R. and Walter, D. R., *Proc. 21st Ann. Power Sources Conf.*, p. 64 (1967); Orsino, J. A. and Jenson, H. E., ibid, p. 60; Malloy, J. P., ibid, p. 68.
27. Andre, H., U.S. Pat. 2,317,711 (1934).
28. Fleischer, A. and Lander, J. J., Eds., *Zinc-Silver Oxide Batteries*, John Wiley, New York (1971).
29. Dirkse, T. P., *Proc. 24th Ann. Power Sources Conf.*, p. 14 (1970).
30. Bowditch, F. T., U.S. Pat. 1,846,246 (1932); Kobe, K. A., and Graham, R. P., *Trans. Electrochem. Soc.*, **73**, 587 (1938); Skinner, W. V., U.S. Pat. 2,199,322 (1940); "How to Charge Flashlight Cells," *Steel*, **113**, 96 (Sept. 1943); "A Method of Recharging Dry Cells for Flashlights," *Gas Age*, **92**, 15 (1943); "Electrical Characteristics of Dry Cells and Batteries (Leclanché Type)," Letter Circular LC 965 (Nov. 15, 1949); Adams, P. H., *Trans. IRE Component Parts*, **5**, 76 (1958); Herbert, W. S., *Electrochem. Tech.*, **1**, 148 (1963); Yeaple, F., *Prod. Eng.*, p. 100–106 (June 7, 1965); ibid. p. 66–70 (Sept. 27, 1965); Lindsley, E. F., *Popular Science*, **198**, 46 (Jan. 1971).
31. Flerov, V. N., *Zh. Prikl. Khim.*, **23**, 1306 (1959); Farr, J. P. G. and Hampson, N. A., *J. Electroanalyt. Chem.*, **13**, 433 (1967).

32. Yeager, E. B. and Schwartz, E. B., in *The Primary Battery*, Eds., G. W. Heise and N. C. Cahoon, Vol. I, Chap. 2, John Wiley, New York (1971).
33. Schumacher, E. A. in ibid, Chap. 4.
34. Foerster, F., *Elektrochemie Waesseriger Loesungen*, 2nd ed., Barth, Leipzig, p. 242 (1922).
35. *American Standard Definitions of Electrical Terms*, American Institute of Electrical Engineers, New York, p. 181 (1941).
36. Casey, E. J., Henderson, H. S. and King, T. E., *Proc. 16th Ann. Power Sources Conf.*, p. 108 (1962).
37. Brodd, R. J. and DeWane, H. J., *Electrochem. Tech.*, **3**, 12 (1965); Falk, S. A., *J. Electrochem. Soc.*, **107**, 661 (1960).
38. *Chemical Engineers Handbook*, 4th ed., Ed. John H. Perry, McGraw Hill, New York, p. 10–13 (1963).
39. Gross, S., *Energy Conversion*, **9**, 55 (1969).
40. Seibert, G., in "Space Power Systems: Energy Storage," European Space Research Organization, ESRO-SP-49, p. 23 (Sept. 1969).
41. Lifshin, E. and Weininger, J. L., *Electrochem. Tech.*, **5**, 5 (1967).
42. Euler, J. and Horn, L., *Electrotech. Z.*, **A81**, 566 (1960).
43. Carson, W. N., "A Study of Nickel-Cadmium Spacecraft Battery Charge Control Methods, NASA Contract NAS 5-9193, Final Report, April 1966.
44. Lifshin, E. and Weininger, J. L., *Electrochem. Tech.*, **5**, 5 (1967).
45. Shephard, C. M., *J. Electrochem. Soc.*, **112**, 657 (1965).
46. Eicke Jr., W. G., *J. Electrochem. Soc.*, **109**, 364 (1962); Salkind, A. J., and Duddy, J. C., ibid, 360.
47. Carson, W. N., *Proc. 20th Ann. Power Sources Conf.*, p. 103 (1966).
48. Amsterdam, R. E., *Proc. 20th Ann. Power Sources Conf.*, p. 105 (1966).
49. Krämer, G., *Electrochim. Acta*, **13**, 2237 (1968).
50. Peukert, W., *Elektrotech. Z.*, **18**, 287 (1897).
51. Selem, R. and Bro, P., *J. Electrochem. Soc.*, **118**, 829 (1971).
52. Selas, S. M. and Russell, C. R., *Electrochem. Tech.*, **1**, 77 (1963).
53. Le Couffee, Y., in *Performance Forecast of Selected Static Energy Conversion Devices*, 29th Meeting of AGARD Propulsion and Energetics Panel, Liege, Belgium (1967).
54. Bauer, P., "Batteries for Space Power Systems," NASA Report, NASA SP-172 (1968).
55. Kober, F. and Charkey, A. in *Power Sources B1*, Ed., D. H. Collins, Oriel Press, Newcastle Upon Tyne, p. 309 (1971).
56. Euler, J., in "Space Power Systems: Energy Storage," European Space Research Organization, ESRO-SP-49, p. 1 (Sept. 1969).
57. Gabano, J. P., Gerbier, G., and Laurent, J. F., *Proc. 23rd Ann. Power Sources Conf.*, p. 80 (1969).
58. Chilton, J. E. and Holsinger, R. W., *J. Electrochem. Soc.*, **111**, 183C (1964).
59. Weber, N. and Kumma, J. T., *Proc. 21st Ann. Power Sources Conf.*, p. 37 (1967).
60. Swinkels, D. A. J., in *Advances in Molten Salt Chemistry*, Vol. I, Eds. J. Braunstein, G., Mamantov, and G. P. Smith, Plenum Press, New York (1971).
61. Cahoon, N. C. and Holland, H. W., in "The Primary Battery", Eds. G. W. Heise and N. C. Cahoon, Vol. I, Chap. 7, John Wiley, New York (1971).

10.

Internal Resistance of Primary Batteries

WALTER J. HAMER

GENERAL CONSIDERATIONS

The measurement and interpretation of internal resistance of primary batteries have attracted the attention of manufacturers and research workers for some time. The manufacturers are interested in the ultimate power derivable from primary batteries and are cognizant of the facts that internal resistance reduces the useful voltage of a cell and leads to internal heat which is lost, a loss that increases with the square of the current. They are also aware that internal resistance indicates the magnitude of the resistance that an external load may have in order to receive maximum power. The research worker, although also aware of the practical aspects, is primarily interested in the measurement of internal resistance as it relates to the battery processes, especially if the resistive component can be separated from the nonresistive part.

The objective of all workers is to obtain an unambiguous value for the internal resistance of a cell. From an electrical point of view, a cell is a combination of an energy source and a resistance. Although it is not difficult to determine the electromotive force (EMF), as with a potentiometer, direct-current methods of measuring resistance, for example, standard direct-current bridge circuits, are generally not applicable because the EMF of the cell directly influences the galvanometer and provides incorrect readings. Thus, it has been difficult until relatively recently to clearly separate the basic factors present in a cell. Furthermore, the internal resistance of any cell is an integral part of the electrical circuit in which the cell is used. As early as 1868, Leclanché(1) recognized the importance of cell resistance since it affected the use of the system in telegraph applications. He reported that an intermediate sized cell, in which the cathode was placed in a porous vessel 15 cm high and 6 cm in diameter, had a resistance equal to that of 550–600 m of iron telegraph wire 4 mm in diameter (approximately 4 ohms).

Measurements of internal resistance are generally approximate in that they include, to various degrees, components other than purely resistive ones. Even so, such measurements are of value for individual classes of cells in giving some indication of the efficiency of cell performance when cells are under various loads. As a first approximation, neglecting electrode polarization, the working voltage of a cell which is discharging is given by

$$V = E - IR \qquad (1)$$

where V = closed-circuit voltage (CCV), measured as near to the instant of closing the circuit as possible, E = open-circuit voltage (OCV), I = current in the circuit including the cell, and R = internal resistance. When E_p, the EMF of polarization is considered, then $V = E - IR - E_p$. (When used subsequently in this chapter these symbols retain the meaning given here.) Thus, it is obvious the higher the internal resistance the lower is the working voltage and the operating efficiency, $V/(V + IR)$, or more generally $[(E - IR)/E]100$ since percent efficiency equals $100 V/E$.

MEASUREMENT METHODS

All of the methods described in this chapter, in one way or another, represent applications of Ohm's law, wherein the change in cell voltage when a known current of electricity is passed through the system, is used as a basis of the calculation of internal resistance.

Various direct-current (dc) and alternating-current (ac) methods have been proposed for the measurement of internal resistance of primary cells. The dc methods may be classed as drop-in-potential, capacitor, bridge, polarization-elimination, and interrupter and pulse methods, although in essence all methods are subclasses of the drop-in-potential method. The ac methods may be classed as the bridge and transient-pulse methods with these also based on a drop in potential. Direct-current interrupter and pulse and ac methods are considered better than other methods in that measurements *are thought to be* made within a time insufficient for electrode polarization to be a factor (see later discussion).

In anticipation of what will be discussed later, it is well to state here that in a number of the methods that follow it is assumed that the EMF of the cell remains constant, that is, there is no back or counter EMF nor chemical changes at the electrodes when electricity flows through a cell or when this flow is altered, that is, there is no electrode polarization. Based on modern understanding of the reaction mechanisms involved, it is now clear that these assumptions are not valid. Nevertheless, a survey of some of the early methods wherein the EMF is assumed to remain constant is pertinent in that these methods approximate conditions encountered in service and tend to give an integrated value of internal resistance and polarization, or the total drop $(IR + E_p)$ in voltage and the overall operating efficiency. The methods chosen for presentation cover fundamental principles and *many are identical in principle and differ only in instrumentation, and modifications in their operational details are, therefore, possible.*

Many of the dc methods which follow are included primarily for historical background; they should not be considered conclusive. In the section of this chapter dealing with the elimination (total or partial) of electrode polarization, as well as the section entitled "Discussion," it will be shown that many dc methods include some electrode polarization in the values obtained. Nevertheless some of the simple dc methods are widely used in practice to provide first approximations of internal resistance levels for many classes of cells.

The methods may be classified as follows:

1. Direct-current methods
   (a) Drop-in-potential
       (1) Voltmeter
       (2) Flash-load
       (3) Volt-ammeter or induced-current
       (4) Flash-current
       (5) Two-step-current
       (6) Half-deflection
       (7) Two-step-resistance
   (b) Capacitor
   (c) Bridge circuits
   (d) Polarization-elimination
       (1) Extrapolation
       (2) Opposition
       (3) Polarity-reversal
       (4) Reference-electrode
       (5) Interrupter and pulse
       (6) Alternating-current (listed below)

2. Alternating-current methods
   (a) Bridge circuits
       (1) Kohlrausch ac-bridge
       (2) Impressed-current
       (3) Substitution
           (a) Wien bridge
           (b) Other bridges including Wayne-Kerr bridge
   (b) Transient-pulse

Direct-current methods

Drop-in-Potential Methods

(1) The simplest dc method of measuring internal resistance is the *voltmeter method* where the OCV and CCV are measured by a voltmeter of the type described in Chapter 8, arranged as shown in Fig. 10.1 with the ammeter, *A*, absent. The internal resistance is given by

$$R = \frac{E - V}{V} \qquad R_e = \Delta V \frac{R_e}{V} \tag{2}$$

where R_e = external or load resistance and ΔV = drop in potential.[1]

(2) By using a load (*flash load*) so that $V = \frac{1}{2}E$, the internal resistance is equal to the known external resistance. This method is sometimes used in the testing of high-potential batteries of high internal resistance, see Chapter 8.

(3) In the *voltmeter-ammeter method* an ammeter is placed in series (Fig. 10.1 with ammeter, A, now present) with R_e permitting a measurement of the current, *I*. Then

$$R = \frac{E - V}{I} = \frac{\Delta V}{I} \tag{3}$$

(It should be noted that R_e/V in equation (2) may be taken as a measure of $1/I$.) The current, *I*, in equation (3) may also be a current passed through the cell from an external source (3) and this method could, therefore, also be called the *induced-current method*.

(4) In a similar method, the *flash current*, I_f, is used; see Chapter 8 for the method of measuring flash current. Then V/I in equation (3) is

[1]This method is used in measuring the internal resistance of standard cells (2) except that a potentiometer is used in the measurement. A 10 megohm resistor is used for R_e and since ΔV is small the current is considered to be given by $I = E/R_e$, hence $R = (E - V)R_e/E$.

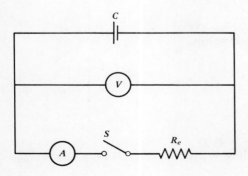

Figure 10.1 The voltmeter or voltmeter-ammeter method. *A*, ammeter; *C*, cell under test; R_e, load resistance; *S*, switch; *V*, voltmeter; *Note.* Ammeter *A* is not included in the voltmeter method.

.01 ohm, the resistance of the ammeter and leads, and, as a result,

$$R = \frac{E}{I_f} - 0.01 \tag{4}$$

The current here is approximately twice that of the flash-load method.

(5) Changes in CCV resulting from changes in I may be used; then, using the arrangement of instruments shown in Fig. 10.2,

$$R = \frac{V_1 - V_2}{I_2 - I_1} = \frac{\Delta V}{\Delta I} \tag{5}$$

where the subscripts 1 and 2 refer to two different but corresponding measurements of V and I. If R_1 were the first resistance and R_2 a second and equal resistance, used as a shunt to R_1, then

$$R = \frac{V_1 - V_2}{2V_2 - V_1} R_e \tag{6}$$

where R_e (the load resistance) $= R_1 = R_2$. Equation (5) may also be written in the form

$$R = \frac{I_1 R_1 - I_2 R_2}{I_2 - I_1} \tag{7}$$

thus, the designation of this method as the *two-step-current method.*

(6) In the *half-deflection method* R_2 is chosen so that $I_2 = \frac{1}{2}I_1$, as indicated by the ammeter; then

$$R = R_2 - 2R_1 \tag{8}$$

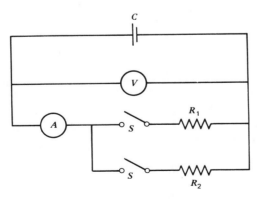

Figure 10.2   The two-step current method. A, ammeter; C, cell under test; R_1, R_2, first and second load resistances; S, switch; V, voltmeter.

(For refinement the resistance of the ammeter is subtracted from the right side of the above equation; this resistance is usually negligible, however. If a galvanometer were used instead of an ammeter, deflections (should be small), d, in the galvanometer scale may be used for I, thus

$$R = \frac{d_1 R_1 - d_2 R_2}{d_2 - d_1} \qquad (9$$

and for the half-deflection method $d_2 = \frac{1}{2} d_1$.

(7) Other modifications of the general dc drop-in-potential method namely, the *two-step-resistance methods* of measuring internal resistance have been proposed by Beetz(4), Fahie(5), and Thomson(6). In the Beetz method a cell C of EMF, E_x, is placed in a closed circuit as shown in Fig. 10.3 and the IR drop in a portion (slidewire or other device) of the circuit measured relative to a known EMF, E_{RC}, of a reference cell RC using a galvanometer as null detector. E_{RC} must be lower in value than E_x. At balance the IR drop across the wire is equal to E_{RC} which is also given by $E_x r_1/(r_1 + R_1 + R)$ where $r_1 =$ resistance of the portion of the slidewire across which IR is measured, $R_1 =$ resistance of the rest of the circuit and R, as above, is the internal resistance of the cell. The resistance R_1 is then increased by resistance AR to R_2 and a new balance obtained for which the IR drop is now given by $E_x r_2/(r_2 + R_2 + R)$. Upon elimination of E_x between these two expressions, the internal resistance is given by

$$R = \frac{r_2 R_1 - r_1 R_2}{r_1 - r_2} \qquad (10$$

in terms of two values of the slidewire and the rest of the circuit, hence the designation as the *two-step-resistance method*. Fahie proposed the same method but modified it by setting $R_2 = 0$ in the second balancing so that the internal resistance is given by

$$R = \frac{r_2 R_1}{r_1 - r_2} \qquad (11$$

In the Thomson method the cell is placed in a circuit with two resistances R_1 and R_2 in parallel. An ammeter is included in the branch with R_2. Since, for a given current, a definite relation exists between R and R_2, if R_1 is removed, R_2 must be changed, say to R_2', to restore the current to its original value. From these measurements, R is given by

$$R = \frac{R_1(R_2' - R_2)}{R_2 + a} \qquad (12$$

Here a, the resistance of the ammeter, is usually negligible. In the original work, Thomson used a galvanometer instead of an ammeter.

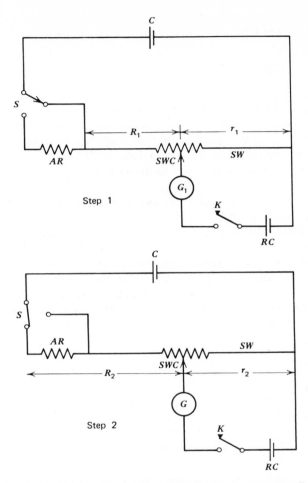

Figure 10.3 The two-step resistance method. AR, added resistance; C, cell under test; G, galvanometer; K, key; S, switch; SW, slide-wire; r_1, r_2, R_1, R_2, are described in text; SWC, slidewire movable contact; RC, reference cell.

CAPACITOR METHOD

In the capacitor method, a somewhat more refined method than the drop-in-potential methods, a precision capacitor of about 1 μF is first charged by a standard cell and then discharged through a ballistic galvanometer with the deflection being d_s. The process is then repeated first with the cell under test and then with the cell when shunted by a precision resistor, R_e, the deflections being d_c^o and d_c^c, respectively. The OCV and CCV are then given, respectively, by $E = (d_c^o/d_s)E_s$ and $V = (d_c^c/d_s)E_s$ wher E_s = EMF of standard cell. The internal resistance of the cell under stud is then given by

$$R = \frac{E - V}{V} R_e = \frac{d_c^o - d_c^c}{d_c^c} R_e \tag{13}$$

R_e is generally kept across the cell terminals for one minute. A modifie procedure would be to make repeat measurements with R_e across the cel terminals for increasing periods of time and then to extrapolate to zer time; a better estimate of R would follow. It can be shown that th internal resistance would be equal to the load resistance, R_e, if the shunt i adjusted so that $d_c^c = \frac{1}{2}d_c^o$.

Kemp(7) who originated this method omitted the step with the standar cell; he measured E independently with a voltmeter. Muirhead(8) an Munro(9) obtained $(d_c^o - d_c^c)$ directly by not discharging the capacito between measurements with the shunted and unshunted cell. They place the galvanometer in the circuit with the cell and capacitor and on closin the circuit the capacitor is charged and the galvanometer needle i deflected by d_c^o or d_c^c depending on whether the cell is shunted or no Then the shunt is either added or removed and the increased or reverse deflection measured, which gives $(d_c^o - d_c^c)$ directly.

Bridge Circuits

Direct-current bridge methods were proposed early by Mance(10) an Siemens(11) for measuring the resistance of primary cells. Mance used modified Wheatstone bridge in which the cell is placed in the arr ordinarily occupied by the unknown resistance and a key for closing th circuit is substituted for the working battery. The galvanometer is shunte and placed in series with a resistance to reduce the magnitude of th deflections produced by the current from the cell. Balance is attaine when the galvanometer deflections remain unaltered when the key i opened and closed. R is then calculated in the usual way from the know resistance of the ratio arms. Lodge(12) modified this method by insertin a capacitor in series with the galvanometer thereby converting the metho to one in which variations in voltage rather than current were detected He also placed a special double key in the circuit so that the galvanomete circuit was broken an instant after the cell was shunted; balance was the less affected by current from the cell. Fahie(13) and Holler(14) placed battery in series with the galvanometer to serve as a counteracting forc to the cell current. Most recently, Sieh(15) using detection instruments o increased sensitivity and a compensation arrangement similar t

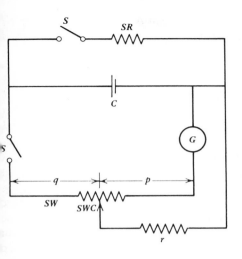

Figure 10.4 A direct-current bridge method. C, cell under test; G, galvanometer; SW, slide wire; SWC, slidewire movable contact; p, q, resistances of portions of the slidewire; r, resistor; SR, shunt resistor; S, switch.

Holler's (14) reinvestigated the application of the Mance method for the determination of the internal resistance of dry cells under various loads.

In the Siemens method shown in Fig. 10.4 the cell is joined in series with a slidewire and a galvanometer. One end of a resistance r is joined at the junction between the cell and galvanometer and its other end to a slider on the slidewire. The slider is moved until the galvanometer deflections are at a minimum; then $R = g + p - q$ where $g =$ galvanometer resistance, $p =$ slidewire resistance between slider and galvanometer, and $q =$ slidewire resistance between slider and cell. This resistance is for the cell when subject to the current given by $E(g + p + r)/[(g + p)(g + p + 2r)]$ where E is measured independently with a voltmeter. The current in the cell is then altered by shunting the cell with a known resistance and the bridge rebalanced. By obtaining R for various values of I, an estimate of R can be obtained for $I = 0$ by extrapolation.

Critique Experience has shown that values of R obtained by the methods outlined above are lower for higher values of I. Hirai, Fukuda, and Amano(16) have most recently confirmed this observation, using a dual purpose dc-ac oscilloscope and an electron-tube timer to read potential drops within 0.1 second. They also reemphasized that in the dc drop-in-potential methods, measurements must be made in as short a time period as possible to avoid effects of changes in concentration polarization with time. Chaney(3) believed that this phenomenon (lower resistance for higher current) arises for dry cells from a contact resistance resulting from a hydrogen-gas film (owing to zinc corrosion) on the zinc anode which is dissipated at higher current densities. However, an alternate explanation is that R, *as mentioned above*, includes polarization which is a function of I.

Polarization-Elimination Methods

In the above methods it is assumed, as was pointed out earlier in this chapter, that the EMF remains constant, that is, there is no back or counter EMF nor chemical changes, when electricity flows through a cell or when this flow is altered, that is, no polarization, assumptions that are not valid. Several methods have been proposed to eliminate or reduce these effects. These are (1) the extrapolation method, (2) the opposition method, (3) the polarity-reversal method, (4) reference-electrode method (5) dc-interrupter and dc-pulse methods, and (6) ac or impedance methods, to be discussed later in this chapter.

(1) In the *extrapolation method* apparent values of R are measured for various values of I and "true" values of R are obtained for $I = 0$ by extrapolation.

(2) In the *opposition method* two similar cells are joined in opposition so that they produce little or no electrical flow of their own. The resistance of one cell is assumed to be equal to half the measured resistance of the combination and to be free of polarization effects. Lawson and Kirkman(17) used this method with an ac bridge.

(3) In the *polarity-reversal method* the cell is first joined to a small storage cell or dc source in series aiding and then in series opposing and the two R's determined in each case by a drop-in-potential method, are averaged to give R "free of polarization."

(4) In the *reference-electrode method*, the potential of each electrode of the primary cell is measured during the passage of electricity against a reference electrode through which there is no current. The differences between the potentials of the operating and reference electrodes are the EMFs of polarization, E_p, which are subtracted from V, the CCV. The internal resistance is then obtained by dividing $E - V - E_p$ by the current I. This method is not readily adapted to routine measurements of the internal resistance of dry cells since the latter must be altered for insertion of the reference electrodes. In essence the method is a destructive one for dry cells although it may be used on a sampling basis. For some wet primary cells the reference electrodes may be inserted without appreciably altering the cell geometry. Even so, E_p may contain a resistive component even though a Luggin capillary is used in its measurement.

(5) *Direct-current interrupter methods* developed from the use of mechanically or electrically-driven units to "interrupt" the dc and thereby

providing a series of current pulses. Early applications of this method prior to 1951 have been briefly reviewed by Ferguson(18) while recent uses have been discussed by Yeager and Schwartz(19). The interrupter method has generally been used in studies of polarization and not of internal resistance; even so it is included here among the methods of measuring internal resistance. *Direct-current-pulse methods* are of more recent origin and utilize commercial pulse generators usually with a "square-wave" setting to furnish square or rectangular pulses of current. The dc-pulse methods are attractive because of several important aspects not inherent in earlier techniques:

(a) The passage of a dc pulse through a cell may represent only a fraction of a coulomb, or at most a few coulombs of energy and thus not polarize the cell electrodes to any appreciable extent.

(b) The use of modern cathode ray oscilloscopes to permit the measurement of the immediate change in voltage of the cell, when the pulse is applied and removed, that is, within a fraction of a millisecond, substantially eliminates interference from concentration polarization at electrode surfaces.

(c) The "instantaneous change" in voltage shown on the oscilloscope when the current pulse is applied or removed represents the voltage change owing to the internal resistance of the cell under study. The resistance may be directly calculated from this voltage change and the current used.

(d) The use of a square or rectangular pulse, characteristic of many pulse methods, makes it possible to readily recognize any marked divergence from this shape caused by electrode operation.

A typical circuit for pulse studies is shown in Fig. 10.5 and a typical single trace is shown in Fig. 10.6.

Steinwehr(20), Glicksman and Morehouse(21), Brodd(22), and Kordesch and Marko(23) have used various pulse methods for the measurement of the internal resistance of dry cells. Glicksman and Morehouse drew or impressed current pulses of 0.2-second duration on or from cells in the forward or reverse direction and measured the instantaneous (10^{-3} to 10^{-4} sec) IR drop (potential decay) or IR rise (potential recovery) with a cathode ray oscilloscope and screen that retains the image for a few seconds. Currents ranging from 5 to 800 ma were used. The IR drop or rise manifests itself as a "vertical" change in the oscilloscope trace, resulting from the resistance of the cell components, as shown in Fig. 10.6. A potentiometer was used to balance out most of the cell voltage thus permitting accurate readings of the potential on the oscilloscope. Their results with Leclanché dry cells were partially in line with the earlier results of Chaney(3), that is, they obtained lower resistance values

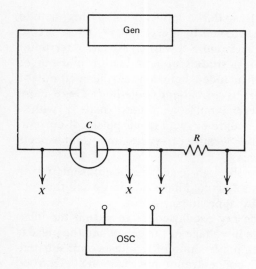

Figure 10.5 A schematic diagram of a pulse circuit. C, cell under test; Gen, pulse generator; R, standard resistance; OSC, oscilloscope; $X-X$ and $Y-Y$, leads to the oscilloscope.

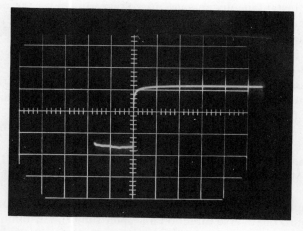

Figure 10.6 An individual oscilloscope trace produced by a pulse circuit.

for higher currents. However, they found that the internal resistance approached a constant value of about 0.3 ohm when $I = 200$ ma for D-size Leclanché dry cells and about 0.7 ohm when $I = 300$ ma for the AA-size cell, for example. On the other hand they found that the internal resistance of mercury dry cells (size M40) was independent of the magnitude of the imposed current. That their values for low current drains also included the effects of polarization is in line with their observation that potential decay invariably gave higher values for the internal resistance than did potential recovery.

Brodd used pulses of shorter duration (1 to 10 μsec) and repetition rate (100 to 5000 pulses per second) and measured the instantaneous (10^{-7} or less) IR drop, produced in the cell by the pulse, at the trailing edge with a Tektronic oscilloscope. The pulses were supplied by a Hewlett-Packard pulse generator, Model 212. The value of I in the circuit was obtained by connecting the oscilloscope across the terminals of a standard 10-ohm resistor in series with the cell, as shown in Fig. 10.5. Representative oscillograph traces are shown in Fig. 10.7. Using currents ranging from 0.008 to 3.96 ampere Brodd found the internal resistance of Leclanché dry cells to be independent of the magnitude of the pulse current although noting(24) that the oscillations or ringing edges of the pulse might conceivably arise from the destruction of an adsorbed hydrogen film (like Chaney's view) or from lead induction of the cell and instrument. However, his later results obtained with DeWane(25)(26) with a Wien substitution ac bridge were consistent with the dc-pulse results. His pulse method, therefore, gives the *resistance of the electronic-ionic carriers*; for Leclanché dry cells these are the zinc can, carbon rod, carbon-MnO_2 matrix, and the electrolyte. Accordingly, these two methods may be considered as reference or referee methods.

Kordesch and Marko(23) used rectified sine-wave pulse currents, taken

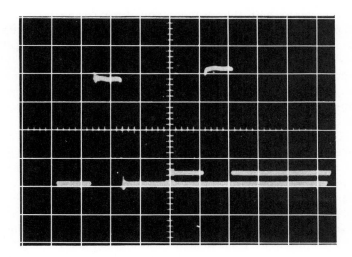

Figure 10.7 Typical oscillographic traces produced with a pulse circuit.

from a 60-Hz line, to discharge a cell under study, and a special test circuit whereby OCV, CCV, and average current could be measured during periodic interrupted discharges. The cell during "off" periods charged a capacitor, the voltage of which was measured with a high-resistance voltmeter. They used their method to measure, in addition to OCV and CCV, the internal resistance, polarization voltage, ohmic-free voltage, and IR drops produced by peak and average pulse currents. The difference between the highest and lowest points on the pulse curve gave the resistance-voltage drop, which when divided by the average current gave the internal resistance. They confined their studies to 500 ma and 60 Hz and consequently gave no data pertaining to the effect of current on internal resistance. They pointed out that the range of their method could be extended by using an audio signal generator whereby frequencies could be varied, and by replacing the voltmeter with a vacuum-tube voltmeter.

Sieh(15) most recently using the Mance-bridge-dc method obtained results that agreed with the pulse methods(21)(22) except for low current drains where the methods are less sensitive.

(6) *Alternating-current methods* (*discussed below*).

ALTERNATING-CURRENT METHODS

Bridge Circuits

Kohlrausch ac Bridge The earliest ac bridge method(27) (in fact the earliest of all ac methods) proposed for measuring the internal resistance of cells is similar to the Kohlrausch method of measuring the resistance of electrolytic solutions except that capacitors are used in the circuit to the ac source and to the detector to confine the dc produced by the cell under study to the bridge arms. The fixed resistor should have a value approximating that of the cell. The cell resistance is calculated in the usual way from the resistance of the ratio arms and corresponds to the resistance for the current given by $E/(R + R_f + R_v)$ where R_f = fixed resistance and R_v = variable resistance; E is measured independently with a voltmeter. The current is then altered by shunting the cell with resistance r; the current is then given by $E/[R + r(R_f + R_v)/(R_f + R_v + r)]$. *By obtaining R* for various values of I, R when $I = 0$ may be obtained by extrapolation. This method suffers from capacitive effects and unless noninductive resistors are used from inductive effects also. Residuals in the arms may be appreciable and complete balance is difficult to achieve. The accuracy of the method is improved by combining it with the opposition method by using two or more cells in series opposing(17). However, R of the individual cells must be taken equal and as the appropriate fraction of the total measured resistance.

Impressed-Current   A common ac method (28)(29) of measuring internal impedance of primary cells is to *impress* an ac signal *current* of low value (about 1 ma) from an audio signal generator across the cell while it is under load and then measure the ac voltage across the cell and load in parallel with a vacuum-tube voltmeter, as illustrated in Fig. 10.8. The method may also be used with cells on open circuit. C is a blocking capacitor of about 1–4 μF and is used to block dc out of the ac circuit so that the impedance of the cell can be read when the cell is on open or closed circuit. Switch S_1 is first turned to position 1 and the IR drop (or V_1) across the precision resistor R measured with the vacuum-tube voltmeter. The ac current is then given by $I_{ac} = V_1/R$. Switch S_1 is then turned to position 2 and switch S_2 is closed to place a load on cell B. An inductor should be used in the dc circuit to prevent the ac measuring circuit from being diverted through R_e which is parallel with the cell; if no inductor is available a correction should be made for the load resistance. Then the ac voltage, V_2, across the cell and load in parallel is measured with the voltmeter. The impedance, Z, of the cell is then given by

$$Z = \frac{V_2}{I_{ac}} = R\left(\frac{V_2}{V_1}\right) \tag{14}$$

If the oscillator output is adjusted so that $I_{ac} = 1$ ma then $Z = V_2 \times 10^3$. This may be done by using as R a 10,000-ohm resistor and setting V_1 at

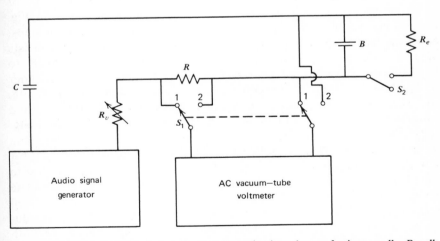

Figure 10.8 Block diagram of ac method of measuring impedance of primary cells. B, cell under test; C, blocking capicator; R, precision resistor; R_e, load resistor for cell; R_v, variable resistor to adjust the level of ac current; S_1, two-position switch; S_2, switch to place load on cell. *Note.* When cell is under load, an indicator should be inserted in the dc circuit.

10 V or using a 50-ohm resistor and setting V_1 at 0.05 V. Although various frequencies may be obtained on audio signal generators it is common practice, for comparative purposes, to select a single frequency (or perhaps 2 or 3 frequencies) for the measurements. Thus, Cahoon(30) using a specially designed instrument (precautions against power-line strays taken) measured the IR drop (impedance) across a cell when it was subjected to 60 Hz. A calibrated noninductive resistance was then adjusted to a value which would give the same IR drop under the same conditions. This value of the known resistance was then taken as the actual resistance of the cell. Walkley(31) used the same approach but used a frequency of 100 Hz. Hübner(32), Fukuda, Hirai, and Manabe(33), and others have shown that the impedance and equivalent IR drop vary with the frequency; this relation is discussed later under the section entitled "Discussion."

Substitution

WIEN BRIDGE. In another ac and more informative method a Wien substitution bridge as described by Grover(34) and Vinal(35) is used. Although the method is a tedious one it yields data on both resistance and capacitance from which the capacitive reactance and the impedance may be calculated. This method has been applied by Brodd and DeWane to various sizes of Leclanché dry cells(25, 26) and sealed nickel-cadmium cells(36) who made use of Cole-Cole(37) semicircle (Argand) diagrams of dielectric theory in their interpretations. Their circuit is shown in Fig. 10.9. R_1 is a variable noninductive resistor, graduated in 0.01-ohm steps, covering a range up to 11,111.1 ohms in six decades. Capacitor C_1 consists of two capacitors in parallel, continuously variable from 50 pF to 1.111 μF and serves to compensate for the capacitance of the cell under study. R_2 and C_2 are a fixed precision resistor and capacitor, respectively, and were selected to cover the range of R and C found for the cells and batteries under study. For single cells R_2 and C_2 were, respectively, 100 ohms and 1 μF. R_3 and R_4 are noninductive precision 1000-ohm resistors. Detector D is a tunable amplifier having a sensitivity of 5 μV for a 10 percent deflection of full scale. The oscillator had a range of 5 Hz to 600 kHz and is coupled to the bridge by an isolation transformer. The bridge is initially balanced with a thick copper bar, then with the cell under study; this substitution technique eliminates the problem of residuals in the balance equations. The resistance and capacitance are both measured on balancing over a wide frequency range (50 to 50,000 Hz).

The resistance, R, and the capacitive reactance, X_c, are obtained at

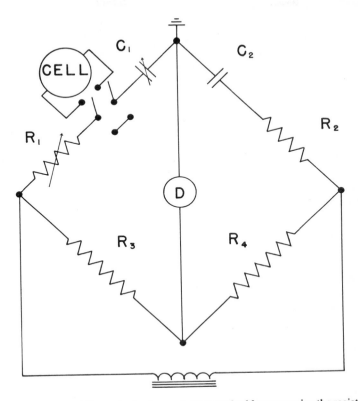

Figure 10.9 Circuit for Wien-substitution-ac-bridge method for measuring the resistance and capacitance of primary cells. C_1 capacitor; C_2, fixed capacitor; R_1 variable resistor; R_2, fixed resistor; R_3 and R_4, precision resistors; D, detector.

each frequency from

$$R = R_i - R_f \tag{15}$$

$$X_c = \frac{1}{\omega}\left[\frac{1}{C_i} - \frac{1}{C_f}\right] \tag{16}$$

where ω is the pulsatance given by $\omega = 2\pi f$, f being the frequency in hertz, and the subscripts i and f refer to the initial (with copper bar) and final (with cell) balances. From these, the internal impedance, Z, is obtained from

$$Z = \sqrt{R^2 + X_c^2} \tag{17}$$

Here inductive effects, as found for nickel-cadmium cells (36) are taken as being negligible. If inductive effects are present X_c is replaced in equation (17) by $X_L - X_c$ where $X_L = \omega L$; L is the inductance in henries.

For illustration, Argand diagrams are given in Figs. 10.10, 10.11, 10.12, 10.13, and 10.14 for general-purpose D-size Leclanché dry cells of a particular brand (26). Here the capacitive reactance is plotted against the resistance where each point corresponds to a definite frequency and is so labeled. *The vector from the origin to where the curve (at the left) cuts the R axis gives the resistive component of the impedance which is frequency independent, that is, the resistance of the electronic-ionic carriers, whereas the vector from the origin to a point on the curve gives the total impedance, including the resistance associated with the electrode proces-ses.* The angle the vector makes with the real axis is the phase angle for that frequency. Figure 10.10 is for a newly manufactured cell (about one month old) while Figure 10.11 is for a cell after storage for six months at 21°C. Figures 10.12, 10.13, and 10.14 are for newly manufactured cells (about one month old) after they were subjected to the general-purpose $2\frac{1}{4}$-ohm intermittent test, the general-purpose 4-ohm intermittent test, and the light-industrial-flashlight-cell test, respectively. These tests are out-lined in Table 1 of Chapter 8 and correspond, respectively, to tests 11, 12, and 14. Comparisons of these diagrams show that the resistance and capacitive reactance both increase slightly on six-month storage while both increase appreciably after the cell has been discharged with the

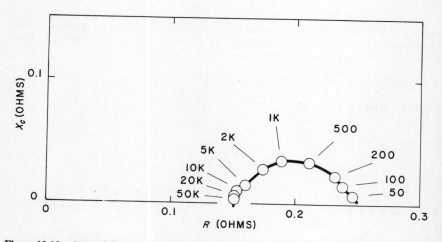

Figure 10.10 Argand diagram for newly manufactured general-purpose D size Leclanché dry cell.

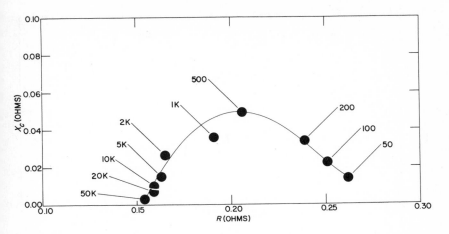

Figure 10.11 Argand diagram for general-purpose D size Leclanché dry cell after six-month storage at 21°C.

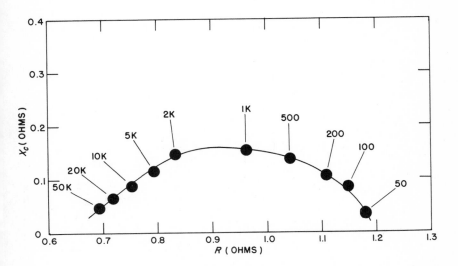

Figure 10.12 Argand diagram for general-purpose D size Leclanché dry cell after discharge on the general-purpose $2\frac{1}{4}$-ohm intermittent test.

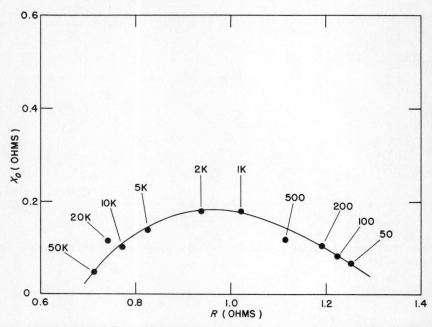

Figure 10.13   Argand diagram for general-purpose D size Leclanché dry cell after discharge on the general-purpose 4-ohm intermittent test.

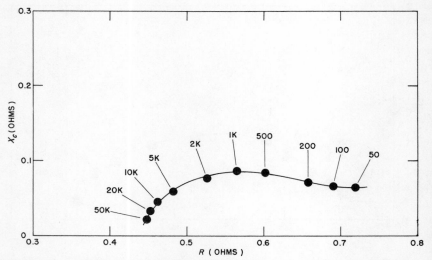

Figure 10.14   Argand diagram for general-purpose D size Leclanché dry cell after discharge on the light-industrial-flashlight-cell test.

increase being more on the general-purpose intermittent tests than on the light-industrial-flashlight-cell test.

The basis for an Argand diagram follows from relaxation theory of dielectrics. In relaxation theory the impedance, Z, is given by

$$Z = R - jX_c \tag{18}$$

where R is the real component of the impedance or resistance, $j = \sqrt{-1}$, and X_c is the imaginary component of the impedance, or the capacitive reactance. It may be shown from relaxation theory[38][2] that

$$Z = R_\infty + \frac{R_s - R_\infty}{1 + j\omega\tau} \tag{19}$$

$$R = R_\infty + \frac{R_s - R_\infty}{1 + \omega^2\tau^2} \tag{20}$$

$$X_c = (R_s - R_\infty)\frac{\omega\tau}{1 + \omega^2\tau^2} \tag{21}$$

where, as shown above, $\omega = 2\pi f$, f being the frequency of the ac field, τ the relaxation time given by $1/\omega_s$ where ω_s denotes a specific value for the frequency of the ac field times 2π, and R_s and R_∞ refer, respectively, to resistances for frequencies far above and far below $1/2\pi\tau$, or at the maximum and minimum values, respectively, of the semicircle of radius $(R_s - R_\infty)/2$. The value of the relaxation time is obtained by plotting values of X_c against values of ω. A maximum value for X_c, according to equation (21), is reached when $\omega\tau = 1$; thus, the value for ω under these conditions is specific and is labeled ω_s and the relaxation time, τ, is given by $1/\omega_s$ or $1/2\pi f_s$ where f_s is the corresponding specific frequency. From equations (20) and (21) it follows that

$$\left(R - \frac{R_s + R_\infty}{2}\right)^2 + X_c^2 = \left(\frac{R_s - R_\infty}{2}\right)^2 \tag{22}$$

By plotting R against X_c a semicircle, the Argand diagram, is obtained with radius $(R_s - R_\infty)/2$ with the center of the semicircle at a distance $(R_s + R_\infty)/2$ from the origin. Very few dry cells[25, 26] yield a semicircle with the abscissa which is a necessary result if, as a function of frequency, only simple relaxation processes prevail for the electrode reactions. Fig. 10.10 represents one of the rare cases where a near semicircle is seen; all others (Figs. 10.11, 10.12, 10.13, and 10.14) show that as the cell ages or is discharged dispersion ensues, or that complex relaxation processes prevail.

---

[2]The impedance, Z, is related to the complex permittivity, ϵ^*, of dielectric theory by $Z = \omega\epsilon^*/j$.

Euler and Dehmelt(39) also discussed data obtained with ac in terms of relaxation theory taking cognizance of the variation of resistance and capacitive reactance with frequency.

Data on the internal resistance, capacitive reactance, and internal impedance of AA-, C-, D-, and F-size Leclanché dry cells, as determined by the above method by Brodd and DeWane(26) are given in Appendix B, for illustration. Except for size F cells the data given are the average for four different brands of cells. In Appendix C similar data are given for four different brands of general-purpose and two different brands of industrial Leclanché dry cells of size D. These latter data show the dispersion in characteristics that may be found between different makes of Leclanché dry cells of the same size.

The data of Appendix B, omitting size F cells, show that the internal resistance of Leclanché dry cells decreases as the volume of the cell is increased. A correlation between R and the cross-sectional area and the distance of electrode separation is not easily made because of the porosity of the positive electrode and since cross-sectional area of each electrode is a function of the electrode separation. Coleman(40) outlined the complexity of this problem especially in regard to the bobbin and discussed the current distribution in dry cells for both ac and dc. Yeager and Schwartz(19), as will be discussed in more detail later, also pointed out the complexities associated with analyses of current distribution for porous electrodes with solution-filled pores and for electrodes with oxide layers.

OTHER BRIDGES INCLUDING WAYNE-KERR BRIDGE.   Various other ac bridges have been used from time to time in studies of primary batteries. Coleman(40) used a Kelvin bridge, Euler and von Lüttichau(41) a Maxwell-Wien RC bridge, and Hübner(32) an impedance bridge. For secondary batteries, Willihnganz(42) used an ac Wheatstone bridge, Vinal(35) a resonance bridge, while Genin(43) gave an extensive review on the subject; these methods could also be applied to studies of primary batteries. In recent years a Wayne-Kerr Universal bridge, Model B221(44) has been used. It is very similar to the Wien bridge described above. It uses fixed transformer ratio arms instead of the fixed resistance ratio arms of the Wien bridge and employs an electronic tuning eye as a detector to balance the phase as well as to measure the actual values of resistance and capacitance. It has the advantage over the Wien bridge in that it may be operated with reasonable speed.

Transient Pulse

The transient-ac-pulse method of measuring the internal resistance of dry cells was advanced by Tvarusko(45); he also gave literature references to its application to systems other than dry cells. His method is a modification to the ac-dc method (not used with primary cells) of Cahan and Rüetschi(46) wherein an ac is superimposed on dc. Tvarusko used the method for cells on open and closed circuit. On open circuit the cell is subjected to an ac pulse from a square-wave generator. Tvarusko observed the potential variation across the cell on a high-sensitivity oscilloscope with differential input. The instantaneous voltage drop in the oscilloscope pattern, caused by the leading edge of the constant-current square wave, indicated the internal resistance of dry cells. On closed circuit the cell under study is paralleled with a discharging resistor and subjected to its own dc current and the impressed ac from the generator, and the internal resistance is calculated using Kirchoff's law. (Cahan and Rüetschi had used a constant-current dc-power supply of high-input resistance for the discharging resistor, thus also imposing a dc on an electrolytic cell.) Currents were varied from 1 to 20 ma and the internal resistance was found by Tvarusko to be independent of frequency over the range of 20 to 10,000 Hz, that is, the method gives the *resistance of the electronic-ionic carriers*, in agreement with the method utilizing dc pulses of short duration(22); this method may, therefore, be considered a third referee method.

In a subsequent paper(47) Tvarusko showed that steady-state ac pulses could be measured in conjunction with the transient ac pulses by determining the steady-state voltage drop attained theoretically after $t = \infty$ but practically within microseconds. This steady-state value gives the sum of the transient resistance and parallel resistance (resistance in parallel with a capacitance, see Fig. 10.16). Knowing the transient value, the parallel resistance, can, therefore, be calculated. The parallel resistance is frequency dependent, that is, is the resistance associated with the electrode processes. Summarizing, then, steady-state ac pulses yield a composite value of internal impedance (transient and parallel resistance) and do not in themselves give values of internal resistance; values of internal resistance are obtained from measurements with transient ac pulses.

Tvarusko(48) later proposed the use of a milliohmmeter, Keithley Type 502, for routine measurements of the steady-state pulse values whereby the tedious analysis of oscilloscope traces is eliminated. The Keithley Milliohmmeter consists of a 100-Hz transistorized constant-current square-wave generator and a highly sensitive ac microvoltmeter which measures the voltage drop on the cell; this voltage drop is displayed on a panel meter calibrated directly in ohms. Tvarusko also modified the assembly so that the output could be recorded on a Texas Instrument recti/riter recorder. A more recent Keithley Milliohmmeter, Type 503, has a factory-installed recorder and no modification is required. The milliohmmeter can be used only for dc voltages less than about 50 mV. This limitation can be overcome by using two good quality 1000 μF blocking capacitors of sufficient dc-voltage rating in proper arrangement with the cell under study. Also, the use of the milliohmmeter does not give directly values of internal resistance except for high discharging rates where transient impedance and internal resistance are nearly equal. It is for this reason that Tvarusko points out that to be certain that correct results are obtained with the milliohmmeter, the system (the instrument may be used with other primary cells besides dry cells) under study should first be studied by the transient and steady-state ac pulse methods (45, 47).

DISCUSSION

That one would expect the internal impedance to vary in value with the frequency, that is, to be of lower magnitude at higher frequencies, may be seen in considerations of the equivalent circuit. At the top of Fig. 10.15

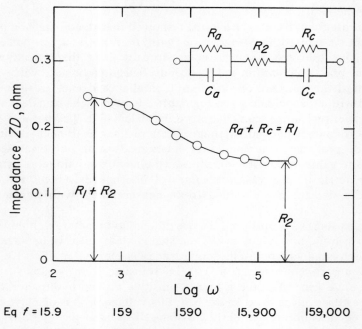

Figure 10.15 Variation of impedance of a D size Leclanché dry cell with pulsatance ($2\pi \times$ cyclic frequency). Equivalent circuit of Leclanché dry cell is also shown.

the ac equivalent circuit[3] is shown for a Leclanché dry cell; here R_a and R_c denote, respectively, the resistance of the anode and cathode related to the electrode processes, C_a and C_c the respective capacitances, and R_2 is the sum of the electronic (within the electrodes) and electrolytic or ionic carriers. At the bottom of Fig. 10.15 data for a typical D-size Leclanché dry cell are given for 26°C as a function of the logarithm of the pulsatance, ω, where $\omega = 2\pi f$, f being the frequency in hertz. The capacitive reactance, X_c, at the electrodes is calculated from the capacitance, C, by $X_c = 1/\omega C$. Since at low frequencies $X_c \cong \infty$, the total resistance R is given by $R_1 + R_2$ where $R_a + R_c$ is represented by R_1, see Fig. 10.15. At the higher frequencies $X_c = 0$ and R_1 is, therefore, short circuited with the result that $R = R_2$. Combining R_1 and R_2 in this way is tantamount to representing the equivalent circuit in a simple way given in Fig. 10.16, a commonly accepted equivalent circuit (see refs. (39) and (47), for example) for a primary cell when treated as an entity. Here R_p, C_p, and R_{se} are used to denote the parallel resistance, parallel capacitance, and series resistance, respectively, without further classification. In the several earlier discussions given in this chapter R_p and C_p are associated with the electrode processes and R_{se} is the resistance of the electronic-electrolytic carriers.

Other equivalent circuits for cells have been considered by Fukuda, Hirai, and Manabe(33), Falk and Lange(49), and Sluyters(50). Falk and Lange considered the parallel resistance, R_p, as consisting of two capacitances in parallel, one representing the electric double layer and the other the Faradaic impedance. Sluyters considered the parallel resistance, R_p, as the sum of a resistance representing activation polarization and a Warburg impedance arising from diffusion polarization and, like Brodd(22), made use of Cole-Cole(37) semicircle (Argand) plots of dielectric theory to interpret impedance. Ukshe(51) and Yeager and Schwartz(19) considered various equivalent circuits for individual electrodes with the latter distinguishing between porous, nonporous, and oxide-layer electrodes. For nonporous electrodes Yeager and Schwartz

[3]For a dc equivalent circuit a voltage would be represented in series with the circuit shown in Fig. 10.15.

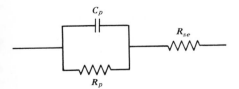

Figure 10.16 A simplified equivalent circuit for a primary cell.

considered the equivalent circuit of an individual electrode to consist of three elements in series: (a) a resistance within the electrode, (b) capacitance and Faradaic (or complex) impedance in parallel, with a pseudo-capacitance associated with the latter, and (c) an electrolytic or ionic resistance; analysis of individual electrodes, especially when used in conjunction with a reference electrode, is facilitated by separating the ionic resistance and the resistance within the electrode. These equivalent circuits, except for interpretations of the source (or cause) of the resistive and capacitance elements all reduce to the simple equivalent circuit given in Fig. 10.16.

For electrodes consisting of electrolyte-filled pores, Yeager and Schwartz(19) point out that the equivalent circuit would consist of resistive elements for porous and nonporous sections in parallel, with the former consisting of a parallel capacitance-resistance element in series with the substrate resistance and the latter of a parallel capacitance-resistance element in series with the electrolytic resistance of the pore. Since current distribution within the system, that is, between the porous and nonporous sections, is not well defined (see ref. (40) also) an integrated value of purely resistive components may not be attainable. Conversely, an integrated value for the various polarizations may also not be attainable. Only if the resistance of the porous and nonporous sections were known a priori could current distribution be ascertained and vice versa, an apparent insolvable problem. As a consequence of these limitations for porous systems, Yeager and Schwartz pointed out that detailed analyses of interrupter measurements with porous electrodes are difficult because some IR components associated with the electronic and electrolytic resistances are included in the potential measured immediately after the interruption of the current because of the two capacitances (in the pores and nonporous section) mentioned above. They also discussed the time factor in switching the current off and its relation to the rise time of the oscilloscope used to follow the potential changes.

For electrodes with oxide layers the situation is even more complex. The equivalent circuit here, as shown by Yeager and Schwartz, differs from that for a nonporous electrode in that the element (b) in the equivalent circuit of a nonporous electrode is replaced by three capacitance-resistance elements in series, one associated with the metal-oxide layer interface, a second with the oxide layer-solution interface, and a third with these two interfaces as a dielectric with the parallel resistance being non-Faradaic. The current distribution between the metal, oxide layer, and interfaces is difficult, if not impossible, to determine. Furthermore, the resistance component of the third element does not disappear instantaneously upon current interruption because of he non-Faradaic capacitance; part of the IR drop, as a consequence, may be included in the polarization measurements and vice versa. With noninterrupter methods, viz., substitution bridges, this problem is largely circumvented.

Although the simplified equivalent circuit of Fig. 10.16 overlooks many of the individual electrode factors discussed immediately above it can be handled mathematically and is the basis of the calculations of the data obtained with the Wien substitution bridge. As stated above, if impedance measurements are to be used to study or characterize electrode reactions, the simple circuit of Fig. 10.16 does not suffice since each electrode reaction has its own specific characteristics, for example, relaxation times (see equation (21) and associated discussion) and distinct frequency characteristics (at least two distinct frequency characteristics are found in battery impedance). If, however, the interest is solely in the internal resistance of the cell itself, as is the case here, the simple circuit of Fig. 10.16 suffices since the internal resistance is identified as R_{se} exclusively. Since some methods yield data for R_{se} alone and others combined data for R_{se}, R_p, and C_p with the latter two being frequency dependent it is necessary to define the method and the frequency used in the measurement of internal resistance or impedance whenever a value is reported.

COMPARISON OF METHODS

The practical use of internal resistance as a laboratory and industrial tool demands that a reasonably precise and rapid method be employed. The dc methods which depend on readings of voltage and current with the usual meters (pp. 430–437) all have the advantage that simple apparatus is needed and readings may be made readily. However, the results obtained, as stated above include components other than purely resistive ones, and tend, therefore, to be high. Of the three referee methods (Wien substitution ac bridge, dc pulses of short duration, and transient ac pulses) described in this chapter the manually operated Wien bridge method is too time consuming for routine use. However, the Wayne-Kerr bridge (44) may be balanced rapidly and offers an alternative to the Wien bridge.

Most of the dc- and ac-pulse methods have the advantage of rapid measurement but require a source of pulsed dc or ac and a calibrated oscilloscope. Although the interrupter methods have generally been confined to polarization studies they could be used in the measurement of internal resistance. A common method of producing interrupted dc is the use of a relay with mercury-wetted contacts driven from an appropriate supply such as a 60-Hz ac line. However, such relays can safely handle currents of only a few amperes and the contacts must be protected by an appropriate capacitance-resistance network when more than 10 ma are used(52). The method is limited, therefore, to studies of cells only at low current drains or to small cells. Most interrupter methods also require a calibrated oscilloscope. The Kordesch-Marko instrument(23), using a rectified sine-wave pulse, has the inherent capability of supplying larger current pulses and feature direct meter readings thereby not requiring a calibrated oscilloscope. The authors(23) have described its use with hydrogen-oxygen fuel cells and refer to its use with nickel-cadmium batteries.

The ac "impressed-current" method may also be used. In contrast to the dc-pulse (see page 438) and Kordesch-Marko methods no net current is passed through the cell. However, corrections for phase angle and inductance are required for precise results.

Table 10.1 presents a comparison of impedance values, R_∞, as defined earlier, with resistance values obtained by the dc-pulse and the flash current methods(26)(53). The cells were approximately one month old and had been stored at 21°C prior to impedance measurements at 26°C about one hour elapsed between removal from storage and impedance measurements. The agreement between the dc-pulse and R_∞ values is excellent. The values of the resistance given by the flash-current method are somewhat higher than those obtained by the other two methods, and as stated above, include values of nonresistive components; in fact, the difference between the R_{dc} and R_∞ values may be considered as giving an estimate of the resistance associated with the electrode processes. The value of R_∞ given in Table 10.1 may be calculated from data such as that given in Appendices B and C using the equations given earlier in this chapter(26). However, R_∞ is most easily found from an Argand diagram by extrapolation of the high frequency end of the circular curve to the point where it crosses the horizontal axis (see p. 466). The value of R at this point gives a close approximation to R_∞.

Additional data(54) extend the comparison of the values of internal resistance and impedance obtained by some additional methods described in this chapter. Table 10.2 presents impedance results on four sizes of Leclanché dry cells using the ac "impressed-current" (cf. p. 443) and the Wayne-Kerr bridge (cf. p. 450) methods. For comparison the internal resistance calculated by the flash-current method (cf. p. 432) is included.

The ac "impressed-current" method was used with a current of 10 ma and a frequency of 1592 Hz, equivalent to a pulsatance, ω, of 10^4 rad sec^{-1} The results were not corrected for phase angle but are reported as ohms impedance. The Wayne-Kerr B221 bridge was used with a parallel-to series low-impedance adaptor Q221 and at the above frequency. The current was varied from 1 to 10 ma. The OCV and flash current of the cells were measured by standard methods already described, see Chapter 8.

TABLE 10.1. A COMPARISON OF INTERNAL RESISTANCE AND IMPEDANCE OF REPRESENTATIVE TYPES AND SIZES OF CELLS MEASURED WITH DC AND AC METHODS AT 26°C[a]

	OCV (volts)	SCC[b] (amperes)	Flash-current R_{dc} (ohms)	dc-pulse dc R_∞ (ohms)	ac-impedance R_∞ (ohms)
Leclanché dry cells					
D size cells					
1.	1.58	7.18	0.220_6 $\pm 0.0006_2$	0.146_4 $\pm 0.004_4$	0.147_8 $\pm 0.006_4$
2.	1.61	6.58	0.245_8 $\pm 0.019_4$	0.186_5 $\pm 0.011_3$	0.191_3 $\pm 0.012_7$
3.	—	—	0.239_8 $\pm 0.025_4$	0.134_6 $\pm 0.007_5$	0.135_0 $\pm 0.008_0$
C size cells					
1.	—	—	—	0.223_0 $\pm 0.007_1$	0.228_0 $\pm 0.005_7$
AA size cells					
1.	1.56	3.53	0.442_0 $\pm 0.006_9$	0.295_0 $\pm 0.007_1$	0.295_5 $\pm 0.003_5$
2.	1.62	6.07	0.267_7 $\pm 0.008_4$	0.169_0 $\pm 0.010_6$	0.170_0 $\pm 0.012_1$
Ni-Cd sealed cells					
Charged	1.27	20	0.063	0.025	0.018
Discharged	1.16	2	0.58^c	0.058	0.056

[a]Each value given for Leclanché dry cells above represents the average value obtained on a group of 3-5 cells. The products of four manufacturers are included in the above data.
[b]Short-circuit current.
[c]This abnormally high value shows the limitations in the SCC method when applied to spent (noncurrent producing) cells.

Groups of 1 to 5 Leclanché dry cells in sizes D, C, AA, and No. 6, obtained from up to five different manufacturers, were measured under open-circuit conditions by the above impedance methods. The cells were six months to one year old and were stored at 21°C for several months prior to impedance measurements at 23°C; cells were removed from 21°C-storage about one to two hours before measurements of impedance were made.

Some important differences appear between the data of Tables 10.1 and 10.2 which warrant mention. The values in Table 10.2 obtained by the two ac methods are in approximate agreement differing only by a few percent. This difference is explained by the fact that no correction was made for the phase angle in the "impressed-current" method whereas this factor

TABLE 10.2. A COMPARISON OF THE INTERNAL RESISTANCE AND IMPEDANCE OF LECLANCHÉ DRY CELLS MEASURED BY ONE DC AND TWO AC METHODS AT 1592 Hz AT 23°C

Cell Size and Manufacturer	OCV (volts)	SCC[b] (amperes)	Internal resistance (R)		Impedance (Z)		
			Flash-current[1] (ohms)	IC[2] at 1592 Hz (ohms)	WK bridge[3] at 1592 Hz (ohms)	Calc. eq. Z^4 at 50 kHz (ohms)	
D size cells							
1[a]	1.594	5.9	0.26	0.234	0.216	0.181	
2	1.641	6.0	0.27	0.258	0.248	0.208	
3	1.653	6.0	0.27	0.356	0.329	0.275	
4	1.624	6.3	0.25	0.290	0.272	0.227	
5	1.601	6.9	0.22	0.244	0.234	0.196	
C size cells							
1	1.622	6.0	0.26	0.246	0.239	0.200	
2	1.613	5.8	0.27	0.289	0.262	0.219	
3	1.632	5.7	0.28	0.360	0.358	0.299	
4	1.592	5.2	0.30	0.294	0.275	0.230	
5	1.564	5.9	0.26	0.295	0.262	0.219	
AA size cells							
1	1.589	4.1	0.37	0.316	0.295	0.247	
2	1.662	3.5	0.48	0.584	0.547	0.459	
3	1.664	4.0	0.41	0.557	0.537	0.449	
4	1.598	5.4	0.29	0.296	0.283	0.237	
5	1.625	4.9	0.32	0.322	0.312	0.261	
No. 6 size cells							
1(Ignition)	1.579	33.2	0.038	0.0396	0.0324	0.0271	
2(Ignition)	1.578	37.5	0.032	0.0382	0.0253	0.0212	
3(Ignition)	1.637	26.0	0.053	0.0518	0.0418	0.0350	

[a]Manufacturer.
[b]Short-circuit current.
[1]Flash-current method given in text, p. 432.
[2]"Impressed-current" method given in text, p. 443.
[3]Wayne-Kerr bridge method given in text, p. 450.
[4]Conversion of data in footnote 3 to 50 kHz (given in text).

was automatically included in the measurements made with the Wayne-Kerr bridge. Thus the latter values are more significant. The Wayne-Kerr values in Table 10.2 are somewhat higher than the R_∞ impedance values in Table 10.1. The main reason for this difference is the use of a lower frequency, 1592 Hz, for the measurements in Table 10.2 as compared to the much higher frequency used to calculate R_∞ in Table 10.1. The column of data in Appendix C headed "mean" which gives the average of the impedance values for four makes of general-purpose D-size cells at various frequencies offers a basis for converting impedance at one frequency to an equivalent impedance at another frequency. Plotting the impedance values in this column against the logarithm of the corresponding frequency provides a curve similar in shape to that given in Fig. 10.15. From this curve it was found that the impedance at 50 kHz is 83.8 percent of that measured at 1592 Hz. Applying this factor to the impedance values obtained by the Wayne-Kerr bridge at 1592 Hz and given in Table 10.2 provides the numbers in the last column of the table. These values are the equivalent impedance at 50 kHz and as such may be compared with the R_∞ values given in Table 10.1. In most instances the equivalent 50 kHz values in Table 10.2 agree satisfactorily with the R_∞ values and the dc-pulses values shown in Table 10.1. There are some exceptions to this generalization, particularly cells of manufactures 2 and 3 of AA-size cells, which show impedance values higher than those of the other manufacturers. It is believed that manufacturing methods and variations in materials influence these results to an unknown extent.

In general the flash-current data on internal resistance given in the two tables agree reasonably well for the same size cell; the difference in measurement temperatures, 23°C and 26°C, and the age of the cells, one and six months, may contribute somewhat to the differences noted. Resistance values for six-month-old cells are somewhat higher than for one-month-old cells, as was shown previously in comparing the data of Fig. 10.10 (one-month-old cells) with that of Fig. 10.11 (six-month-old cells). Furthermore it should be recognized that the cells involved in the two series of measurements were made at different times; Table 10.1 data were obtained on cells made in 1958 whereas Table 10.2 data were obtained on cells made in 1971–1972. However the internal-resistance levels given in Table 10.2 are not always higher than the impedance values given in Table 10.1. The probable reason for this situation lies in the changing manufacturing methods and materials used by the various manufacturers. It is well known that the output of cells of the same manufacturer changes (usually rises) from year to year (55), and accordingly the internal resistance changes also. The change would not necessarily be a decrease (in fact some of the data of Tables 10.1 and 10.2 show the opposite) since quality, density, and porosity of cell materials and cell geometry are critical factors in comparing cell output from one year to another.

The simplest equivalent circuit for a cell given in Fig. 10.16 and the explanation in the text emphasize that the series resistance R_{se}, the internal resistance, is the basic quantity that is being sought without the interfering effects of electrode polarization which influence the paralleled parts of the circuit, R_p and C_p. To eliminate the electrode phenomena it seems clear that a dc or ac method utilizing very rapid pulse measurements is necessary. However, if the experimenter selects the ac bridge or ac "impressed-current" method he should, in either case, use a high frequency ac, for example, 50,000 Hz to assure that the final impedance results are not significantly affected by R_p and C_p. If he selects the ac "impressed-current" method he should be aware that the results, owing to the lower accuracy inherent in this method (phase angle shifts) will be slightly higher than those given by the ac-bridge method unless corrections are made for phase angle.

CONCLUSIONS

Although each of the various methods, outlined in this chapter, for measuring the internal resistance of primary cells and batteries has unique features of merit, if a choice had to be made of a method most suited for rapid testing purposes one would probably select the ac "impressed-current" method, outlined in connection with Fig. 10.8, or the dc method of Kordesch and Marko. The first is convenient and suitable for comparative tests and is free of the uncertainties of the earlier methods. The method of Kordesch and Marko is also relatively simple and has wide applicability not only to Leclanché and similar cells but to those that show electrode passivity or high internal resistance. For research purposes, however, either the Wien or Wayne-Kerr substitution ac method or the composite transient and steady-state ac-pulse method should be chosen since they afford information on the purely resistive elements (electronic-ionic carriers) of primary cells as well as on the resistance associated with the electrode processes which are frequency dependent.

REFERENCES

1. Leclanché, Georges, Les Mondes 16, 532 (1868).
2. Hamer, W. J., "Standard Cells," chap. 12, p. 464, The Primary Battery, Vol. 1, Ed. George W. Heise and N. Corey Cahoon, John Wiley, New York (1971).
3. Chaney, N. K., Trans. Am. Electrochem. Soc. 29, 183 (1916).
4. Beetz, W., Ann. Physik. (2) 142, 573 (1871).
5. Fahie, J. J., Tel. J. Elec. Rev. 12, 203 (1883).
6. Thomson, Sir William, J. Soc. Teleg. Eng. 1, 399 (1872).
7. Kempe, H. R., J. Soc. Teleg. Eng. 1, 419 (1872).
8. Kempe, H. R., A Handbook of Electrical Testing, 6th ed., p. 339, E. & F. N. Spon. Ltd., London (1900).
9. Id, p. 334.
10. Mance, Sir Henry, Proc. R. Soc. 19, 248 (1871).
11. Siemens, Werner, J. Soc. Teleg. Eng. 1, 407 (1872).
12. Lodge, O. J., Phil. Mag. (5) 3, 515 (1877).

13. Ref. (8), p. 194.
14. Holler, H. D., *J. Electrochem. Soc.* **97**, 271 (1950).
15. Sieh, H. W., *Electrochimica Acta* **13**, 2139 (1968).
16. Hirai, T., Fukuda, M., and Amano, Y., *Nat. Tech. Rep.* **8**, No. 2, 160 (1962).
17. Lawson, H. E. and Kirkman, F. J., *Trans. Electrochem. Soc.* **70**, 44 (1936).
18. Ferguson, A. L., chap. 23, *Electrochemical Constants*, Ed. W. J. Hamer, Natl. Bur. Standards Circular 524, U.S. Government Printing Office (1953).
19. Yeager, E. B. and Schwartz, E. P., chap. 2, *The Primary Battery*, Vol. 1, Ed. George W. Heise and N. Corey Cahoon, John Wiley, New York (1971).
20. Steinwehr, H., *Elektrotechnik* **2**, 91 (1948).
21. Glicksman, R. and Morehouse, C. K., *J. Electrochem. Soc.* **102**, 273 (1955).
22. Brodd, R. J., *J. Electrochem. Soc.* **106**, 471 (1959).
23. Kordesch, K. and Marko, A., *J. Electrochem. Soc.* **107**, 480 (1960).
24. Brodd, R. J., *J. Electrochem. Soc.* **106**, 1083 (1959).
25. Brodd, R. J. and DeWane, H. J., *J. Electrochem. Soc.* **110**, 1091 (1963).
26. Brodd, R. J. and DeWane, H. J., *NBS Tech. Note* 190, 1963.
27. Northrup, E. F., *Methods of Measuring Electrical Resistance*, p. 226, McGraw-Hill, New York (1912).
28. Mundel, A. B., Sonotone Corp., Elmsford, N.Y., private communication.
29. Potter, N. M., Union Carbide Consumers Product Company, Cleveland, Ohio, private communication.
30. Cahoon, N. C., *Trans. Electrochem. Soc.* **92**, 159 (1947).
31. Walkley, A., *Australian J. App. Sci.* **3**, 324 (1955).
32. Hübner, W., *Elek. Tech. Z.* **61**, 149 (1940).
33. Fukuda, M., Hirai, T., and Manabe, H., *Denka* **27**, 247 (1959).
34. Grover, F. W., *Natl. Bur. Standards Bull.* **3**, 278 (1907).
35. Vinal, G. W., *Storage Batteries*, 4th ed., pp. 327, 329, John Wiley, New York (1955).
36. Brodd, R. J. and DeWane, H. J., *Electrochem. Tech.* **3**, 12 (1965).
37. Cole, K. S. and Cole, P. H., *J. Chem. Phys.* **9**, 341 (1941).
38. Böttcher, C. J. F., *Theory of Electric Polarization*, pp. 363–374, Elsevier, New York (1952).
39. Euler, J. and Dehmelt, K., *Z. Electrochemis* **61**, 1200 (1957).
40. Coleman, J. J., *Trans. Electrochem. Soc.* **90**, 545 (1946).
41. Euler, J. and von Lüttichau, H. G., *Z. Electrochemis* **61**, 1196 (1957).
42. Willihnganz, E., *Trans. Electrochem. Soc.* **79**, 253 (1941).
43. Genin, G., *Rev. Gen. Elec.* **56**, 159 (1947).
44. Wayne-Kerr Universal Bridge Model 221 with Q221 adaptor, manufactured by the Wayne-Kerr Laboratories, Ltd., Roebuck Road, Chessington, Surrey, England.
45. Tvarusko, A., *J. Electrochem. Soc.* **109**, 557 (1962).
46. Cahan, B. D. and Rüetschi, P., *J. Electrochem. Soc.* **106**, 543 (1959).
47. Tvarusko, A., *J. Electrochem. Soc.* **109**, 881 (1962).
48. Tvarusko, A., *Electrochem. Tech.* **1**, 354 (1963).
49. Falk, G. and Lange, E., *Z. Elektrochem.* **54**, 132 (1950).
50. Sluyters, J. H., *Rec. Trav. Chim. Pays-Bas* **79**, 1092 (1960).
51. Ukshe, E. A., *Elektrochimiya* **4**, No. 9, 1116 (1968).
52. Ref. (19), p. 139.
53. Brodd, R. J., private communication of unpublished data (1972).
54. Dereska, J. S., private communication of unpublished data (1972).
55. Hamer, W. J., *Electrochem. Tech.* **5**, 490 (1967).

APPENDIX A
LIST OF SYMBOLS

V = closed-circuit voltage (CCV)

E = open-circuit voltage (OCV)

I = current

R = internal resistance, resistance, total resistance, real component of resistance

E_p = electromotive force (EMF) of polarization

dc = direct current

ac = alternating current

R_e = external resistance, load resistance, precision resistor

ΔV = drop in potential

I_f = flash current

V_1, V_2 = different values of voltage, ac voltages

I_1, I_2 = different values of current

R_1 = first resistance, sum of parallel resistance of anode and cathode, variable noninductive resistor

R_2 = second resistance, sum of resistance of electronic and ionic carriers, fixed precision resistor, variable resistor

d = deflection of galvanometer needle

d_1, d_2 = different deflections of galvanometer needle

E_x = EMF of cell

E_{RC} = EMF of reference cell

r_1, r_2 = different resistances of portions of slidewire

R_2' = new value for R_2 when R_2 is a variable resistor

a = resistance of ammeter

d_s = deflection of ballistic galvanometer from a standard cell

d_c^o = deflection of ballistic galvanometer from cell under test on OCV

d_c^c = deflection of ballistic galvanometer from cell under test on load

E_s = EMF of standard cell

r = a resistance

g = galvanometer resistance

p = part of slidewire resistance

q = remainder of slidewire resistance

R_f = fixed resistance

R_v = variable resistance

I_{ac} = ac current

Z = impedance in ohms

C = capacitor, capacitance

C_1 = two capacitors in parallel

C_2 = fixed precision capacitor

R_3 = noninductive precision 1000-ohm resistor

R_4 = noninductive precision 1000-ohm resistor

R_i = initial resistance

R_f = final resistance
X_c = capacitive reactance, imaginary component of impedance
ω = pulsatance
f = frequency
X_L = inductive reactance
L = inductance
C_i = initial capacitance
C_f = final capacitance
$j = \sqrt{-1}$
ϵ^* = complex permittivity
R_s = resistance for frequency far above $\frac{1}{2}\pi\tau$, i.e., maximum value
R_∞ = resistance for frequency far below $\frac{1}{2}\pi\tau$, i.e., minimum value
τ = relaxation time
ω_s = specific value of frequency of ac field times 2π
f_s = specific frequency of ac field
R_a = resistance of anode in parallel with a capacitance
R_c = resistance of cathode in parallel with a capacitance
R_p = resistance in parallel with capacitance in analog circuit
C_p = capacitance in parallel with resistance in analog circuit
R_{se} = series resistance in analog circuit identified as internal resistance
Note: Several other symbols are defined in the notes to the figures.

APPENDIX B

INTERNAL RESISTANCE, CAPACITIVE REACTANCE, AND INTERNAL IMPEDANCE E FOR VARIOUS SIZES OF CYLINDRICAL LECLANCHÉ CELLS AT 26°C

Frequency Hz	(Mean[a] of four different brands of new cells)			
	AA	C	D	F[b]
Resistance, ohm				
50	1.237	0.649	0.333	0.441
100	0.681	0.590	0.305	0.305
200	0.444	0.486	0.262	0.245
500	0.331	0.371	0.226	0.175
1,000	0.302	0.321	0.205	0.143
2,000	0.287	0.296	0.190	0.131
5,000	0.278	0.280	0.175	0.123
10,000	0.271	0.271	0.170	0.111
20,000	0.267	0.267	0.167	0.107
50,000	0.262	0.262	0.164	0.104
Capacitive Reactance, ohm				
50	1.651	0.099	0.034	0.102
100	1.018	0.144	0.051	0.118
200	0.580	0.168	0.058	0.115
500	0.263	0.141	0.050	0.084
1,000	0.145	0.099	0.039	0.055
2,000	0.078	0.061	0.030	0.035
5,000	0.038	0.034	0.017	0.022
10,000	0.023	0.021	0.013	0.018
20,000	0.015	0.014	0.008	0.021
50,000	0.009	0.008	0.006	0.022
Impedance, ohm				
50	2.063	0.657	0.334	0.453
100	1.225	0.607	0.309	0.327
200	0.731	0.513	0.269	0.271
500	0.423	0.397	0.232	0.194
1,000	0.335	0.336	0.208	0.153
2,000	0.298	0.302	0.192	0.136
5,000	0.280	0.282	0.176	0.125
10,000	0.272	0.272	0.171	0.112
20,000	0.268	0.267	0.168	0.109
50,000	0.262	0.262	0.164	0.106

[a] $R_m = (R^2/n)^{1/2}$; $X_m = (X_c^2/n)^{1/2}$; $Z_m = (Z^2/n)^{1/2}$ where $n =$ number of brands.

[b] Average of four cells of same make.

APPENDIX C

INTERBAK RESISTANCE, CAPACITIVE REACTANCE, AND INTERNAL IMPE-
DANCE OF 4 DIFFERENT BRANDS OF GENERAL-PURPOSE AND 2 DIFFERENT
BRANDS OF INDUSTRIAL D-SIZE LECLANCHÉ CELLS AT 26°C

Frequency Hz	(New cells)							
	1^d	2	3	4	Mean[b]	1	2	Mean[b]
	Resistance, ohm							
50	0.246	0.330	0.335	0.401	0.333	0.314	0.363	0.339
100	0.238	0.303	0.314	0.354	0.305	0.301	0.336	0.319
200	0.232	0.263	0.241	0.307	0.262	0.285	0.294	0.290
500	0.211	0.214	0.238	0.241	0.226	0.254	0.242	0.248
1,000	0.189	0.196	0.224	0.208	0.205	0.231	0.222	0.227
2,000	0.174	0.183	0.207	0.193	0.190	0.219	0.208	0.214
5,000	0.160	0.171	0.184	0.185	0.175	0.207	0.197	0.202
10,000	0.153	0.164	0.181	0.181	0.170	0.199	0.192	0.196
20,000	0.151	0.159	0.180	0.178	0.167	0.196	0.187	0.192
50,000	0.151	0.155	0.175	0.175	0.164	0.191	0.183	0.187
	Capacitive Reactance, ohm							
50	0.007	0.045	0.006	0.049	0.034	0.018	0.045	0.034
100	0.015	0.064	0.005	0.077	0.051	0.028	0.066	0.050
200	0.023	0.068	0.011	0.090	0.058	0.036	0.073	0.058
500	0.034	0.053	0.021	0.074	0.050	0.040	0.057	0.049
1,000	0.035	0.038	0.028	0.051	0.039	0.034	0.041	0.038
2,000	0.028	0.028	0.030	0.034	0.030	0.026	0.028	0.027
5,000	0.015	0.018	0.022	0.017	0.017	0.018	0.016	0.017
10,000	0.011	0.013	0.015	0.011	0.013	0.013	0.012	0.013
20,000	0.007	0.008	0.010	0.007	0.008	0.007	0.007	0.007
50,000	0.004	0.004	0.008	0.005	0.006	0.006	0.005	0.006
	Impedance, ohm							
50	0.246	0.333	0.335	0.404	0.334	0.315	0.366	0.341
100	0.238	0.310	0.314	0.362	0.309	0.302	0.342	0.323
200	0.233	0.272	0.241	0.320	0.269	0.287	0.303	0.295
500	0.214	0.220	0.239	0.252	0.232	0.257	0.249	0.253
1,000	0.192	0.200	0.226	0.214	0.208	0.234	0.226	0.229
2,000	0.176	0.185	0.209	0.196	0.192	0.221	0.210	0.215
5,000	0.161	0.171	0.185	0.186	0.176	0.208	0.198	0.203
10,000	0.153	0.165	0.182	0.181	0.171	0.199	0.192	0.196
20,000	0.151	0.159	0.180	0.178	0.168	0.196	0.187	0.192
50,000	0.151	0.155	0.175	0.175	0.164	0.191	0.183	0.187

[a]Numbers refer to manufacturer; thus, general-purpose cell 1 and industrial cell 1 were produced by the same manufacturer (No. 1).
[b]$R_m = (\Sigma R^2/n)^{1/2}$; $X_m = (\Sigma X_c^2/n)^{1/2}$; $Z_m = (\Sigma Z^2/n)^{1/2}$ where n = number of brands.

11.

Energy Conversion

R. C. Shair

The objective of this chapter is to present, in perspective, the position of the primary battery in the broad spectrum of sources and converters of energy. This involves a comprehensive look at the conversion picture as a whole, with emphasis on systems in the same realm of application as the battery itself. Batteries convert primary chemical energy into an electrical output. There are other primary sources of energy and many other types of converters are known. Some of these converters are quite old; others are of relatively recent origin. Many have been the subject of study and the more promising ones have been the subject of considerable development effort.

In this comprehensive study of energy conversion, we include all the energy sources and conversions that have had significance. It should be kept in mind, however, that batteries have a specific area of applicability, and probably of most interest to electrochemists and battery engineers today are those energy conversion systems that are in the same realm of application as batteries. Batteries have been used where portability is a prime consideration; their continuous mission duration has usually ranged from a few seconds to many days; their power level has been from milliwatts up to hundreds of kilowatts. As secondary batteries coupled to an electrical recharging source, they can be recycled many times for a total of many years.

Let us begin by examining the various existing sources of energy and the techniques available for energy conversion. The most significant primary energy sources available to man are:

a. Chemical
b. Natural energy sources (hydroelectric, winds, waves, tides, lightning, and cosmic energy)
c. Nuclear
d. Solar

By far, the most widely used chemical sources of energy are the foss·
fuels. About 80 percent of the world's electric power is produced b·
burning fuels to produce heat which is converted to electrical energ
through a heat-engine driving a generator. The internal combustion engin
converts chemical energy into a mechanical output.

Among the natural energy sources, hydroelectric power is well de
veloped and much used; windmills of many varied designs have bee
fabricated and used; and, to a lesser extent, the other sources have bee
tapped for practical use and/or for scientific curiosity. In genera
however, the natural energy sources are not as versatile, universall
available, or as easily portable as either chemical, solar, or nuclear energ
sources, This is particularly true with regard to space applications, whic
have provided much of the current interest and impetus in the develor
ment of energy conversion techniques.

Nuclear sources are growing in application for large stationar
electrical-energy power plants. These are fission plants, and considerabl·
development effort is underway to improve their technolog>
Radioisotope energy-sources are used in smaller applications and hav
been applied in aerospace power plants.

A general outline for the treatment of the subject of this chapter i
provided by Tables 11.1 and 11.2. Table 11.1 lists the more importar
energy sources and the conversion means whereby they are eithe
converted to electrical energy directly or are converted to heat energy o·
to chemical energy as an intermediate step in the process leading to a·
electric output. Table 11.2 lists the processes for the conversion o
thermal and mechanical energy into an electrical output. The numbere
subjects are discussed in further detail in the text. The lettered topics ar
listed only and are referred to in Fig. 11.1 which shows schematically th·
various energy sources and the paths of conversion to electrical energ>
The numbers and letters are the items listed in Tables 11.1 and 11.2.

It is interesting to note the multiplicity of paths whereby electric;
energy may be produced. As an example, consider the case of sola
energy which, as shown in Fig. 11.1, can be converted directly t
electrical energy by utilizing a photovoltaic device such as a silico·
photodiode. The current level of silicon solar cell conversion efficiency i
12 percent. A much more complex and considerably less efficient path i
the conversion and storage of solar energy via photosynthesis to
chemical form such as an algae culture. The algae, for example chlorell;
can convert solar energy at an efficiency of about 8 percent and the energ
can be stored before use. The chemical energy in chlorella can in turn b
converted t to heat by burning the algae, the heat can be converted t
mechanical energy, and finally, mechanical energy can be converted t

TABLE 11.1. ENERGY SOURCES AND CONVERSIONS

Source and Conversion Process	Conversion Mechanism	Output
Chemical Sources		
(1) Electrochemical	(a) Primary battery	Electrical
	(b) Secondary battery	Electrical
	(c) Fuel cell	Electrical
(2) Thermochemical	(a) Burning of fossil fuels	Heat
	(b) Exothermic reactions	Heat
Solar Sources		
(3) Photogalvanic	Photosensitive oxides or dyes	Electrical
(4) Photovoltaic	Solar battery	Electrical
(5) Photoelectric	Photoelectron emission	Electrical
(6) Photochemical	(a) Photosynthesis	Chemical
	(b) Photodecomposition	Chemical
(7) Conversion of radiant energy to heat	Solar furnace	Heat
Nuclear Sources		
(8) Radioisotope decay	(a) Beta current battery	Electrical
	(b) Contact potential	Electrical
	(c) p-n junction	Electrical
	(d) Absorbtion of radiation by container	Heat
	(e) Radiochemical regeneration of fuel cell reactants	Chemical
(9) Fission	(a) Reactor, thermopile	Heat
	(b) Plasma–electromagnetic coupling	Electrical
(10) Fusion	(a) Contained explosion	Heat
	(b) Plasma–magneto-hydrodynamics	Electrical
Natural Mechanical Sources		
(A) Elevated water	Water turbine	Electrical
(B) Winds	Windmill	Electrical
(C) Tides	Fluid mill	Electrical
(D) Waves	Fluid mill	Electrical

TABLE 11.2. CONVERSION OF THERMAL AND MECHANICAL ENERGY
TO ELECTRICITY

Static Thermal Processes	
(11) Thermoelectric	(a) Metals
	(b) Intermetallic compounds
	(c) Plasmas
(12) Thermionic	
(13) Thermomagnetic	
(14) Pyroelectric	
(E) Thermogalvanic	Thermal regeneration of fuel cell reactants.
Dynamic Thermal Processes	
(15) Dynamic heat engine	(a) Rankine
	(b) Brayton
(16) Magnetohydrodynamics	
(F) Internal combustion engine	
Mechanical Processes	
(17) Piezoelectric	
(G) Electromagnetic	
(H) Hydroelectric	
(I) Windmill	

electrical energy. Total overall conversion efficiency from solar energy to electrical energy by this method is well below 1 percent.

CONVERSION MECHANISMS

In the following paragraphs, the various energy sources and conversion processes are more fully discussed. Among the energy conversion processes, several have emerged as significant contenders for useful application as power supplies. These, other than electrochemical, are photovoltaic (solar cells), thermoelectric, thermionic, Rankine and Brayton dynamic heat engines, and magnetohydrodynamics (MHD). Accordingly, they are covered in greater detail.

ELECTROCHEMICAL

In primary batteries, continuous feed and fuel cells, and secondary batteries, energy in a chemical source is converted directly to electricity by way of an electrochemical reaction. The conversion process is usually very efficient, and a high percentage (60 to 80 percent) of the total free energy may be realized. Additional advantages for batteries include portability, operation at ambient temperatures, simplicity of construction,

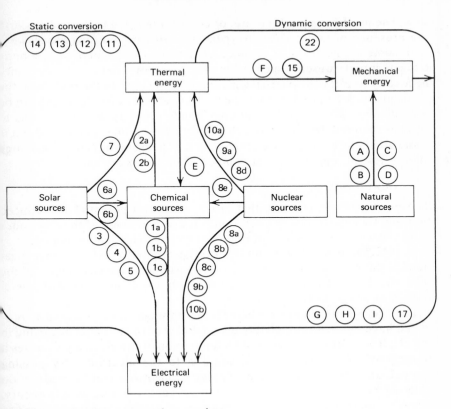

Figure 11.1 Energy sources and conversions.

freedom from moving parts, reliability and minimum maintenance, and immediate availability of full power after periods of idleness. Continuous feed and fuel cells are more complex than conventional batteries and require auxiliary plumbing and pumping accessories. Some even operate at elevated temperature. They are, however, operable for thousands of hours and have been considered for multikilowatt power sources; though high cost and low power density at present limit their usefulness for such applications.

THERMOCHEMICAL

By far the largest amount of electrical energy produced today involves the burning of elements or compounds in air to produce heat which is then converted to electrical energy through a heat engine-generator combina-

tion. The most common fuels are, of course, the fossil fuels, and on eart
the presence of atmospheric oxygen facilitates the burning process.

In some applications in confined environments or where atmospheri
oxygen is not present, such as in aerospace or undersea, it become
necessary to carry the oxidizing medium for the fuel. In the pure form th
oxidizer can be carried in containers and released as needed, or it can b
provided combined in chemical compounds properly dispersed so as t
react as required. In many cases the oxidizing medium is not oxygen but i
some other electronegative element (such as halogen) having a combinin,
affinity for an organic carbon compound or metallic element.

PHOTOGALVANIC

The photogalvanic effect is that effect wherein a potential is produce
between two electrodes immersed in solution when one of the electrode
or the electrolyte adjacent to one of the electrodes is illuminated.

In 1942, Copeland, Black and Garret(1) presented a review of investiga
tions in this field, and in 1955 Sancier(2) presented a concise review of th
field. In 1958, Rosenberg(3) discussed photogalvanic processes in
review of photochemical processes.

The precise mechanisms whereby the photogalvanic potential is pro
duced have not been completely elucidated. For the reaction occurring i
the electrolyte near one electrode, illumination of the electrolyte causes
transfer of electrons from the solution to the electrode. By passin
through an external circuit, the transferred electrons can be made to d
external work. When illumination ceases, reverse reaction usually occurs
releasing thermal energy into the solution. If any storage features are t
be realized from such a cell the rate of the reverse thermal reaction mus
be very slow.

Several types of photogalvanic cells are known(3): (a) metal electrode
immersed in electrolytes, (b) metal electrodes coated with inorgani
compounds immersed in electrolytes, (c) metal electrodes coated wit
dyes immersed in electrolytes, and (d) metal electrodes immersed i
organic liquids.

One system which has been extensively studied is the iron-thionin
system(4). The potential of the cell has been reported to be 0.4 V, and a
overall efficiency of 1 percent was realized, but the reverse reaction wa
too rapid to permit energy storage.

PHOTOVOLTAIC

When a potential is produced as a result of the effect of illumination on
surface material the effect is referred to as the photovoltaic effect. Th
barrier-layer photovoltaic cell was first developed around 1876 by Adam

and Day(5). The light-sensitive material utilized, selenium, still used today in instruments and electronic devices, converts light energy to electrical energy with an efficiency of about 0.6 percent. In 1954, Chapin, Fuller, and Pearson(6) of Bell Telephone Laboratories announced the development of a silicon photovoltaic device having a conversion efficiency of 6 percent. Rappaport(7) has reviewed progress in the field up to 1958. Crossley, Noel, and Wolf(8) prepared an extensive review in 1968.

Considerable development has taken place in the technology of silicon solar cells, spurred on by their widespread use in aerospace applications. Current commercial product has a conversion efficiency of 12 percent while laboratory samples have achieved 15 percent. The theoretical ultimate has been calculated to be 22 percent. The cost for bare silicon solar cells is about $60 to $80 per watt but mounted arrays have cost $300 to $1600 per watt. At the mean earth orbital distance from the sun the thermal radiation level is 1.4 kW per square meter, and therefore at 12 percent efficiency the specific power for solar cells is 168 watts per square meter. Solar cell arrays can be built with specific power of 4 to 20 W/lb. By using roll-out arrays it is expected that 30 W/lb can be achieved.

The key element of the solar cell is the p-n junction close to the surface of a crystal of silicon. Pure silicon is a semiconductor; the element has four valence electrons. By introducing small amounts of impurities into the silicon crystal its semiconductor properties are enhanced. If an atom of silicon is replaced by a five-valence electron element such as arsenic, antimony or phosphorus, there is a loosely bound extra electron in the lattice structure which is free to migrate and impart conductivity. Such a semiconductor is called n-type. If an atom of silicon is replaced by a three-valence electron element such as boron, indium or gallium there is a hole in the lattice structure which can move about and impart conductivity. Such a semiconductor material is called p-type. At the boundary between p- and n-type materials there is a field which can be utilized.

Consider the potential energy diagram of a silicon solar cell as shown in Fig. 11.2. Since the average energy of the carriers, the Fermi level, must be equal in both the p-type and n-type regions, a potential barrier is set up at the p-n junction. Light causes electron-hole pairs to form in the crystal by moving electrons from the valence band up into the conduction band. If these are formed within a diffusion distance of the p-n junction, the potential of the junction causes holes to flow across the junction into the p-type and electrons to flow across the junction into the n-type. These pairs can be recombined through an external circuit to do useful electrical work.

Figure 11.3 illustrates the construction of a solar cell. The bulk of the silicon is n-type typically doped with arsenic. A layer of boron is diffused

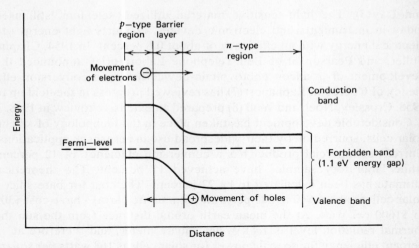

Figure 11.2 Potential energy diagram for a silicon photovoltaic converter.

into the surface of the crystal to convert it to p-type. The p-layer, located on top so that it can absorb the photon energy readily, is very thin so that the electron-hole pairs formed will come under the influence of the junction before they have a chance to recombine in the p-layer.

The electrical characteristics of solar cells are shown in Fig. 11.4. Current density is directly proportional to radiation intensity while the open circuit voltage tends to approach a constant value at high radiation

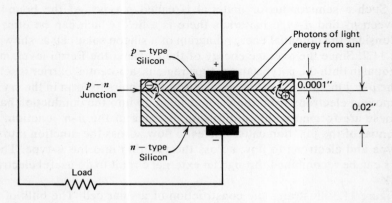

Figure 11.3 Schematic diagram of a silicon solar cell.

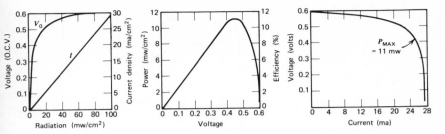

Figure 11.4 Electrical characteristics of solar cells (7).

values. Power output is a maximum at an intermediate voltage value. This maximum power point corresponds to the point where the maximum-area rectangle can be drawn under the voltage current characteristic.

Silicon has an energy gap (see Fig. 11.5) of 1.1 eV. By comparison, metallic conductors have a zero energy gap and insulators have an energy gap greater than 3 eV. Silicon thus is responsive to radiant energy above 1.1 eV, that is, all shorter wavelengths, which is very desirable because this includes a good fraction of the sun's visible and infrared radiation. For energies below 1.1 eV the low energy photons cannot create electron-hole pairs. On the other hand, energies much above 1.1 eV are not useful either since the electron-hole pairs are formed quickly and so far away from the p-n junction that they recombine before reaching the junction, and no useful energy is released to the external circuit. Since the high energy radiation is not too readily used by silicon, other materials such as gallium arsenide, indium phosphide, or cadmium telluride which have a higher energy gap might be used to more efficiently utilize the energy in the ultraviolet region. Figure 11.5 shows the theoretical efficiency of various solar cell materials exposed to the solar radiation above the earth's atmosphere as a function of the energy gap of the semiconductor material. A tandem combination of several materials might be able to utilize a wider total spectrum of energy.

Solar cells have been used extensively in conjunction with nickel-cadmium batteries in orbiting aerospace power supplies. During the sunlit period of the orbit the solar cell array provides the electrical power needs plus the power to recharge sealed nickel-cadmium batteries. During the dark period of the orbit the nickel-cadmium battery carries the electrical load. The largest solar array-battery power system built to date is 10.5 kW for NASA's Apollo Telescope Mount. A solar array-battery power system of 100 kW size is being considered for the Manned Orbiting Space Station to be launched in the late 1970s.

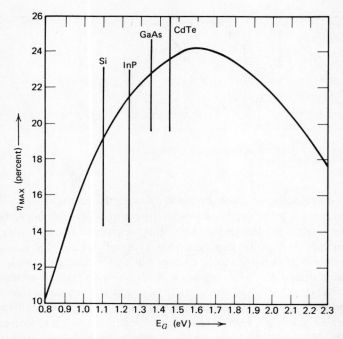

Figure 11.5 Theoretical efficiency of various solar cell materials as a function of semiconductor energy-gap (6).

PHOTOELECTRIC

The photoelectric effect is that effect whereby radiant energy acts on a cathode material to cause the emission of electrons. Einstein deduced the equation for the photoelectric effect to be $h\upsilon = \frac{1}{2} m\upsilon^2 + P$ where $h\upsilon$, the energy of the incident photon, is equal to the kinetic energy of the emitted electron, $\frac{1}{2} m\upsilon^2$, plus the energy P required to release the electron from the metal (the work function). The required energy and the efficiency for the conversion depend on the work functions of the materials used for cathode and anode.

In a speculative paper by Castruccio(9) in 1958 there was described a proposed photoelectric power plant for use on the moon, utilizing the vacuum environment of the moon as the dielectric. A photocathode constructed of a thin plastic sheet covered with a thin film of a photosensitive compound would be stretched out over a large area of the moon's surface. The photoanode would be a thin wire mesh stretched out just above the plastic film but separated from it a fixed small distance by insulating spacers. For high voltage, sections of the device would be connected in series. The concept was never pursued.

PHOTOCHEMICAL

A molecule exposed to radiation may absorb a photon of energy. As a result of this absorption of energy, a change may occur which is capable of yielding a product which can subsequently be reacted to yield heat, or products may be formed which can subsequently be recombined electrochemically to yield useful electrical energy.

The most familiar photochemical process is photosynthesis wherein chlorophyll brings about the reaction of carbon dioxide and water in sunlight to produce complex carbohydrates. In the process photon energy is converted to chemical energy and can be stored in that form. The photosensitive reactions in the presence of chlorophyll are complex and still not completely elucidated. In general the efficiency for the conversion of solar energy is about 1 percent, but in some laboratory experiments using monochromatic light in the red region, where absorption by chlorophyll is strongest, efficiencies as high as 70 percent have been reported. Many studies have been made of photosynthesis and several reviews have been published(3)(10)(11).

As a conversion process from solar energy to electricity the species of algae known as chlorella has been studied as the chemical intermediate and energy-storage medium(12). The actual efficiency of solar energy conversion by chlorella is 1 percent although the theoretical value is 8 percent. The energy chemically stored by algae is converted to electricity by burning the algae as a fuel and then converting the heat to electricity by a heat engine. The efficiency of the latter step may be about 10 percent, and the overall efficiency for the process would thus be about 0.1 percent.

An interesting consideration in the culture of algae is their response to illumination. Much of the intensity of solar radiation is wasted on chlorella, resulting in conversion to heat and reduced efficiency. An effective way of more efficiently utilizing solar energy is to diffuse it over a large area of algae culture. This is an interesting contrast to the philosophy of most solar collectors where the intent is to concentrate solar energy and raise the temperature at the focal point of the solar collector.

In the energy conversion process utilizing photodecomposition, solar energy is absorbed by a chemical reaction which splits a chemical compound into several constituent compounds. These can be recombined when desired to release heat or electrical energy. The photodecomposition of nitrosylchloride is of interest as a solar-energy conversion system, since the decomposition is sensitive to almost the whole visible spectrum, and the quantum yield is high. An efficiency of 10 percent has been reported for the decomposition. The reverse reaction under controlled conditions, such as in a fuel cell, would yield electrical energy.

CONVERSION OF RADIANT ENERGY TO HEAT

The amount of solar energy incident on the surface of the earth is approximately 1.4 kW per square meter. The surface area required for collection of useful amounts of power is fairly large, and, although the energy is free, the capital cost of the collecting device can be quite high. An important consideration is collector efficiency which can be as high as 65 percent. Work done in this field has been reviewed by Duwez(13) and Telkes(14).

Incident solar energy must be absorbed by an opaque surface which converts the radiant energy into thermal energy. The surface may be flat, or it may be curved or augmented with mirrors and lenses to concentrate the solar energy. In either case the area of surface required per unit of energy absorbed is the same, but the use of concentrators permits the working fluid to be brought to a higher temperature which increases the efficiency of the heat engine into which the collector may be working.

Flat plate collectors usually are fabricated of a metallic surface behind which a working fluid circulates. High absorption by the plate is essential and means must be provided to prevent heat losses by re-emission from the black plate.

The simplest type of concentrating collector is one in which the edges of a flat plate collector are boxed in with mirrors which reflect solar energy to the flat absorbing plate. This simple technique can double the absorption capability of the collector.

More complex concentrating collectors utilize absorbing surfaces having optical shapes, either spherical, parabolic or cylindrical. The maximum theoretical concentration ratio achievable with a parabolic collector is about 46,000 times that received by a flat surface. Extreme geometric perfection is not possible in building these collectors, and concentration ratios of 10,000 are achievable practically. The maximum efficiency of conversion of these devices is around 50 percent and a temperature of 3000° to 4000°C can theoretically be attained at the focus.

Where only moderate temperatures of the order of 1000°C to 2000°C are required, optical perfection in the collector is not necessary and collector efficiency can be maintained at high values. The foregoing is the requirement for the conversion of heat to electrical energy using heat engines since the upper temperature limit for the common materials used in heat engines is about 2000°C.

RADIOISOTOPE DECAY

The specific power release associated with the decay of radioisotopes is many orders of magnitude greater than from chemical reactions. During the spontaneous decay of radioisotopes the kinetic energy of the emitted radiation can be absorbed by an enclosure and converted to heat, which can then be converted to an electrical output. The energy conversion process involving the intermediate generation of heat is the most important and most widely used process. There are, however, several other processes which have been reported wherein radioisotope decay is converted directly to electricity.

Three types of radiation emanate from isotopes; alpha, beta and gamma. The alpha radiation is safest, and the least penetrating. Gamma radiation is the most damaging both biologically and to materials of construction.

Radioactive isotopes are available today from two sources: waste products from fission reactors and artificially produced materials prepared in an atomic pile. In selecting an isotope as a primary energy source several factors are of importance; (a) specific power, (b) half-life, (c) availability, (d) cost, and (e) biological and radiation damage hazards and shielding requirements. Table 11.3 presents data on some available radioisotopes which are used as power sources.

The technology of radioisotope-fueled power supplies coupled to a heat-to-electrical energy converter has advanced significantly in the past decade. Much of the work has been directed at developing aerospace power supplies under the SNAP (Systems for Nuclear Auxiliary Power) program. In these systems, the radiation from the decay of the radioactive isotope is first absorbed in a metal container and converted to heat. The selection of the isotope, the amount of isotope and the size and design of the container are dependent on the temperature desired which is in turn dependent on the heat to electricity conversion device which is used.

TABLE 11.3. PROPERTIES OF RADIOISOTOPES FOR POWER SOURCES

Isotope	Half-Life Years	Principal Radiation	Shielding Required	Watts/ gram	Watts/ cm^3	Cost Range $/ Thermal Watt
Strontium ^{90}Sr	28	β, χ	Heavy	0.22	0.82	25–35
Promethium ^{147}Pm	2.6	β	Minor	0.17	1.8	200–600
Cerium ^{144}Ce	0.78	β, γ	Heavy	3.8	25.3	1
Curium ^{242}Cm	0.45	α, η	Minor	98	882	—
Polonium ^{210}Po	0.38	α	Minor	82.4	824	10–25
Plutonium ^{238}Pu	87.4	α	Minor	0.36	3.6	500–700

Reference: *Isotope and Radiation Technology*, Vol. 6, No. 2, Winter 1968–1969.

In 1959, design considerations for isotope power sources were reviewed by Kittl(15). He presented detailed equations based on the power density of the nuclear fuel, the power decay law and the heat flow in thermoelectric and thermionic converters. Knowing the voltage output desired from the device, the desired volume of the device and the operating temperature of the heat junction (1000°K for thermoelectric devices and 2500°K for thermionic devices), the derived functions indicate the specific power required of the source. Considerable further work has been reported, which Corliss and Harvey(16) have presented in a text.

For the subsequent generation of an electrical output from the radioisotope generated heat, the use of thermoelectric converters has predominated. A series of SNAP isotope-thermoelectric systems has been built and used in aerospace missions(17)(18). Table 11.4 presents data on some of these systems(19). Currently a specific power slightly better than 1.0 W/lb has been achieved. The lead-telluride materials are the most fully developed thermoelectrics. When in a cascade arrangement with higher temperature silicon-germanium materials, currently in advanced development, it is expected that 2.0 to 2.5 W/lb will be achieved. Efficiencies of present systems are 4 or 5 percent but 8 to 9 percent is anticipated. Increased efficiency is significant since it will result in reduced fuel inventory thereby reducing costs and potential hazards. Plutonium appears to be the most attractive fuel at the present time but curium is attractive for future development because of its high specific power, suitable for higher temperature converters such as thermionic diodes, and its cost, may be significantly lower.

TABLE 11.4. SNAP RADIOISOTOPE THERMOELECTRIC SYSTEMS

System	Application	Isotope	Power Level (watts)	Weight (lbs)	Mission Duration
SNAP-3A	Satellite power	^{238}Pu	2.5	4	90 days
SNAP-7A-F	Buoys, beacons and weather stations	^{90}Sr	7.5–60	600–4600	2 yr
SNAP-9A	Satellite power	^{238}Pu	25	27	5 yr
SNAP-11	Moon probe	^{242}Cm	21–25	30	90 days
SNAP-17	Communications satellite	^{90}Sr	25	30	5 yr
SNAP-19B	Weather satellite	^{238}Pu	30	30	5 yr
SNAP-21	Deep sea	^{90}Sr	10	500	5 yr
SNAP-23	Terrestrial	^{90}Sr	60	900	5 yr
SNAP-27	Lunar landings	^{238}Pu	60	30	5 yr
SNAP-29	Various future missions	^{210}Po	500	500	90 days

Over the years various devices have been reported which directly convert radioisotope energy to an electrical output.

In 1913 Moseley(20) constructed an isotope converter using radium in a thin sphere which radiated beta particles (i.e., electrons) through a vacuum dielectric to the inner walls of a second larger collecting sphere surrounding the inner emitter. A high potential of 150 kV was realized. The beta current type nuclear batteries in use today utilize the principles applied by Moseley. Such a battery is shown in Fig. 11.6. The dielectric is either vacuum or a plastic or some other type insulator. Developments in isotope devices producing electrical energy were reviewed by Coleman(21). Strontium 90, krypton 85 and tritium have been used as the energy source. One of the first prototype batteries used 10 mCi of strontium 90 and developed 50 $\mu\mu$a at 14 kV in a volume of 1 in.^{3}. The characteristic of this type of device is very low current output at a high potential, limiting its use as a power source.

A contact potential device has been described(22)(23), in which a radioactive source irradiates a low-pressure gas contained in an enclosed space between two electrodes fabricated of dissimilar materials. Such a device is shown in Fig. 11.7. The irradiation causes the gas to ionize and the positive and negative ions are attracted to the oppositely charged electrodes where they give up their charge causing electrons to flow through an external electrical circuit. The potential of this cell is of the

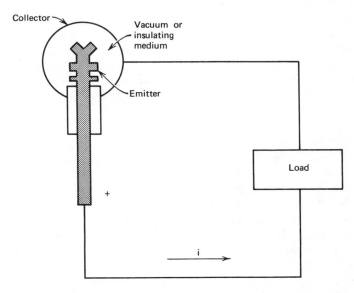

Figure 11.6 Schematic diagram of a beta current radioisotope converter.

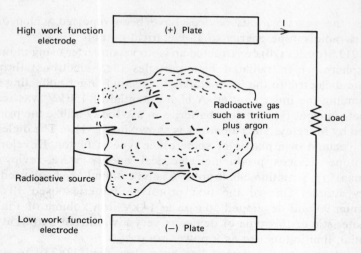

Figure 11.7 Schematic diagram of a contact potential radioisotope converter (23).

order of a fraction of a volt to several volts and is governed by the work function of the materials used as electrodes. Current output is of the order of 10^{-10} amp. Tritium has been used as the radioactive source.

Work has also been done on p-n junction devices to convert nuclear radiation directly to electrical energy (22). A silicon p-n junction subjected to the radiation from strontium 90 yielded 0.2 V and current in the microampere range. One major shortcoming of this device is the destruction of the p-n junction by the radiation. The destructive action of the p-n junction was eliminated in one device wherein the radioactive decay energized a phosphorescent material, and the light generated therefrom was converted to electrical energy in a photovoltaic p-n junction device. Overall efficiency for the multiconversion device was less than 1 percent.

FISSION

In a fission reaction the large energies released are due to the conversion of a small portion of mass (m) into energy (E) in accordance with the relationship $E = mc^2$ in which c is the velocity of light. E then becomes the kinetic energy of the fission fragments as they electrostatically repel each other. When the uranium fission process occurs the uranium nucleus is excited and two bundles of protons begin to oscillate toward and away from each other. The two bundles are caused to separate due to the strong electrostatic repulsive forces which come into play at distances of 10^{-12} cm. The fragments which fly apart have very high kinetic energy. The energy released by a fissioning uranium atom is 10^7 times the energy stored in the valence electrons of the atom, which is the energy normally available from a chemical reaction.

The most widely utilized fission process is the light water reactor process which is fueled with uranium-235 and uranium-238. When a free slow neutron penetrates the ^{235}U nucleus it causes it to fission. The fission splits the nucleus into smaller fission fragments, releases energy as gamma ray emission and produces two new fast neutrons. One of these fast neutrons is moderated by collision with the hydrogen nuclei of the water present in the reactor, and when sufficiently slowed down will enter another ^{235}U nucleus, thus sustaining the fission reaction. The second fast neutron is 40 percent wasted as heat and 60 percent absorbed by ^{238}U which is converted to plutonium.

The emitted radiation, generated in the core of the nuclear reactor is removed by the water which is used as the heat transfer fluid circulating in the jacket surrounding the core. The isotope ^{235}U occurs naturally only to the extent of 0.71 percent of the uranium element as found in nature. During the fission process there is always a net consumption of fissionable ^{235}U.

More efficient use of uranium can be achieved in the liquid metal fast breeder reactor which is fueled with plutonium and ^{238}U. In this type of reactor a fast neutron is caused to penetrate the plutonium nucleus causing it to fission. The fission releases energy and produces 2.5 new fast neutrons. One of these fast neutrons enters another Pu nucleus and sustains the fission reaction. The remaining 1.5 fast neutrons convert ^{238}U into more plutonium. Thus a fast reactor breeds more plutonium than it consumes. Since the reactor design must favor the interaction of fast neutrons with the plutonium fuel, this type of reactor is much smaller and more difficult to control and to cool.

Fission reactors yielding thermal energy for conversion to electrical energy are attractive primarily where large amounts of power (above 3 kW) are desired. A fission reactor requires considerable shielding, and rejection of the waste heat also poses a considerable problem. Heat engines are readily coupled to fission reactors and these usually are the dynamic turbine types. The high temperatures produced yield fairly high heat-conversion efficiencies. Direct conversion of the fission-generated heat can also be achieved by the use of thermoelectric or thermionic converters, which are discussed later in this text. Still another direct conversion method embodies the use of a fission reaction to produce a plasma which by magnetohydrodynamic techniques results in an electrical output.

A more speculative energy conversion scheme described by Colgate and Aamodt(24) is a device utilizing a high temperature-fissionable uranium plasma contained in a cylindrical container. An initial excitation causes the uranium gas to go critical and fission. The shock wave travels down the cylinder, compresses the plasma at the other end and fission again occurs reversing the shock wave. The plasma oscillating between the two ends of the container is electromagnetically coupled to coils outside the container and an electric potential is induced in these coils.

FUSION

In the fusion process deuterium reacts with itself or with tritium, accompanied by a large release of energy. The attractiveness of fusion power lies in the essentially inexhaustible and cheap supply of fuel— seawater (1 gallon of seawater yields the energy-equivalent of approximately 300 gallons of gasoline). The big technical problems associated with the controlled use of fusion power are the extremely high temperature (200 million °K), at which the thermonuclear reactions occur, and the attendant difficulty of containing them. Colgate (25) and Werner and his co-workers (26) have summarized these problems and approaches pertinent to the controlled utilization of thermonuclear reactions.

In the AEC's project Plowshare program, consideration was given to the contained thermonuclear explosion as a means of generating power. A conceptual design shown in Fig. 11.8 consists of an underground cavity with a steam line leading to a turbogenerator plant and a lock and passage for the introduction of thermonuclear bombs into the cavity. A plant capability of 2000 MW at an assumed thermal efficiency of 25 percent would require a 200-kiloton explosion every 30 hours.

The objective of another project, known as the Sherwood project, has for its objective the attainment of steady-state fusion power by magnetic field containment of an ionized plasma. The heat content of the plasma is not very large or difficult to handle, but it is difficult to prevent the plasma from cooling down. If the plasma cools, the temperature-catalyzed thermonuclear reaction stops. In a plasma the high temperature electrons and ions readily transfer their heat to a wall because they can strike the wall in a linear path with very little collision among themselves. To contain the plasma, strong magnetic fields are applied to it which react on the charged particles, causing them to assume a circular motion which keeps them away from the container walls. The magnetic field must exert a force on the particles great enough to contain the pressure of the plasma. A magnetic field of 5000 gauss will contain a plasma at one atmosphere. Typical conditions in a "magnetic bottle" are 100 atmospheres pressure at 100 million °K, and a field of 50,000 gauss is used to contain the plasma. The main problem is heating the plasma initially and preventing it from leaking out of the magnetic bottle and cooling. Thermal

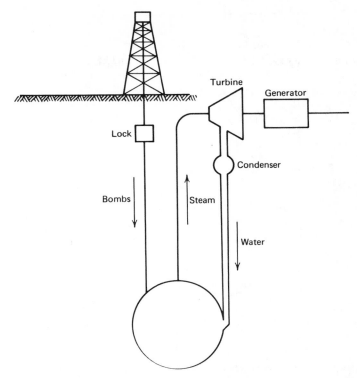

Figure 11.8 Schematic design for a power plant to utilize thermonuclear explosions for the generation of electrical power (25).

losses are present due to radiation from the plasma and diffusion of the plasma out of the magnetic bottle.

Several different techniques for the magnetic confinement of thermonuclear plasmas are being investigated. These include the stabilized pinch, the magnetic mirror, the Stellerator and the Astron, which are described in the literature of this field.

In Fig. 11.9 is plotted the fusion power generated per unit of volume as a function of plasma temperature for the reaction between deuterium and tritium (DT) and deuterium and deuterium (DD). Also plotted is the radiation loss at each temperature. A self-sustaining reaction must be at a higher temperature than the intersection of the radiation loss line and the power generation lines. This is called the ideal or critical ignition temperature.

The method for converting thermonuclear energy to electrical energy is

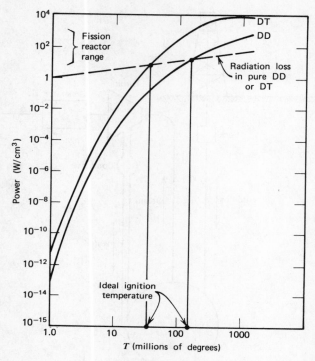

Figure 11.9 Fission power generation as a function of temperature (25).

dependent on the thermonuclear reaction. Where the kinetic energy is converted to heat by absorption at container walls, conversion to an electrical output is by a heat engine or by thermoelectric or thermionic direct energy converters. Where, however, the bulk of the energy released is in the form of charged particles, the coupling of plasmas to electric coils is being considered, but this art is still in its early stages of development. For the present, most effort is being devoted to the problem of controlling and containing the thermonuclear reaction itself.

THERMOELECTRIC

In 1822, the German physicist, Thomas Seebeck, discovered that a voltage was produced when he heated one junction of a loop containing two dissimilar materials while the other junction was kept cold. The Seebeck coefficient "S" is the coefficient of proportionality between generated voltage and temperature difference. In 1834 a French watchmaker, Jean Peltier, reported that a current passing through a junction of two conductors caused the junction to become heated or cooled depending on the direction of the current. The Peltier coefficient, π, relates the heat effect to the current. In 1854, Lord Kelvin applied thermodynamic theory to confirm Seebeck's discoveries and theoretically related the Seebeck and Peltier coefficients. The early history and work on thermoelectric effects have been reviewed by Ioffe(27) and Telkes(28).

Early studies of thermoelectricity concentrated on the use of metals as the thermoelectric materials. Conversion efficiencies of about 1 percent or less were attainable. In 1947 Telkes in the United States reported on the use of new materials, PbS-ZnSb, which gave efficiencies of 5 percent. Ioffe(27) in Russia has also been predominant in the development of semiconductor technology. Several summary articles have codified the technology and reviewed progress in the field(29)(30)(31).

A thermoelectric device for power generation is depicted schematically in Fig. 11.10. For such a device, efficiency, "η", is independent of physical size and geometry, and may be defined as the ratio of electric power output in a load P_l to the rate of heat flux Q entering the thermoelement hot junction:

$$\eta = \frac{P_l}{Q} \tag{1}$$

The thermocouple efficiency, η_{tc}, is expressed as a fraction which multiplied by the theoretical maximum Carnot cycle efficiency, η_{carnot}, gives the device efficiency

$$\eta = \eta_{carnot} \, \eta_{tc} \tag{2}$$

$$\eta_{carnot} = \frac{T_h - T_c}{T_h} \tag{3}$$

in which T_h and T_c are the temperatures of the hot and cold junctions respectively.

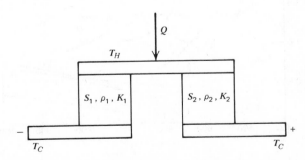

Figure 11.10 Schematic diagram of a thermoelectric element.

Thermoelectric materials have been the subject of considerable research in this field. The three parameters of a material which govern the efficiency of a thermoelectric device are the Seebeck coefficient S, the electrical resistivity ρ, and the thermal conductivity K. Ioffe has related these as shown below to yield the figure of merit Z;

$$Z = \frac{S^2}{\rho K} = \frac{(\text{volts/°C})^2}{(\text{ohm-cm})(\text{watts/cm-°C})} \tag{4}$$

For metals the Seebeck coefficient S is very small, (a few microvolts per degree centigrade) and K and ρ are related by the Wiedemann-Franz-Lorenz law ($K\rho = LT$) where L is the Lorenz number (a constant) and T is the absolute temperature. Semiconductors have values of S which are as high as 200–300 μV/°C but the thermal conductivity of semiconductors is high because of conduction both by the charge carriers and the crystal lattice. Ioffe reasoned that the use of mixed crystals would provide superior materials because the charge mobility could be kept substantially unchanged while the crystal disorder introduced would reduce thermal conductivity through the crystal lattice. Ternary compounds of antimony, arsenic, bismuth, indium, lead, phosphorus, selenium and tellurium are today among those having the highest Z values known. Figure 11.11 shows the figure of merit for several polycrystalline materials. Semiconductors of germanium or selenium crystals have the shortcoming that at elevated temperatures the intrinsic conductivity of the material increases to an intolerable value.

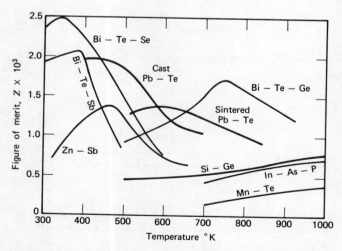

Figure 11.11   Figure of merit of several thermoelectric materials.

While many semiconductors have been extensively investigated during the past 10 years, lead-telluride and silicon-germanium have emerged as the two most significant materials. At low temperatures, lead-telluride has the higher figure of merit but when the temperature approaches 1000°K, the silicon germanium material begins to have a higher figure of merit. Assuming S, ρ, and K to be temperature independent, Ioffe has derived an expression for the efficiency of a thermoelectric couple consisting of materials 1 and 2.

$$\eta = \frac{T_h - T_c}{T_h} \left(\frac{\sqrt{1 + ZT_{av}} - 1}{\sqrt{1 + ZT_{av}} + \dfrac{T_c}{T_h}} \right) \tag{5}$$

where

$$T_{av} = \frac{T_h + T_c}{2}, \qquad Z = \frac{(s_1 - s_2)^2}{(\sqrt{\rho_1 K_1} + \sqrt{\rho_2 K_2})^2} \tag{6}$$

Z is the figure of merit for thermoelectric materials.

The cascading of thermoelectric elements of different materials in series permits the optimum matching of materials and their effective temperatures and increases the overall efficiency of the device. Generally the number of stages needed is quite small; four or five.

Early work with metallic couples such as chromel-alumel yielded efficiencies of about 1 percent. Semiconductor compounds such as PbTe give 5 to 6 percent efficiencies. Some recent materials have yielded 10 to 20 percent in laboratory measurements.

Figure 11.12 shows the thermoelectric efficiency attainable with various different materials as a function of temperature. The maximum theoretical efficiency calculated for thermoelectric materials is 36 percent. In working devices semiconductor materials have yielded 4 to 5 percent overall efficiency. By cascading high and low temperature devices and operating between 1000° and 30°C, an overall efficiency for a working device is expected to be 10 percent.

The technology of complete thermoelectric power supplies has advanced considerably in the past decade. The thermoelectric converter has been coupled to such sources as radioisotope fueled, solar fueled, fossil fueled and nuclear fueled. Many radioisotope-thermoelectric power supplies have flown in aerospace applications as SNAP systems. They have achieved specific powers of 1 W/lb using PbTe and it is anticipated that 2 to 2.5 W/lb will be achieved with higher temperature SiGe converters. Many thermoelectric systems have also found use in terrestrial and undersea applications. In 1965 a 500-W thermoelectric converter with a nuclear reactor source was test flown successfully by the Air Force.

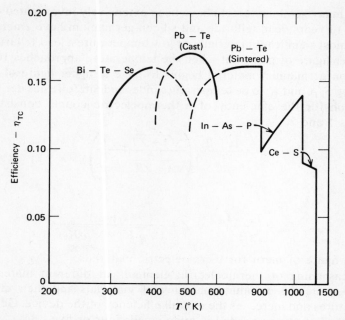

Figure 11.12 Thermoelectric efficiency of several thermoelectric materials (30).

THERMIONIC

Thermionic energy conversion is based on the Edison effect patented in 1883 which describes the flow of current from the surface of a heated element in a vacuum. In 1953 a systematic study of this phenomenon was begun to utilize it as a technique for the direct conversion of heat to electrical energy. Moss (32) in 1957 described the use of thermionic diodes as energy converters. Detailed studies of the thermionic converter and analysis of its performance have been published by Hatsopoulos and Kay (33), Wilson (34), Hernqvist, Kanefsky and Norman (35) and others.

The thermionic converter is a true heat engine. Heat is supplied at a hot cathode where electrons are boiled off to form an electron gas which is the working fluid. This electron gas passes through a vacuum and condenses on an anode which is kept at a lower temperature. The basic principle of the device is that a specific number of electrons emitted by a hot cathode have sufficiently high emission velocities to overcome a retarding electrostatic potential barrier between a cathode and an anode in a vacuum. The high initial kinetic energy of these electrons lifts them to a high potential energy position and if they fall back into another surface

having a different work function from the emitting surface, the difference
between the two work functions provides an emf which can be utilized to
do work in an external electric circuit connecting anode to cathode.

The potential energy diagram of the thermionic converter is shown in
Fig. 11.13. The cathode is a material having a high work function ϕ_c, and it
is heated to emit electrons which have sufficient energy to overcome the
space charge barrier δ. The electrons pass to the anode or collector which
is a material of lower work function. The net potential energy V_o is
available as an electric potential. Normally a low work function surface
will emit electrons more readily than a high work function surface but in
this device the high work function cathode is heated and electrons are
emitted according to the Richardson equation:

$$J_s = AT^2 \exp\left(-\phi/kT\right) \tag{7}$$

where J_s = saturation current, amps/cm^2;
    A = constant, amp/°K . cm^2;
    ϕ = work function, volts;
    k = Boltzmann constant, 0.861×10^{-4} eV/°K;
    T = absolute temperature, °K.

In order to achieve a useful conversion efficiency, it is necessary to
provide a means to cancel the effect of the space charge. Four methods
are available: (a) place the electrodes very close together to reduce space
charge formation(33), (b) use positive ions to neutralize it(34), (c) use
electric or magnetic fields to conduct electrons from cathode to

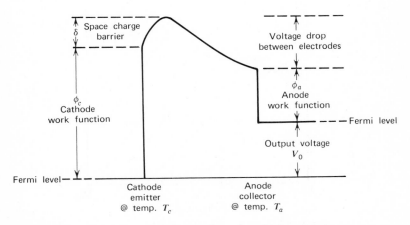

Figure 11.13 Potential energy diagram for a thermionic converter.

anode(35), and (d) use a grid to accelerate the electrons past the space charge. The first two techniques have been most successfully utilized.

In the approach taken by Hatsopoulos(33), precisely machined emissive surfaces were utilized such that very precise small distances of the order of 0.001 inch or less were provided between anode and cathode. In addition, very careful heat-transfer designs were employed to greatly reduce back-current losses from the anode to the cathode. Figure 11.14 shows the relation between maximum power output and spacing between electrodes for a range of work function differences from 0 to 1 V.

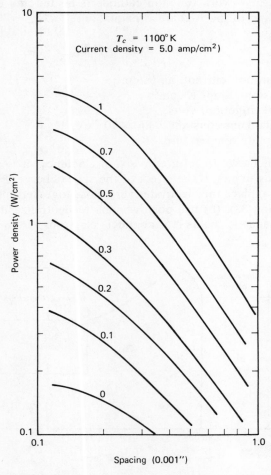

Figure 11.14 Maximum power output as a function of interelectrode spacing for various values of work-function difference between electrodes (34).

Wilson's(34) work described the use of cesium vapor in the space between electrodes to: (a) neutralize the space charge and (b) control the work functions of the electrode surfaces. If a tungsten wire is heated in a low pressure cesium vapor atmosphere, Cs is adsorbed on the wire, lowers its work function and permits greater emission. Emission increases with temperature until Cs is boiled off, at which time tungsten becomes exposed and the average work function of the wire increases. As the temperature is raised and more tungsten becomes exposed, the average work function increases further and emission decreases. Thus for maximum power output a compromise between voltage and current is demanded. On the cold anode the presence of adsorbed cesium lowers the work function and tends to increase the output potential, V_o, of the device. The lowest work-function materials available are oxide surfaces with adsorbed cesium.

In the application of thermionic energy converters, the heat source should be at a value between 2000° and 3500°K in order to realize high efficiencies. These high temperatures cause deterioration of the cathodes, and effort is currently being directed at finding and developing cathode materials which have a lower work function and would permit high efficiency at low temperatures. At the present time at an operating temperature of 2000°K, efficiencies of about 15 percent have been realized although theoretical efficiencies of over 50 percent have been calculated. Figure 11.15 shows theoretical conversion efficiency as a function of cathode work function for various values of cathode temperature.

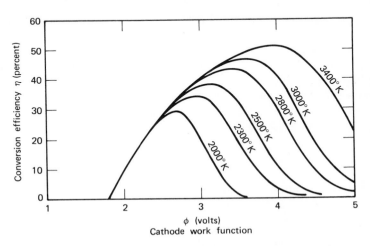

Figure 11.15 Conversion efficiency of a thermionic converter as a function of cathode work-function for various values of cathode temperature (35).

Considerable effort has been devoted to thermionic converter systems in the past two decades(36). In working thermionic converter systems tungsten emitters and niobium collectors have been used. Close spacing of electrodes and the use of cesium vapor have been applied. For the primary energy source, fossil fuels, solar energy and nuclear reactors have been used.

Thermionic conversion systems appear capable of yielding specific powers of 4 to 6 W/lb. The technology for the converter itself requires continuing effort to achieve long life and reliability. Significant promise appears to be ahead for the use of "in-pile" thermionic systems whereby the thermionic device directly surrounds a nuclear core. Ultimately thermionic diodes working at high temperature may be cascaded with thermoelectric converters to achieve system efficiencies of 20 percent and higher.

THERMOMAGNETIC

There are two thermomagnetic effects which have been considered for energy conversion. One of them is based on the change in magnetic properties of a material caused by a change of temperature. The magnetic properties of a material are lost when it is heated to the Curie temperature. If a magnetic material in a magnetic flux circuit is alternately heated to the Curie point and allowed to cool the resulting decrease and increase in its permeability will cause a change in flux density in the magnetic flux circuit. A coil wrapped around any portion of the flux path will have an emf induced in it. The thermomagnetic generator is thus a low-frequency ac source. A mechanical means is required to control the heat flow and the heating and cooling cycle of the generator. Early work on this device was briefly summarized by Betts(38).

The second effect which is also commonly referred to as the thermomagnetic effect is the Ettingshausen-Nernst effect. If a magnetic field is applied to a conductor and heat is caused to flow through the conductor at right angles to the magnetic field then a potential is developed along the third mutually perpendicular axis of the material. The effect is described by an equation of the type

$$E_y = QB_z \frac{dt}{dx} \tag{8}$$

Where E_y is the potential developed;

Q is the heat term;

B_z is the magnetic field term;

$\frac{dt}{dx}$ is the temperature gradient;

x, y, z are the respective perpendicular axes.

The magnitude of the potential developed is only about one-tenth of the potential developed by thermoelectric devices and therefore this method of energy conversion is not particularly attractive.

PYROELECTRIC

When an oriented crystal is subjected to heat, a potential is generated between the two faces of the crystal. When properly oriented these crystals can detect temperature variations on the order of millionths of a degree. Tourmaline has been the subject of most of the study of the pyroelectric effect. Current output is of the order of 5×10^{-9} amp. Betts(38) briefly reviewed this effect and concluded that due to the low magnitude of current produced the phenomenon was not considered a significant energy conversion method.

DYNAMIC HEAT ENGINE

Dynamic heat engines have been the longest known and most developed means of converting energy. Two broad categories of engines exist; reciprocating engines and turbine engines. Much written material and many texts have been devoted to the subject, and no attempt will be made herein to cover the field. Rather an effort will be made to outline recent developments and concepts pertinent to compact, lightweight, maintenance-free engines some of which may be useful in aerospace applications.

Currently interest is limited to the turbine-type power plants. In the turbine power plant area the Brayton gas turbine and the Rankine vapor turbine are important.

An understanding of the features of the two systems can be obtained by examining the temperature-entropy diagrams for the two cycles as shown in Fig. 11.16(39), and the schematic representations of the Rankine cycle system in Fig. 11.17 and the Brayton cycle system in Fig. 11.18. In the Rankine cycle, a pump is used to pump the condensate returning from the radiator from a pressure corresponding with the condensing temperature up to a pressure corresponding with the boiling temperature. The liquid at high pressure is heated to boiling temperature, boiled along a constant temperature line, and expanded down to condensing pressure and temperature in a turbine that drives a generator. The vapor is then condensed and returned to the pump. It is important to note that the pressures in this cycle are determined by the vapor pressure-temperature relation of the working fluid and are not independent of the temperatures.

In the Brayton cycle, a compression process is followed by heating an inert gas at approximately constant pressure, first in the recuperator and finally in the heat source where the gas is heated to the turbine inlet-temperature. The gas is expanded through the turbine and is then

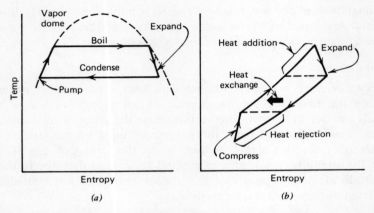

Figure 11.16 Temperature-entropy diagrams for (*a*) Rankine and (*b*) Brayton systems (39).

cooled along an essentially constant pressure path, first through the recuperator and then through the radiator where the waste heat is rejected. The cycle is completed as the cooled gas from the radiator is returned to the compressor. Because this is a gas cycle rather than a vapor cycle, the cycle pressure level is independent of the temperature.

In a comparison of the characteristics of the two cycles, it is first noted that the working fluids of interest in the Rankine cycle systems are liquid metals and involve a phase change. The working fluids in the Brayton

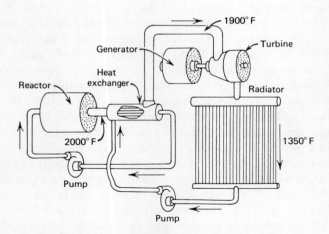

Figure 11.17 Schematic diagram of a Rankine cycle system.

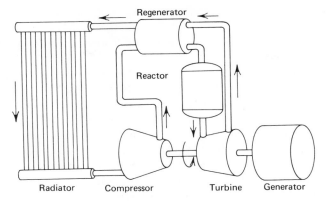

Figure 11.18   Schematic diagram of Brayton cycle system.

cycle systems, on the other hand, are inert gases and involve no phase change. Certain problems or characteristics appear as a result of the nature of the working fluids involved in the two cycles.

An extremely important consideration is overall efficiency since it regulates the size of the energy source and the heat rejection sink. Figure 11.19 presents a comparison of the efficiencies of several types of turbine

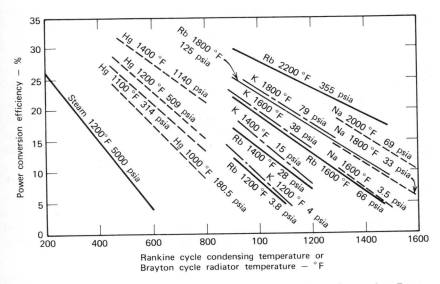

Figure 11.19   Comparison of efficiencies of several types of turbine cycles. Brayton: $N_T = 87\%$; $N_{ALT} = 90\%$; $N_{MECH} = 95\%$. Rankine: $N_T = 80\%$; $N_{ALT} = 90\%$; $N_{MECH} = 95\%$.

cycles with various working fluids operating over a wide temperature range.

Shure and Schwartz(19) have compared Rankine and Brayton system for aerospace power systems. The mercury Rankine cycle has been coupled to solar, radioisotope and nuclear energy sounds for application up to 60 kW. Cycle efficiency of 8 to 10 percent has been achieved. Features of the Rankine cycle are: (1) potential applicability over a wide range of powers up to megawatts, (2) small radiator area and weight, (3) moderate thermal efficiency, (4) adaptability for use with a diversity of heat sources, (5) wealth of background related to terrestrial technology and (6) difficult technology because of high temperatures, two-phase flow problems and corrosive nature of working fluids.

The Brayton cycle has the potential of achieving conversion efficiencies of from 20 to 30 percent. Its features are: (1) high efficiency, (2) large radiator area and weight, (3) applicability over a wide range of power levels, but not competitive with the Rankine cycle at very high powers, (4) adaptability for use with a diversity of heat sources, (5) large background of related terrestrial technology (gas turbines), and, (6) considerably less difficult technology because of inert, single-phase working fluid.

MAGNETOHYDRODYNAMICS

The magnetohydrodynamic generator is based on long known principle of the interaction between ionized or conducting fluids and magnetic fields. Faraday, ca. 1830, showed that power could be generated by substituting a flowing conductive liquid such as mercury for the copper wire conductors of a generator.

In the magnetohydrodynamic generator, hot ionized gases travel through a magnetic field which is at right angles to the fluid flow. The motion of the gas through the field induces an emf in it which is picked up by two electrodes in the fluid, and current flows in an external circuit. Voltage output of an MHD generator is proportional to the intensity of the magnetic field, the velocity of the gas and the distance between electrodes. A schematic representation of a linear MHD generator is shown in Fig. 11.20. Other geometrics have also been considered including a vortex generator, a radial outflow generator and an annular Hall generator.

Rosa and Kantrowitz(40) in 1960 reported on the engineering aspects of magnetohydrodynamic energy conversion. The process for this heat engine device is described by the following equations:

$$\text{Current density: } j = \sigma(uB - E) \tag{9}$$

$$\text{Power output: } P = \rho = \frac{dh}{dx} = jE \tag{10}$$

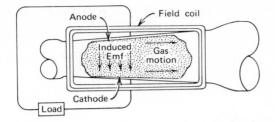

Figure 11.20 Schematic diagram of a magnetohydrodynamic generator (40).

where σ is gas conductivity, u is gas velocity, B is magnetic field strength, E is polarization, or back emf, ρ is density, and h is enthalpy per unit mass.

The conductivity of the gas is one of the most important considerations. Temperatures of the order of 5000°F are necessary to ionize a pure gas. If, however, the gas is seeded with potassium or cesium it will ionize at temperatures around 2000°C to yield a plasma which can interact with the magnetic field. At these high temperatures, materials of construction become important and presently limit the sustained use of MHD generators. Zirconia has been used to line the inside walls.

A discussion of recent engineering aspects of MHD is presented by Sutton(41). Under proper conditions it is estimated that an MHD generator can convert approximately 50 percent of its heat input to electricity. The device is inherently a large power device and high efficiency is only attainable for designed outputs of megawatts and for operating temperatures in excess of 2000°C. Fossil fueled MHD converters have been built and tested, the fuel being burned with oxygen and the working fluid seeded with potassium. There still exist many problems such as erosion and corrosion due to the combustion gas. The most promising use for MHD appears to be as a topper in fossil fueled generating plants. Ultimately it is anticipated that fission of fusion plasmas will be the source of the high temperature working fluid.

PIEZOELECTRIC

The piezoelectric effect was first discovered by Hauy in 1782, but Pierre and Jacques Curie have been associated around 1880 with the early discussions of the phenomenon. When a dielectric crystalline material is subjected to a properly oriented strain, an electric potential is developed at the surfaces which is proportional to the applied strain. Mason(42) has reviewed the field of energy conversion in the solid state and discusses the piezoelectric effect in some detail. The equations governing piezoelectric behavior for a single mode of motion along a single axis in the same direction as the applied field are of the form:

$$S = S^E T + dE$$

and

$$D = \epsilon^T E + dT$$

where S is the strain;
         T is the stress;
         E is the applied field;
         D is the electric displacement;
         S^E is the compliance modulus or ratio between S and T when the applied field E is constant;
         D is the piezoelectric constant or the ratio between S and E;
         ϵ^T is the dielectric constant or the ratio between D and E when the stress T is constant.

The electromechanical coupling factor, k, is a dimensionless quantity indicating the ability of a piezoelectric material to convert energy from mechanical to electrical. It is given by

$$k = \frac{d}{\sqrt{S^E \cdot \epsilon^T}} \tag{11}$$

Power conversion efficiency is proportional to the fourth power of k.

Quartz was one of the first piezoelectric materials investigated and has a value of k of about 0.1. Rochelle salt is also well known and has a high value of k of about 0.8. Other salts which have been studied are ammonium dihydrogen phosphate, potassium dihydrogen phosphate, dipotassium tartrate, lithium sulfate and ethylene diamine tartrate.

Most of the early salt crystals were temperature sensitive and affected by humidity. They have been replaced by more advanced types of piezoelectric material which do not have these shortcomings. These are the ferroelectric ceramic materials, such as barium titanate which is made by sintering together powders of BaO and TiO_2, and the lead titanate zirconates.

As a mechanical-to-electrical energy converter a crystal can be operated at its resonant frequency, powered by a vibrating driver mechanism such as a vibratory gas engine. Operating into a tuned circuit and a matched impedance output, the piezoelectric crystal can convert 90 percent of the mechanical power to electrical power. Power density of piezoelectric materials is 10 to 50 W/cm^2. Mason described and analyzed the performance of a vibratory gas engine coupled to a ferroelectric ceramic transducer. A 1000 W engine was estimated to weigh about 5 lb.

SUMMARY

A wide variety of energy sources and energy conversion methods is available to yield an electrical output. Of most interest in a book devoted to batteries are those sources and conversions that have the same realm of applicability as do batteries. Among the primary energy sources the chemical, solar and nuclear (radioisotope and fission) are most important. Of the many conversion systems known, the most pertinent ones are photovoltaic, thermoelectric, thermionic, and the Rankine and Brayton dynamic heat engines.

Photovoltaic systems are widely used coupled with nickel-cadmium batteries. Efficiency has not improved beyond 12 percent in almost two decades and costs are still high. A breakthrough is needed for solar array-battery plants to find widespread terrestrial use.

Thermoelectric converters are widely coupled to fossil fueled burners, solar energy concentrators, and radioisotope and nuclear fission sources. The semiconductor lead-telluride has emerged as the major material for temperatures of a few hundred degrees while silicon-germanium shows promise for use at 1000°C. Efficiencies of about 5 percent have been achieved. Cascading of the two materials to cover a wide temperature range will result in better efficiency for the overall power conversion system.

Thermionic conversion systems operate at temperatures of 2000°C and are potentially more efficient than thermoelectrics. They appear to have the potential to yield 15 percent efficiency, and because of their high temperature of application, appear to be well suited for use in "in-pile" nuclear reactor installations. Ultimately, thermionic diodes working at high temperature may be cascaded with thermoelectric converters to achieve system efficiencies of 20 percent and higher.

Among the dynamic heat engines, Rankine cycles have moderate efficiencies and are more competitive at high power than are Brayton engines. The Brayton gas turbine is more efficient and has less difficult technological problems at high temperature but it is larger and heavier. Both systems can extend to powers as high as megawatts.

REFERENCES

1. Copeland, A. W., Black, O. D., and Garrett, A. B., *Chem. Reviews* **31**, 177 (1942).
2. Sancier, K. M., Conference on Solar Energy; The Scientific Basis, Tucson, Arizona (1955).
3. Rosenberg, N., *Proc. Seminar on Advanced Energy Sources and Conversion Techniques*, A.S.T.I.A. No. AD209301, Dept. of Commerce, O.T.S. No. P.B.151461, p. 139 (Nov. 1958).

4. Rabinowitch, E. I., *J. Chem. Phys.* **8**, 551 (1940).
5. Adams, W. G. and Day, R. E., *Proc. R. Soc.* **A25**, 113 (1877).
6. Chapin, D. M., Fuller, C. S., and Pearson, G. L., *J. Appl. Phys.* **25**, 676 (1954).
7. Rappaport, P., A.S.T.I.A. No. AD209301, p. 127 (Nov. 1958).
8. Crossley, P. A., Noel, G. T., and Wolf, M., Final Report "Review and Evaluation of Past Solar-Cell Development Efforts," NASA Contract NASW-1427 (June 1968).
9. Castruccio, P. A., "Proposed Moon Power Station", Westinghouse Technical Paper A-5946 (September 10, 1958).
10. Rabinowitch, E. I., *Photosynthesis and Related Processes*, Interscience, New York (1945, 1951, 1956).
11. Hill, R. and Whittingham, C. P., *Photosynthesis*, John Wiley, New York (1955).
12. Meyers, J., *Proc. World Symposium on Applied Solar Energy*, Stanford Research Institute, Menlo Park, California (1956).
13. Duwez, P., A.S.T.I.A. No. AD209301, p. 123 (Nov. 1958).
14. Telkes, M., *Proc. 13th Annual Power Sources Conf.*, U.S. Army Signal Research and Development Laboratory, Fort Monmouth, N.J., p. 69 (April, 1959).
15. Kittl, E., *Proc. 13th Annual Power Sources Conf.*, p. 22 (April 1959).
16. Corliss, W. R. and Harvey, D. L., *Radioisotope Power Generation*, Prentice Hall, Englewood Cliffs, N.J. (1964).
17. Mead, R. and Corliss, W. R., "Power from Radioisotopes", U.S. Atomic Energy Commission, Division of Technical Information, Series on Understanding the Atom (1964).
18. Schulman, F., "Isotopes and Isotope Thermoelectric Generators", Space Power Systems Advanced Technology Conference Proceedings, Cleveland, Ohio, p. 7 (1966), NASA SP-131.
19. Shure, L. I. and Schwartz, H. J., "Survey of Electric Power Plants for Space Applications" *Chemical Engineering Progress Symposium Series 75*, **63**, 95 (1967).
20. Moseley, *Proc. R. Soc.* **A88**, 471 (1913).
21. Coleman, J. H., A.S.T.I.A. No. AD209301, p. 159 (Nov., 1958).
22. Linder, E. G., Rappaport, P., and Loferski, J. J., *International Conference on the Peaceful Uses of Atomic Energy*, p. 283 (June, 1955).
23. Thomas, A., "Nuclear Batteries: Types and Possible Uses," *Nucleonics* **13**, 129 (Nov. 1955).
24. Colgate, S. and Aamodt, R. L., *Nucleonics* **15**, 50 (Aug. 1957).
25. Colgate, S., A.S.T.I.A. No. AD209301, p. 151 (Nov. 1958).
26. Werner, R. W., Meyers, B., Lee, J. D., and Mohr, P. B., "Controlled Thermonuclear Power" 4th Intersociety Energy Conversion Engineering Conference, Washington, D.C., p. 176 (1969).
27. Ioffe, A. F., *Semiconductor Thermoelements and Thermoelectric Cooling*, Infosearch Ltd., London (1957).
28. Telkes, M., *J. Appl. Phys.* **18**, 116 (1947).
29. Zener, C., *The Impact of Thermoelectricity on Science and Technology*, Symposium Ed. Paul Egli, McGraw-Hill, New York (1960).
30. *Elec. Eng.* **79**, 353 (1960)—special issue on thermoelectricity.
31. Fuschillo, N., Gibson, R., Eggleston, F., and Epstein, J., *Advanced Energy Conversion* **6**, 103 (1966).
32. Moss, H., *Brit. J. Electronics* **2**, 305 (1957).
33. Hatsopoulos, G. N. and Kay, J., *J. Appl. Phys.* **29**, 1124 (1958) and *Proc. IRE* **46**, 157 (1958).
34. Wilson, V. C., *J. Appl. Phys.* **30**, 475 (1959).

5. Hernqvist, K. G., Kanefsky, M., and Norman, F. H., *R.C.A. Review* **XIX**, 244 (June 1958).

36. Studies of Thermionic Materials for Space Power Applications, NASA Contractor Report CR-54980 (May 1966).

37. Research Program for the Long Term Testing of Cylindrical Diodes, NASA Contractor Report CR-54656 (April 1965).

38. Betts, A. L., Unconventional Electrical Power Sources, Dept. of Commerce, O.T.S. No. PB131411 (1954).

39. Stewart, W. L. et al., "Brayton Cycle Technology," Space Power Systems Advanced Technology Conference Proceedings, Cleveland, Ohio, p. 95 (1966), NASA SP-131.

40. Rosa, R. J. and Kantrovitz, A., "MHD Energy Conversion Techniques" in *Direct Conversion of Heat to Electricity*, Ed. Kaye and Welsh, John Wiley, New York (1960).

41. Sutton, G. W., *Engineering Magnetohydrodynamics, McGraw-Hill, New York* (1965).

42. *Mason, W. P., Piezoelectric Crystals*, Van Nostrand, New York (1950).

Author Index

Subject Index

514